Fungal Genetics

MYCOLOGY SERIES

Editor

Paul A. Lemke

Department of Botany and Microbiology
Auburn University
Auburn, Alabama

1. *Viruses and Plasmids in Fungi*, edited by Paul A. Lemke
2. *The Fungal Community: Its Organization and Role in the Ecosystem*, edited by Donald T. Wicklow and George C. Carroll
3. *Fungi Pathogenic for Humans and Animals* (in three parts), edited by Dexter H. Howard
4. *Fungal Differentiation: A Contemporary Synthesis*, edited by John E. Smith
5. *Secondary Metabolism and Differentiation in Fungi*, edited by Joan W. Bennett and Alex Ciegler
6. *Fungal Protoplasts*, edited by John F. Peberdy and Lajos Ferenczy
7. *Viruses of Fungi and Simple Eukaryotes*, edited by Yigal Koltin and Michael J. Leibowitz
8. *Molecular Industrial Mycology: Systems and Applications for Filamentous Fungi*, edited by Sally A. Leong and Randy M. Berka
9. *The Fungal Community: Its Organization and Role in the Ecosystem, Second Edition*, edited by George C. Carroll and Donald T. Wicklow
10. *Stress Tolerance of Fungi*, edited by D. H. Jennings
11. *Metal Ions in Fungi*, edited by Günther Winkelmann and Dennis R. Winge
12. *Anaerobic Fungi: Biology, Ecology, and Function*, edited by Douglas O. Mountfort and Colin G. Orpin
13. *Fungal Genetics: Principles and Practice*, edited by Cees J. Bos

Additional Volumes in Preparation

Fungal Genetics

Principles and Practice

edited by

Cees J. Bos
Wageningen Agricultural University
Wageningen, The Netherlands

CRC Press
Taylor & Francis Group
Boca Raton London New York

CRC Press is an imprint of the
Taylor & Francis Group, an informa business

First published 1996 by Marcel Dekker, Inc.

Published 2019 by CRC Press
Taylor & Francis Group
6000 Broken Sound Parkway NW, Suite 300
Boca Raton, FL 33487-2742

First issued in paperback 2019

ISBN 13: 978-0-367-44864-6 (pbk)
ISBN 13: 978-0-8247-9544-3 (hbk)

Visit the Taylor & Francis Web site at
http://www.taylorandfrancis.com

and the CRC Press Web site at
http://www.crcpress.com

Library of Congress Cataloging-in-Publication Data

Fungal genetics : principles and practice / edited by Cees J. Bos.
p. cm. — (Mycology series ; 13)
Includes biblographical references and indexes.
ISBN 0-8247-9544-X (hardcover : alk. paper)
1. Fungi—Genetics. I. Bos, Cees J. II. Series:
Mycology series ; v. 13.
QK602.F85 1996
589.2'0415—dc20 95-52699
CIP

Preface

Fungi play an important role in biotechnological processes, phytopathology, food technology, and biomedical research. For more than 50 years these lower eukaryotes have been useful in genetic and other fundamental biological research. In the first few decades of this research, fungal genetic studies were done mostly by geneticists and mycologists who were already familiar with the main aspects of these organisms. In recent years, biochemists and other scientists (who may have little genetic background) have used molecular genetic analysis and gene manipulation. Just as the molecular approach can profit from formal genetics, classical genetic analysis can profit from molecular techniques. Since fungi are often very suitable as hosts for heterologous genes, and since molecular genetic analysis techniques are becoming more widespread, there was a need for a book on fungal genetics covering formal and molecular genetics and techniques.

The first part of the book deals with genetic principles. The second part contains case studies on the genetics of specific fungi, indicating what the interesting features of each organism are for genetic studies. The aim is to cover the main aspects of fungal genetics using a practical approach. The book gives a broad view on fungal genetics from numerous specialists on particular subjects or organisms. The chapters are suitably formatted for use in courses and as a practical reference. Each chapter starts with a treatise of the basic concepts, followed by an experimental design that illustrates the practical approach. Several experimental protocols contain an illustrated outline, discussion points that focus the reader on special aspects, and suggestions for further reading. This organization facilitates an understanding of the practical approach without the reader having to do the experiment. However, the experiments are reproducible without requiring very much experience, and the strains are commonly available for general use. This reference on the genetics of fungi demonstrates the concerted use of classical and molecular methods and techniques for genetic analysis and breeding of fungi—two aspects of fungal genetics that make it such a fascinating field of science.

Cees J. Bos

Contents

Contributors

María I. Alvarez Department of Genetic Microbiology, University of Salamanca, Salamanca, Spain

José Arnau* Norwich Molecular Plant Pathology Group, School of Biological Sciences, University of East Anglia, Norwich, England

Kirk A. Bartholomew Pulp and Paper Research Institute of Canada, Pont Claire, Quebec, Canada

Cees J. Bos Department of Genetics, Wageningen Agricultural University, Wageningen, The Netherlands

Jon J. P. Bruchez Department of Genetics, The University of Leeds, Leeds, England

**Current affiliation*: Department of Research and Development, Biotechnological Institute, Lyngby, Denmark.

Fons Debets Department of Genetics, Wageningen Agricultural University, Wageningen, The Netherlands

Arturo P. Eslava Department of Genetic Microbiology, University of Salamanca, Salamanca, Spain

Karl Esser Allgemeine Botanik, Ruhr-Universität, Bochum, Germany

Mark L. Farman Department of Plant Pathology, University of Wisconsin, Madison, Wisconsin

Theo Goosen Department of Genetics, Wageningen Agricultural University, Wageningen, The Netherlands

Rolf F. Hoekstra Department of Genetics, Wageningen Agricultural University, Wageningen, The Netherlands

James Robertson Kinghorn School of Biological and Medical Sciences, University of St. Andrews, St. Andrews, Fife, Scotland

Sally Ann Leong Disease Resistance Research Unit, Agricultural Research Service, U.S. Department of Agriculture, and Department of Plant Pathology, University of Wisconsin, Madison, Wisconsin

Benjamin C. K. Lu Department of Molecular Biology and Genetics, University of Guelph, Guelph, Ontario, Canada

P. T. Magee Department of Genetics and Cell Biology, University of Minnesota, St. Paul, Minnesota

Amy L. Marion Department of Biological Sciences, Purdue University at Fort Wayne, Fort Wayne, Indiana

Charles P. Novotny Department of Microbiology and Molecular Genetics, The University of Vermont, Burlington, Vermont

Richard P. Oliver Norwich Molecular Plant Pathology Group, School of Biological Sciences, University of East Anglia, Norwich, England

Heinz D. Osiewacz Botanisches Institut, Johann Wolfgang Goethe-Universität, Frankfurt am Main, Germany

Patricia J. Pukkila Department of Biology, The University of North Carolina at Chapel Hill, Chapel Hill, North Carolina

Alan Radford Department of Genetics, The University of Leeds, Leeds, England

Maureen B. R. Riach* School of Biological and Medical Sciences, University of St. Andrews, St. Andrews, Fife, Scotland

Stewart Scherer Department of Microbiology, University of Minnesota, Minneapolis, Minnesota

D. S. Shaw School of Biological Sciences, University of Wales, Bangor, Gwynedd, Wales

David Stadler Department of Genetics, University of Washington, Seattle, Washington

Paul J. Stone Department of Genetics, The University of Leeds, Leeds, England

Klaas Swart Department of Genetics, Wageningen Agricultural University, Wageningen, The Netherlands

Fawzi Taleb Department of Genetics, The University of Leeds, Leeds, England

Robert C. Ullrich Department of Botany, The University of Vermont, Burlington, Vermont

Klaus Wolf Institute for Biology IV (Microbiology), Rheinisch-Westfälische Technische Hochschule, Aachen, Germany

**Current affiliation*: Blackwell Science Ltd., Edinburgh, Scotland

1

Biology of Fungi

Cees J. Bos
Wageningen Agricultural University, Wageningen, The Netherlands

1. INTRODUCTION

Why do we study fungal genetics? Many fungi are excellent objects to study genetics, and genetics plays an important role in various fields of fundamental and applied mycology. In this chapter we consider some aspects of the biology of fungi that are essential for understanding genetic features and processes. In addition, various aspects of fungal genetics are reviewed in order to elucidate how they play a role in fundamental biology, biotechnology, plant pathology, and other fields of applied biology.

Fungi are lower eukaryotes, and most of them can be grown on defined media. That was the basis for the work of Beadle and Tatum [1], who made a new approach in the early 1940s to study genetic control of metabolism. Some fungi are obligate biotrophs (obligate parasites), such as the rust fungi and mildews. They can be grown only in conjunction with their host plant, but some stages can be studied in vitro. Genetic studies with these fungi have been done, although they are very laborious. They enable, however, the study of the genetics of such close host–parasite relations. In fact the rusts (*Puccinia, Melampsora*) provided the first proofs of a gene-for-gene relationship in host–parasite systems [2].

The most characteristic aspect of fungi is that they grow as a thread of cells (hypha) and are propagated by generative and/or vegetative spores. A spore germinates with a germ tube that grows only at the tip. Transverse cell walls (septae) are formed at some distance behind the tip except in Phycomycetes, which are coenocytic. In general, the septae have a pore that allows transport of cytoplasm, mitochondria, and even nuclei.

The top cell of a hypha often has more than one nucleus located at some distance from the tip. Several interesting studies on the growth of hyphal tips have been done, and this subject has an old tradition [3]. Soon after germination, hyphae will form branches, and the result is a network of hyphae (mycelium). Hyphal fusions (anastomoses) can occur at some distance from the tips. In principle they allow exchange of cytoplasm and also nuclei between hyphae of different origin leading to heterokaryons (Chapter 4). A heterokaryon is a dynamic system of fusing and segregating hyphae. Even in a very well balanced heterokaryon, only a fraction of the hyphal tips is heterokaryotic. On a solid surface fungi grow radially and in a vigorously shaken liquid culture as compact spherical colonies. The compact spheres may consist of an aggregation of different mycelia or may contain ungerminated spores. Most yeasts do not have a predominant mycelial growth, although cells may stick together. Growth proceeds by budding (*Saccharomyces*) or by fission of cells (*Schizosaccharomyces*). The cells are mostly uninucleate and haploid, but diploid strains do exist. Some fungi are dimorphic—i.e., they can grow yeastlike as well as with hyphae. The best-known example is the smut fungus *Ustilago maydis*. One form is haploid and unicellular, divides by budding, and is nonpathogenic. The filamentous form is dikaryotic and pathogenic to maize. The yeast form can be grown on synthetic media and is very suitable for genetic studies [4].

Four main groups of fungi are distinguished on basis of the structures for sexual reproduction:

1. Phycomycetes. The sexual spores arise in a sporangium. They may be uninucleate or multinucleate and provided with flagella (e.g., zoospores of *Phytophthora*) or be nonmotile (*Phycomyces*). These fungi are in general coenocytic.

2. Ascomycetes. The sexual spores are formed within an ascus. The simplest form is found in yeasts. Two cells fuse and karyogamy is immediately followed by meiosis, resulting in four spores without cell division. Such a cell with four meiotic products is called an ascus. In many Ascomycetes the tetrad cells undergo an additional mitotic division, resulting in eight ascospores having pairwise the same genotype. The yeasts are called hemiascomycetes. The eu-ascomycetes have specialized fruiting bodies (ascocarps) that may contain hundreds to thousands of asci. The main forms are:

Cleistothecium: closed spherical ascocarp (*Aspergillus*)
Perithecium: a spherical ascocarp with a pore for the release of the spores (*Sordaria, Neurospora*)
Apothecium: the ascocarp is open at the upper side (Discomycetes as *Helotium, Sclerotinia*).

In some fungi the ascospores are situated linearly in the ascus. Others have unordered asci.

3. Basidiomycetes. In the Basidiomycetes the generative spores are formed on the meiocyte (basidium). The holobasidiomycetes have an undivided basidium. They are the higher fungi known as mushrooms (*Agaricus, Cantharellus* etc.). The hemibasidiomycetes have a divided basidium. The main groups are the rusts (Uredinales) and the smut fungi (Ustilaginales).

4. Fungi Imperfecti. This group of fungi consists of fungi of which no sexual stage is known. Most of them are grouped in form genera on the basis of their structures for vegetative reproduction and the form of the vegetative spores. Many Fungi Imperfecti are related to Ascomycetes.

2. GENETIC PROCESSES

A main point in the life cycle of any organism is the alternation of haploid and diploid "generation." In the diploid phase an organism has two sets of chromosomes (two genomes), and the homologous chromosomes differ when the parents contained different mutations (alleles) on homologous loci. In animals the haploid phase consists only of the gametes, and in higher plants the haploid phase is restricted to few divisions. In lower plants (mosses, ferns) the haploid phase is much longer, and in fact the moss plant is haploid and only the sporangium and the sporangium stalk are diploid. The haploid phase that ultimately produces the gametes is called the gametophyte, and the diploid phase the sporophyte. The phase alternation consists of plasmogamy–karyogamy–meiosis. In both phases the nuclei of the somatic cells divide in a characteristic way (mitosis). In the life cycle only one cell (the meiocyte) has a different nuclear division (meiosis) by which the homologous chromosomes segregate into the daughter cells. Meiosis consists of two divisions (the second resembles a mitotic division) and results in four haploid cells (a tetrad) that can grow out to a gametophyte. The processes are illustrated in Figures 1 and 2. Two different recombination processes occur during meiosis: reassortment of homologous chromosomes, and exchange of genetic material between nonsister chromatids of homologous chromosomes (crossing over). The genetic consequences and the use for genetic analysis are discussed in Chapter 3. The organization of the genetic material and the meiotic process are discussed in Chapters 6 and 18.

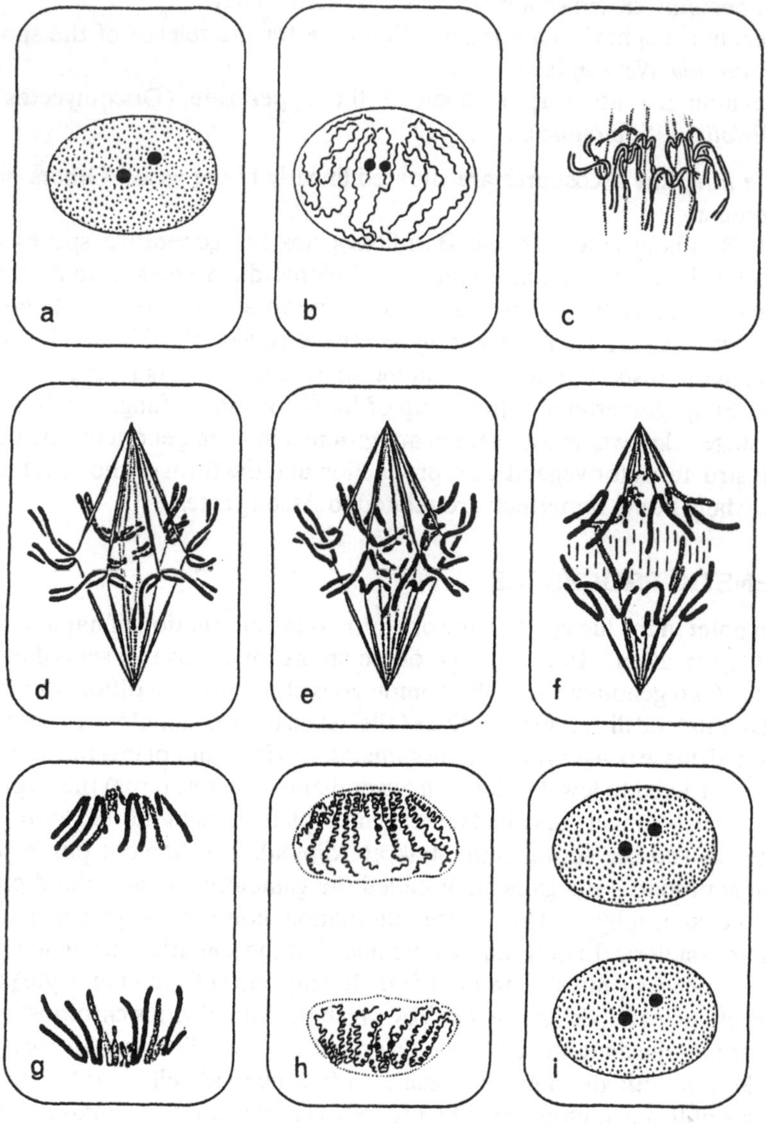

Figure 1 Mitosis in a diploid cell: interphase (a); prophase (b, c); metaphase (d), in which the centromeres are in the equatorial plane due to the forces of the spindle; anaphase (e–g), during which the sister chromatids separate; telophase (h); interphase (i). The result is that each daughter cell has the same genetic constitution. DNA is replicated during interphase, and the chromosomes consist of one chromatid before and of two after replication.

Fungi have also extrachromosomal genetic elements of which the mitochondria are the most important. There is interaction between extrachromosomal elements and the genome by transposable elements. Chapters 7, 10, and 16 deal with these phenomena. The transmission of extrachromosomal elements is also interesting on population level and can be used to study horizontal gene flow in view of risk assessment of genetically manipulated fungi.

In fungi the alternation of the haploid and diploid phase can be complex, because there can be a long period of time between plasmogamy and karyogamy. In general, plasmogamy between different strains results in a heterokaryon. In case of the transition of the gametophytic to the sporophytic phase, plasmogamy results in a dikaryon that is maintained in a specific way of concerted nuclear divisions. In Ascomycetes as well as in Basidiomycetes this is accompanied by a special mechanism of cell division to ensure the dikaryotic situation (see Chapter 19).

A. Vegetative Reproduction

Vegetative reproduction is the most important way for dispersion in many fungi. Some fungi form specialized organs for vegetative reproduction as a peritheciumlike pycnidium (*Phoma* sp.) or an apotheciumlike acervulus (*Gloeosporium* sp.). The rusts have a complicated life cycle with nearly closed or open sori in which the vegetative spores are formed. In some fungi the vegetative spores arise at the top of normal hyphae, and others have specialized conidiophores. These fungi and also the Phycomycetes with simple sexual organs are very suitable to study differentiation. In the Aspergilli the conidiophore has a swollen top cell bearing sterigmata and on them phialides in which the conidiospores are formed in a chain, with the oldest at the end of the chain. The conidiospores of a chain are of the same genotype, and even on a heterokaryon the various chains on a conidiophore (spore head) usually have a similar genotype. This facilitates the isolation of mutants (Chapter 2) or vegetative recombinants (Chapter 4).

B. Generative Reproduction

In fungi we find not only a wide variety of reproduction structures, but also a variety of reproduction processes. The generative process starts with plasmogamy (somatogamy or gametogamy), and the result is a dikaryotic phase. In yeast, cell fusion is immediately followed by meiosis. It is the simplest form of the alternation of the haploid and diploid phases. This is much more complex in many fungi. In the Phycomycetes the meiosis seems to occur just before the formation of the gametes. In such a situation the fungal mycelium is diploid and more difficult to study genetically. The genetics of *Phycomyces* is discussed in Chapter 20.

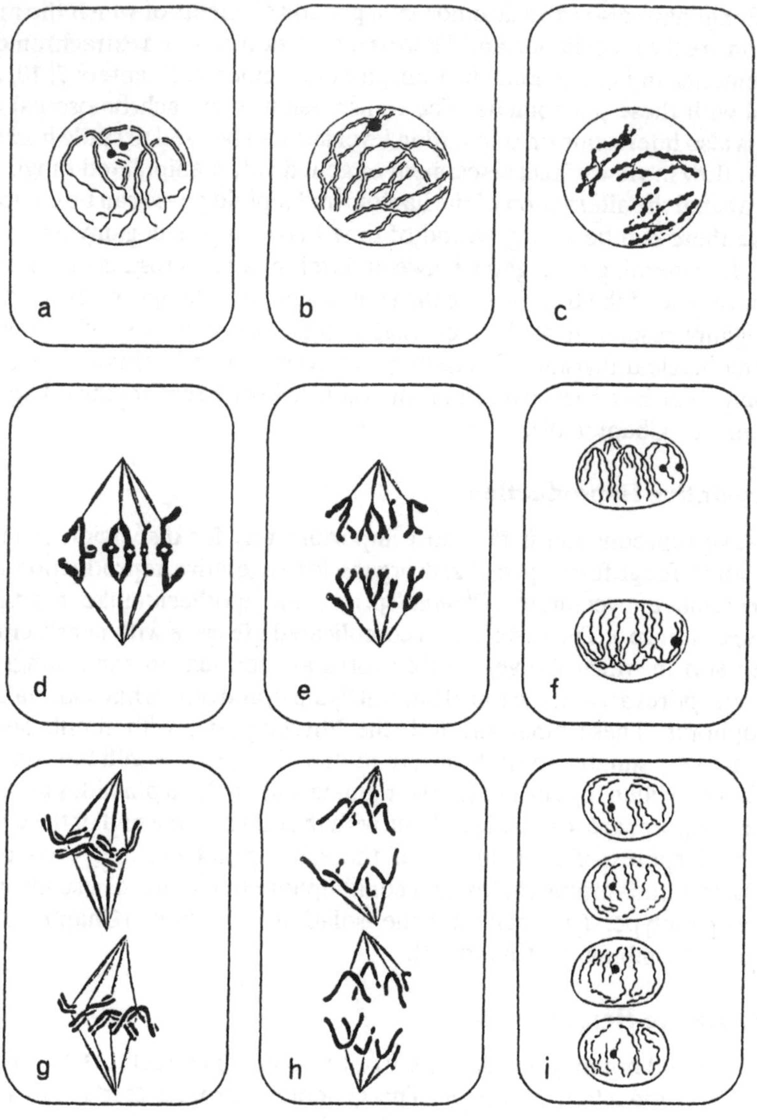

Figure 2 Meiosis consists of two divisions. The cell that undergoes meiosis is called a meiocyte. During MI the homologous chromosome pair (a) can be seen, and in the prophase chiasmata can be seen (b, c). The chiasmata are the result of exchange between homologous nonsiter chromatids (cross-over). During the metaphase the centromeres are not in the equatorial plane. The homologous chromosomes stick together owing to the chiasmata which terminalize due to the forces of the spindle (d, e), and the result is two haploid nuclei (f). (*legend continues on facing page.*)

In many Basidiomycetes there is a space of time between plasmogamy and karyogamy. The dikaryon that is established can grow for a long time. In the Holobasidiomycetes it makes a fruiting body, and then cells differentiate to basidia in which karyogamy takes place. The basidia lie in a layer (hymenium) that covers a side of the fruiting body. The four basidiospores that are formed on the basidium are often vigourously dispersed. In the rust and the smut fungi (both hemi-Basidiomycetes) the dikaryon grows on a special host, where they form abundant resting spores (teliospores). These teliospores are incapable of vegetative growth, but upon germination they form a basidium consisting of four cells on which the basidiospores bud off (often forming chains of cells with similar genotype).

In the Ascomycetes the dikaryotic phase is found only within the ascocarp. The dikaryotic hypha is also maintained by a special way of cell growth and is called ascogenous hypha, because the asci arise at the tips of the branched ascogenous hyphae. Karyogamy occurs in the young ascus cells and is immediately followed by meiosis. These meiocytes develop into asci with four or eight (some species even 16) ascospores. In some fungi the asci can be isolated, and the four meiotic products that belong together (a tetrad) can be studied. In some species the ascospores stay in the linear order in which they have been formed. In such a situation we can infer what happened at the first meiotic division, and we can distinguish between prereduction and postreduction (see Chapter 3).

C. Parasexual Processes

It has been shown for many fungi that recombination can occur also in somatic cells. Diploid nuclei can arise in a heterokaryon by somatic karyogamy at very low frequency. The diploid nuclei can lose chromosomes during mitosis, resulting in new haploid sets of chromosomes. Moreover, mitotic crossing-over can occur in diploid nuclei. The occurrence of mitotic crossing-over had already been detected in *Drosphila* by Stern in 1936. In genetic studies with fungi, somatic recombination turned out to be a real alternative to sex [5]. It encouraged genetic studies and it has also been used in genetic analysis and breeding. The importance of the parasexual mechanisms in natural situations is still unknown. The simplest form of somatic recombination is the formation

---

The first meiotic divison is immediately followed by second one (MII) that resembles a mitotic division (g–i). During the anaphase of MII (h), the centromeres divide and the chromatids of a chromosome separate. The result is four haploid nuclei (i, tetrad). For some loci the alleles (the homologous genes) separate during MI, but for other loci the homologous genes separate during MII due to one or more crossovers between locus and centromere. The effect is that genes that are located close together show linkage whereas others may inherit independently.

and segregation of heterokaryons, but in various fungi somatic karyogamy can also occur, giving rise to somatic diploids. Chapter 4 deals with somatic recombination. The use in genetic analysis and breeding is also discussed in Chapter 9.

D. Mating Types and Incompatibility

Both vegetative and generative cell fusions are restricted to compatible strains. The compatibility factors for sexual fusion are known as mating types. In *U. maydis* both partners must have different alleles on two loci (e.g., one strain *a1 b1* and the other *a2 b2*). Otherwise no filamentous growth occurs and the fungus will be nonpathogenic. Somatic diploids can be used to study the various aspects of the incompatibility system. In this fungus one gene (*a*) is involved in a pheromone response pathway and to establish filamentous growth [6]. The *b* gene is involved in tumor formation and is a prerequisite for pathogenic development. As probably at least 25 *b* alleles exist in nature, it is a complex system [see 6]. Also in other fungi the mating-type systems seem to be complex, and pheromones are involved. Chapter 19 deals with a case study on mating type genes in *Schizophyllum commune*. In general the possibility of sexual contact is a complex process depending on several genes. One defective gene can already block sexual contact between two strains. Wheeler detected such pheromones in *Glomerella cingulata* in 1954 [7]. Fungi that can form reproductive organs without participation of another strain are called homothallic (vs. heterothallic). But a single mutation can change a strain in partner dependence.

Besides mating-type (mt) genes, there are also incompatibility genes that function at the vegetative level. Vegetative incompatibility genes do not usually interfere with generative reproduction, but mt genes might have a function during the vegetative phase. Vegetative incompatibility inhibits the formation of plasmogamy, but sometimes stable heterokaryons can be obtained by protoplast fusion (Chapter 4). In contrast to the mating-type genes, in the case of the incompatible genes, the allele on the corresponding *het* loci should be identical. Vegetative incompatibility genes limit horizonal transfers of extrachromosomal elements and viruses that could be deleterious. In the chapter on population genetics (Chapter 16) attention is paid to the role that vegetative incompatibility might play in nature.

3. SPECIAL FIELDS OF INTEREST

A. Metabolic Studies and Biotechnology

Many fungi are very suitable for genetic studies because of their haploid nature (when the colonies are gametophytes) and because of their growth on defined media. Beadle and Tatum laid a basis for studies on the relation between genetic material and metabolism in the 1940s with *Neurospora crassa* [1]. Fungi also served to study gene expression and gene manipulation in eukaryotes (see Chapter 5). In general, fungi can express heterologous genes very well, so they are very suitable for the production of primary gene products (proteins, peptides) that are of pharmaceutical importance. The growth of genetically modified fungi is ethically easier to accept than the manipulation of animals because of the interference with the integrity of the animal. Various metabolic pathways in fungi have been studied and several groups have concentrated on regulation of nitrogen, sulfur, or carbon metabolism [see e.g., 8,9]. Around 1970 several textbooks were published [10–13], and 11 fungal systems were described in the *Handbook of Genetics* in 1974 [14]. Recently several monographs and handbooks have again reviewed the state of art [15–18]. Now, the concerted use of formal and molecular techniques is a very useful tool to knock down a metabolic pathway and to restore it again ultimately in breeding production strains. Chapters 5, 8, and 9 discuss the possibilities and strategies for genetic manipulation and breeding. Now even fungi that were not accessible to classical genetic analysis can be analyzed genetically (Chapter 11).

B. Plant Pathogenic Relationships

The fungi turned out to be very suitable for the study of symbiotic relationships. The variation in fungi and their complex variability have drawn much attention from the beginning of the 20th century, starting with Erikson and Barrus, whereas Stakman and Christensen recognized the genetic basis of the phytopathological relationships in the 1920s in studies with *Puccinia* and *Helminthosporium*, respectively [19,20]. Flor [2,21] came with a gene-for-gene concept studying flax rust, *Melampsora lini*, on flax. In principle there are two processes: the fungus must be able to use the plant as host, and the symbiotic relation must become pathogenic. Assuming that in the initial situation the plant has not been killed, a mutation from avirulence to virulence will lead to a pathogenic attack. This probably will involve the loss of a function in the fungus. Therefore the fungal genes are described as avirulence genes. Briggs and Johal [22] argue for the use of the terms compatibility factor and incompatibility factor. Compatibility factor causes general changes in the host's physiology so that the fungus can invade the host. In an in-principle compatible situation a second pattern of avirulence or incompatibility factors might exist. These factors change the host's physiology in such a way as to prevent infection. This system is regarded typically unstable because loss of function (due to resistance of the host) can be compensated by a mutation in an avirulence gene of the fungus [22]. Molecular genetic techniques allow the study of the molecular mechanisms involved in a gene-for-gene relationship (Chapter 14). Certain races of *Cladosporium fulvum* produce a protein that is toxic only for tomato cultivars that carry the Cf9 resistance gene (see Chapter 17). On the other hand the use of gene disruption techniques (see Chapter 8) has revealed that several fungal genes that were believed to play an important role in host–parasite relationships, had no effect on pathogenicity [23].

Resistance that prevents the action of compatibility factors is regarded to be rather stable, whereas resistance based on an incompatibility interaction is likely to be unstable, because loss of a function is easier to achieve. It is, however, questionable whether a totally new compatibility factor is needed to overcome this and, on the other hand, to what extent an incompatibility gene (avirulence gene) contributes to the fitness of the fungus. Physiological races have been found in many plant pathogenic fungi. In the case of gene-for-gene relationship it is often easy to find new resistance genes, but the fungus often will find an answer as soon as the resistant host variety is grown on a large scale. That provides an excellent selection system for new virulence genes. It has been recognized that the more durable field resistance mostly has a more complicated genetic background and that quantitative genetic characters are often involved. Gene disruption can be a tool to identify whether a gene plays a role in the pathogenicity process. Combination with physical karyotyping (Chapter 5) has shown that pathogenicity genes may be located on chromosomes that can be lost without lethal effect [24]. Also from population genetic studies it is obvious that fungal species may show a great variation in karyotype. The gene manipulation techniques open possibilities for biological control of plant pathogens. Genetics may also prove to be needed for breeding fungi for biological control of insects. For plant breeders it will stay necessary to understand and to consider how plant pathogens might answer and therefore it is essential to study genetics of pathogenic fungi.

C. Perspectives

Fungal genetics is still in the picture and so is the use of molecular techniques to study diversity in fungal species (e.g., mating types and vegetative incompatibility, phylogenetic trees, pathogenic characters). Perhaps not only plant–parasite relationships, but also pathogenic processes for human and animal pathogens will be studied [25]. Fungi have an important role in genetic research. Several genetic processes have been extensively studied in fungi, and new insights have been obtained. Examples include the "one-gene-one-enzyme" hypothesis, allelic complementation, tetrad analysis, process of crossing-over, gene conversion, gene expression, and gene manipulation. In the future fungi will continue to have their place in research to study genetic mechanisms and fundamental biological problems, because especially fungi have various features in common with other eukaryotes, while at the same time they are accessible metabolically. Fundamental research directed to application will perhaps prevail. There are challenges in the search for new ways to antibiotics as resistance makes many known antibiotics unuseful. Fungi are also excellent organisms to produce pharmaceuticals and other high-value products. Fungi can perhaps also be more adapted to food technology processes especially in view of food supply in developing countries. Many fungi of industrial interest have been studied genetically and are accessible to breeding. The main application will, however, still be in biotechnology, where a concerted use of mutation, gene manipulation, and classical recombination has the best perspectives.

REFERENCES

1. Beadle, G. W., E. L. Tatum (1941). Genetic control of biochemical reactions in *Neurospora. Proc. Natl. Acad. Sci. U.S.A. 48*:400.
2. Flor, H. H. (1942). The inheritance of pathogenicity in *Melampsora lini. Phytopathology 32*:334.
3. Holliday, R. (1974). *Ustilago maydis*. In: King, R. C. (ed.). *Handbook of Genetics*. Plenum Press, New York.
4. Köhler, E. (1930). Zur Kentnisse der vegetativen Anastomosen der Pilze. II. *Mit. Planta 10*:3.
5. Pontecorvo, G. (1958). *Trends in Genetic Analysis.* Columbia Univ. Press, New York.
6. Banuett, F. (1994). *Ustilago maydis*, the delightful blight. *TIG 8*:174.
7. Wheeler, H. E. (1954). Genetic and evolution of heterothalism in *Glomerella. Phytopathology 44*:342.
8. Smith, J. E., D. R. Berry (1975). *Filamentous Fungi* (Vols. I–III). Edward Arnold, London.
9. Marzluf, G. A. (1993). Regulation of sulfur and nitrogen metabolism in filamentous fungi. *Ann. Rev. Microbiol. 47*:31–55.
10. Fincham, J. R. S., P. R. Day, A. Radford (1979). *Fungal Genetics.* Blackwell, Oxford.
11. Esser, K., R. Kuenen (1965). *Genetik der Pilze*. Springer, Berlin.
12. Sermonti, G. (1969). *Genetics of Antibiotic-Producing Micro-Organisms*. Wiley, London.
13. Burnett, J. H. (1975). *Mycogenetics.* Wiley, London.
14. King, R. C. (ed.) (1974). *Handbook of Genetics.* Vol. I. *Bacteria, Bacteriophages and Fungi.* Plenum Press, New York.
15. Kinghorn, J. R., G. Turner (1992). *Applied Molecular Genetics of Filamentous Fungi*. Blackie, London.
16. May, G., W. E. Timberlake (eds.) (1994). Aspergillus *Handbook* (*A Methods Manual*). Springer-Verlag, New York.
17. Martinelli, S. W., J. R. Kinghorn (1994). Aspergillus: 50 years on. *Progress in Industrial Microbiology*, Vol. 29, Elsevier, Amsterdam–New York.

18. Kück, U. (ed.). *The Mycota*. Vol. III: *Genetics and Biotechnology*. Springer, Berlin.
19. Stakman, E. C., F. J. Piemeisel, M. N. Levine (1918). Plasticity of biologic forms of *Puccinia graminis. J. Agr. Res. 15*:221.
20. Christensen, J. J. (1922). Studies on the parasitism of *Helminthosporium sativum. Univ. Minn. Agr. Expt. Sta. Bull. 11.*
21. Flor, H. H. (1971). Current status of the gene-for-gene concept. *Annu. Rev. Phytopathol. 9*:275.
22. Briggs, S. P., G. S. Johal (1994). Genetic patterns of plant host–parasite interactions. *TIG 10*:12.
23. Tudzynski, P., H. van den Broek, C. A. .M. J. J. van den Hondel (1994). Molecular genetics of phytopathognic fungi: new horizons. *TIM 2*:429.
24. VanEtten, H., D. Funell-Baerg, C. Wasmann, K. McCluskey (1994). Location of pathogenicity genes on dispensable chromosomes in *Nectria haematococca* MMVI. *Antonie van Leeuwenhoek 65*:263.
25. Holden, D. W., C. M. Rang, J. M. Smith (1994). Molecular genetics of *Aspergillus* pathogenicity. *Antonie van Leeuwenhoek 65*:251.

2

Mutation

Cees J. Bos
Wageningen Agricultural University, Wageningen, The Netherlands

David Stadler
University of Washington, Seattle, Washington

1. INTRODUCTION

In genetic and physiological research in the breeding of strains, we need mutants. In formal genetics, we can only recognize a gene after we have detected a mutant allele of it and observed the segregation in the progeny of the heterozygote. With molecular methods, it is now possible to identify a gene even when no mutant allele is known. Mutants are not only important for genetic analyses. Metabolic pathways can be blocked by mutations or opened by reverse mutations. Mutations are also important tools in molecular genetic studies and gene manipulation.

Some key concepts of this chapter are:

Spontaneous mutation occurs at random at low frequency and in principle at any site of a gene.

Mutations often result in a change in the metabolism of an organism, but some mutations have no effect (silent mutations), and others may have different consequences for the phenotype.

The effect of mutagenic treatment on survival is complex, and there may be no simple causal relationship between mutant yield and survival. Moreover, multiple mutants may arise in the same individual. Mutants should, in general, be isolated only after low-dose mutagenesis.

Some types of mutants can be selected directly but, especially for auxotrophic mutants, enrichment procedures are needed. It is usually desirable to include a propagation step between mutagenesis and isolation of mutants.

The isolation of mutants is a random process, and what is obtained depends mainly on the success of selection, whereas many potentially valuable mutants escape detection.

Directed mutagenesis can be achieved in certain cases by means of gene manipulation techniques such as gene disruption or even site-directed mutagenesis.

2. SPONTANEOUS MUTATION

A mutation is a permanent change in a heritable trait. For any given gene locus, it happens very rarely under normal conditions; this is spontaneous mutation. Radiations or chemicals that damage DNA can cause more frequent mutation. This is induced mutation, and the causal agents are called mutagens.

The study of spontaneous mutation is made difficult by the rarity of the event. For a given genetic locus, only one individual in 10^5 may carry a new mutation. For some genes the frequency is one in 10^6 or even lower. Another problem is that the mutant gene is likely to be recessive to the wild-type allele it replaces. A diploid individual that is heterozygous for this new mutation will still show the normal (wild-type) trait, so the mutation will not be detected. It may be some generations later that an individual occurs that is homozygous for the mutant gene. Only then will the mutation be detected. This makes it very difficult to do quantitative studies (such as measuring mutation rates) in natural populations.

However, in diploid organisms direct counts can be made for dominant mutations. Thus the data collected at large maternity hospitals in various parts of the world have given us mutation rates for the gene for achondroplasia (dwarfism). Mutation rates for sex-linked recessive traits in humans can be *estimated* by their frequencies in male progeny of families with no history of the trait. Even autosomal-recessive mutants can be detected and counted in experimental studies with fruit flies or other suitable species. A wild-type parent is crossed to a mutant (homozygous recessive). All progeny are heterozygous wild-type in the absence of mutation in the gamete from the wild-type parent. Mutations are signaled by mutant progeny.

The above examples deal with *germinal* mutations. In cultures of microorganisms (bacteria, yeasts, and fungi) we study *somatic* mutations. In a growing culture of cells, spontaneous mutations may occur at any time. A mutation that takes place early in the growth of the culture results in a large number (clone) of mutant cells in the final population. A mutation occurring late produces only one or a few mutant cells. Thus the relationship between mutant frequency and mutation rate is not simple. If a series of parallel cultures of the same strain are grown under identical conditions and then assayed for mutants, there may be large fluctuations in the mutant frequencies. It was this property of mutant distribution that Luria and Delbrück [1] used to show that spontaneous mutations occurred in bacterial cultures.

Lederberg and Lederberg [2] used a replica-plating technique to show that mutants were present previous to the application of selective conditions. They plated *Escherichia coli* bacteria that were sensitive to bacteriophage T1 on nutrient medium in a confluent layer. Replicas were made onto medium to which the phage was added. Resistant colonies arose on the replica plates, and it could be shown that there were already some resistant cells on the corresponding spots on the master plate. The conclusion was that they were selecting mutants that had arisen spontaneously.

Mutation rates at specific loci have been measured in a variety of organisms. These data generally show that forward mutations (from wild-type to mutant) are more frequent than reverse mutations. This makes sense because forward mutation may occur at any of the many sites in a gene that code for the required amino acid sequence. Reverse mutation, on the other hand, may only be achieved by a change at the precise site of the original mutation. There is little information in the published literature about frequencies of spontaneous forward mutation in fungi, although reverse mutation rates have been reported for a number of genes for nutritional requirements. In a study of spontaneous mutation at a specific locus in *Neurospora* [3], the forward mutation rate was 2×10^{-7}, while reverse mutations were at least an order of magnitude less frequent (D. Stadler, unpublished). In *Aspergillus nidulans* acriflavine-resistant mutants (often dominant) were much more frequent than *pabaA1* revertants [4]. When measuring mutation rates one must remember that a distinct mutant phenotype can often be caused by mutations in various genes.

In principle, mutational lesions may occur in only one of two DNA strands, and after replication this should result in a mutant and a nonmutant double helix. Consequently, upon mitosis only one of two daughter nuclei are expected to carry the mutation. In general, starting with uninuclear cells, only very few heterokaryotic colonies are found. This means either that mutations directly involve both strands, or that single-strand mutations by some mechanism (such as DNA repair) lead to a mutation in the complementary strand. A mutation arising in the G2 phase (post DNA replication) of a uninuclear cell will be carried in only one of the two daughter cells.

A. Spectrum of Spontaneous Mutations

What types of mutation can occur? In principle, all metabolic functions can be changed. Each metabolic step is controlled by an enzyme, and the structure and function of that enzyme depend on its amino acid sequence. In turn, the amino acid sequence of a gene product depends on the genetic information of the gene in the form of the base sequence, in which each triplet of bases encodes one amino acid. So, a gene has three times as many sites for point mutation as the enzyme has amino acids. However, the code is degenerate, so most changes in the third base of the triplet are silent (cause no change in the amino acid sequence). Even some of the mutations that cause an amino acid substitution may have little or no effect on the enzyme function. Mutations often result in a defect, but rarely do they give an enzyme with new characteristics, as the potential to attack new substrates. This is of practical importance for host–parasite relationships (see Chapter 14).

For any given gene locus, the molecular spectrum of spontaneous mutations runs from simple base substitutions to deletions and insertions as big as the whole gene. There are also smaller deletions and insertions, with the lower limit being those we call frame-shift mutations. A study of spontaneous mutations in a specific gene in the fungus *Neurospora* [5] showed that the great majority of the mutations were either base substitutions or deletions, with much smaller numbers of insertions (tandem duplications) and frame shifts.

The first effective mutagen to be discovered was x-irradiation. H. J. Muller and his colleagues did many experiments in the 1930s on the kinetics of x-ray–induced mutations in *Drosophila* sperm. They concluded that there was a simple, first-order relationship between dose of x-ray and amount of mutation. Thus 1000 roentgens resulted in sex-linked recessive lethal mutations in 3% of the treated X chromosomes. This was true whether the x-ray was administered in a short acute treatment of 1 minute or in a low-level chronic exposure requiring 100 minutes.

The combination of Muller's demonstration of one-hit kinetics for mutation with Luria and Delbrück's demonstration of the distribution of mutants in bacterial populations led to a compelling picture of spontaneous mutation as a sudden, permanent change in a gene, taking place without influence by the environment. The two steps in evolution, mutation and selection, were seen as completely separate and independent of each other.

B. Genetic Repair

This simple picture of mutation was complicated by the discovery of the systems for DNA repair. All organisms are endowed with enzymatic systems for the repair of genetic damages. The repair is usually perfect, restoring the gene to its previous (undamaged) form. It is the occasional mistake in repair that produces a mutant gene.

Thus mutation is a two-step process. The first step is a lesion in the double-stranded structure of the DNA—e.g., a break caused by x-ray or a pyrimidine dimer produced by UV. The second step is the processing of that lesion by the repair systems. Physiological studies of repair have been done using induced mutations, due to the low frequency of spontaneous mutations.

Experiments on radiation-induced mutations in mice [6,7] gave a different result from that found earlier by Muller in his *Drosophila* studies. The mouse experiments showed dose-rate effects. That is, the same total dose of radiation gave different amounts of mutation, depending on how it was administered. Chronic or interrupted exposure resulted in less mutation than the same dose given as one acute exposure. This seems logical in relation to what we know about genetic repair. The repair systems can keep up with the low level of damage produced by chronic treatment, but they are overwhelmed by the high level of damage present after acute exposure. The reason that Muller did not get this result was that *Drosophila* sperm (the cells he was monitoring for radiation-induced mutation) are specialized cells that do not have repair enzymes. The mouse experiments were designed so that the cells being monitored for mutation were at earlier stages of gametogenesis at the time of treatment, stages at which repair enzymes were present.

Spontaneous mutation also involves two steps. In this case the initial step is a sequence mistake made in replication. Here we find that nature has prepared us with a system called mismatch repair, which screens the newly synthesized DNA strand for mispaired bases and replaces them. If a mismatch is not corrected before the next round of replication, it is too late. There is now a homoduplex with the mutant sequence, and it cannot be detected by the mismatch repair system. Thus the processing step in spontaneous mutation is a race between repair and replication.

Mutation, whether spontaneous or induced, involves two steps: initiation and processing. The initiation step appears to have the properties first proposed by Luria and Delbrück: a random change, unaffected by the environment. But the processing step (repair and/or replication) is an aspect of metabolism, and its rate and efficiency can certainly be influenced by the environment.

C. "Adaptive" Mutation

In 1988, Cairns et al. [8] reported that nongrowing bacteria, starving on a deficient medium, were rescued by what appeared to be adaptive mutations. The medium contained all the nutrients needed by wild-type cells, but the bacteria were of a strain that carried a mutant gene which blocked their growth.

The "adaptive" mutation was the reversion of this mutant gene. But the authors did not find evidence of mutations at other loci—mutations for which the medium was not selective.

This work has attracted great interest, and some observers are skeptical of this kind of "adaptive" mutation, believing that it requires some supernatural "awareness" of the plating conditions by the mutating organisms. But perhaps there is a logical explanation. The experiment was designed so that the plated cells could grow for a period before being arrested: the medium contained a small amount of the nutrient needed for the growth of the mutant (unreverted) cells. It seems likely that when growth was arrested, the last round of replication may have produced errors at any site in any gene. A tiny fraction of the cells had replication errors of the right kind at the right site to restore the function of the mutant gene. Perhaps this altered DNA sequence is able to serve as template for the synthesis of mRNA which can be translated to produce a functional gene product. If so, these particular cells may gain the energy required to carry out the processing step (in this case another round of DNA replication) to produce a completed back-mutation. Cells with errors at other sites would not gain this energy, and their DNA would remain unreplicated long enough for repair to reverse the sequence change. Thus the adaptive mutations would remain while those at other loci were negated.

This is not to say that the process discovered by Cairns et al. is trivial. On the contrary, it appears to be an important way in which populations of organisms can reverse harmful mutant traits. It may be especially important in microorganisms that are subject to abrupt changes in the nutritional environment.

3. EFFECT OF MUTAGENIC TREATMENT ON SURVIVAL AND MUTANT YIELD

As the yield of mutants per surviving cell in general increases with increasing dose of mutagen, it has often been concluded that it is most efficient to apply high mutagen doses for the induction and isolation of mutants. In this section we discuss the relationship between mutant yield and survival.

A. Survival Curves

Examination of dose-response survival curves provides information on the process of cell-killing itself, and indirectly on the choice of the optimal mutagen dose for obtaining mutants without too heavy a load of genetic damage in the genetic background.

A suspension of unicellular uninucleate haploid conidiospores is expected to behave as a population of single target cells. When the survival is plotted against irradiation dose, such curves become linear on a log survival scale. They often have an initial shoulder at low doses and tend to become flat at high doses.

The linear relationship can be explained by assuming that the hits are randomly distributed both over the cells and over the targets within a cell. The number of hits per target follows a Poisson distribution where the fraction of targets that receives no hits equals e^{-kt} (kt is the effective mutagen dose). When the number of hits (h) needed to kill a target (n) is 1 and when the cells are unicellular and uninucleate (n = 1), the survival function becomes $S = e^{-kt}$ and then $\log S = -kt \log e$, a straight line through the origin (Fig. 1). When h and/or n are greater than 1, the curve will show a shoulder. The log S intercept is called the extrapolation number [9]. The meanings of the terms (multi-)hit and (multi-)target, which sometimes are wrongly used interchangably, are discussed elsewhere [10]. In terms of point mutations one hit may be sufficient to kill a uninucleate cell. The extrapolation number has been used to estimate the number of targets and as an indication for the presence of multinucleate cells. A change in hit number has, however, a much greater effect on the shoulder (extrapolation number) than has a corresponding change in target number [10].

Several studies on survival curves were done around 1950. Atwood and Norman [11] found a shoulder upon UV irradiation of *Neurospora crassa* macroconidia, but Giles [12] and Norman [13] found a linear log survival curve for *N. crassa* microconidia. The cause seems to be that the microconidia are uninucleate and the macroconidia are not. Later, however, Chang and Tuveson [14] found a shoulder also for a wild-type microconidial *N. crassa* strain, but not for a UV-repair–deficient strain. We tend to suppose that the strains used by Giles and Norman might have been UV-repair deficient. The experimental design has a great effect on what we observe. It is very difficult to detect the effect of small doses of mutagen, because it is a compound genetic effect and low doses also have a physiological effect. Moreover, in several cases very low doses of mutagen seem to stimulate germination of conidiospores [15].

A third factor that may contribute to a shoulder is the dark repair mechanism [16]. Haynes [17] suggested that the cell may have effective repair mechanisms that can cure UV lesions at lower doses but that become saturated or inhibited at higher doses. It has been shown experimentally that this initial repair causes an obvious shoulder in the survival curves of uninuclear unicellular haploid *Aspergillus nidulans* conidiospores [10]. Haynes and Eckardt [18] made a theoretical study of dose-response relations in view of mutant yield.

We will have a look at some aspects of survival curves in Figures 1 and 2 (this topic is discussed in [10]):

1. The factors that influence the shoulder are an initial repair capacity, the number of hits needed to kill one target, and the target number per cell.

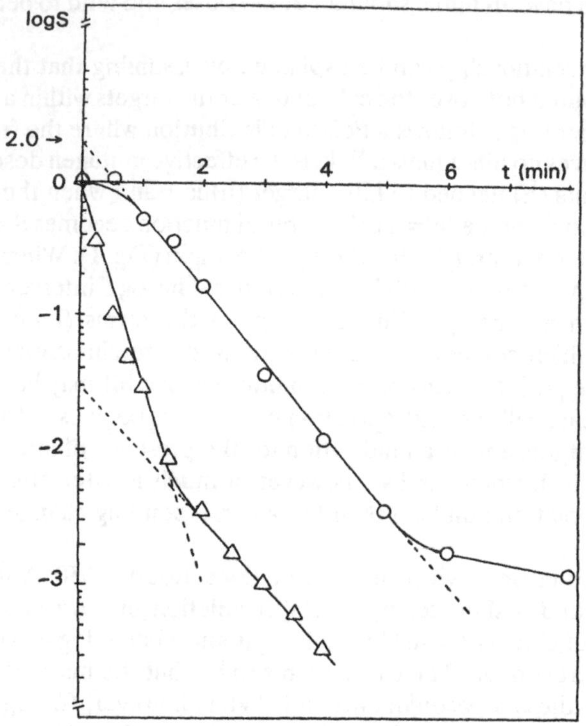

Figure 1 Survival curves of *A. nidulans*: a wild-type strain (○) and UV-repair–deficient (*usvD*51) strain (Δ). On the Y axis the logarithm of the survival: the wild-type strain shows an extrapolation number of 2.0. The survival curve of the mutant strain suggests that the shoulder of the wild-type strains is caused by an initial repair capacity. Furthermore the slope is steeper (more sensitive due to less repair), and the strain seems to consist of about 3% of revertants. UV dose rate of the Philips TUV tube (30 W) was about 120 J/m^2/min.

In different systems (organisms) the relative importance of these factors may be different. Usually repair has a greater effect than target number.

2. The slope of the curve depends on the effective dose (kt). The dose depends on the duration of the treatment (t) and on a compound factor k. Differences in k may result from differences in inherent sensitivity between cell populations (physiological state), from differences in penetration (dose rate received in the cell), or from shelter effects (e.g., density of the suspension).

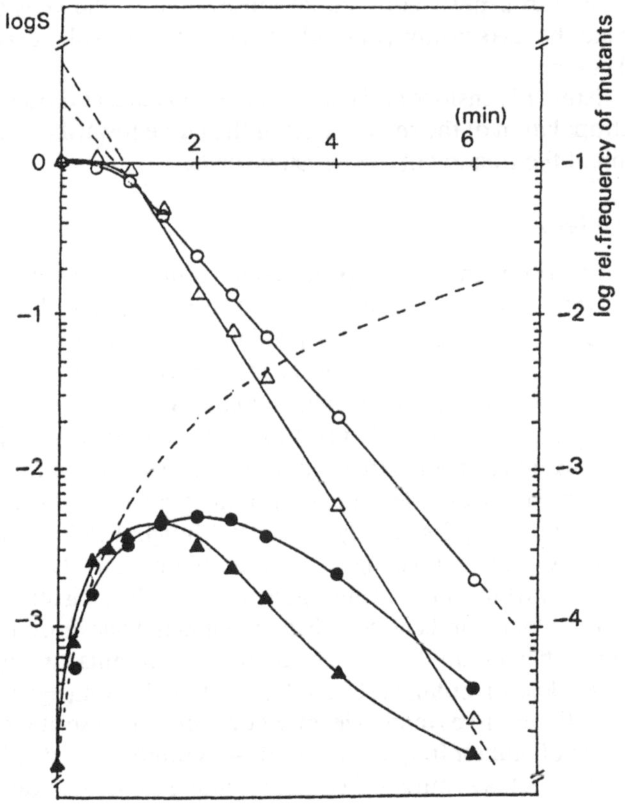

Figure 2 Mutation experiment with a *metG*1 strain of *A. nidulans*. Survival curves for conidiospores without (○) and with (Δ) 3 h preincubation in liquid supplemented medium. MetG+ revertants were isolated on minimal medium, and the mutant yield (solid marks) can be read on the right Y axis. The solid line with solid marks represents the frequency of the revertants among all conidiospores. The logarithmic relative frequency of the mutants among the survivors is presented as a dashed line. Preincubation (DNA replication) makes the strain more sensitive to UV.

3. A change in k as a result of initial repair capacity has a strong effect on the extrapolation number. The sensitivity to a mutagen increases upon preincubation when the cells come into the S phase [19], but the repair capacity also increases strongly.

4. The log-survival curve is often S-shaped with a tail at higher mutagen doses. A (very) small fraction of the cells seem to be more resistant to killing.

That can be due to experimental conditions. Some of the cells surviving a higher dose may be less sensitive, and subcultures of them will show a less steep log-survival curve.

5. A mixture of sensitive and less sensitive cells causes the survival curve to bend. Extrapolation of the second part of the curve toward the Y axis gives an estimation of the fraction of less sensitive cells.

B. Mutant Yield

In industrial practice high-yielding production strains have been obtained by treating cells with a mutagen until a certain "desired" kill is obtained (often >>99%). The survivors are tested for production characteristics, and those that are better in this respect are preserved. However, high mutation doses can result in chromosome rearrangements [20,21], and in general they disturb the genetic background by an enhanced load of unnoticed mutations, especially when recurrent mutagenic treatments are used, as is often the case with strains of industrial interest. The genealogy of the penicillin-producing strains is a good illustration [22]. Especially in asexual fungi, which cannot be outcrossed, genetic background damage can accumulate easily. Upon treatment of conidiospores with high doses of mutagen, many small and irregular growing colonies are often found, indicating a distorted genetic background.

It has often been found that the frequency of mutants among survivors is linear with the dose of mutagen [12,23,24]. Witkin [25] long ago pointed out that, in general, these curves tend to level off or even decrease at higher doses. The relationship of mutant frequency with dose is often presented by plotting the untransformed relative mutant frequency among survivors against log S or log t [16,26–29]. For short intervals this often results in a straight line, suggesting that it is profitable to screen for mutants in cell suspensions that have received a high dose of mutagen. The steep rise in these two-sided logarithmic survival graphs, however, is mainly a result of a scaling effect. Curves in which the log frequency of mutants among survivors is plotted against the dose (on linear scale) give a more realistic picture. In the examination of the efficiency of a mutagenic treatment it is better to focus on mutant yield (mutants as fraction of the total amount cells) than on the mutant fraction (relative frequency of mutants among survivors).

As mutations (forward mutants as well as revertants) may concern different loci, complex dose response relationships can be expected [4]. That is all the more reason why the value of graphs of the mutant fraction is questionable. There are several systems in which we can study the mutant yield–mutagen dose relationship [30]. One of them, the *metG1* system of *A. nidulans* [31], is described in the experimental section of this chapter. Such experiments show that the maximum yield of mutants is at relatively high survival (>50%). Obviously mutants are also subjected to kill (!) by additional mutagenic treatment (Fig. 2). So, our conclusion is that low mutagen doses are more efficient. In order to avoid unwanted background damage, which is especially important if we want to establish a collection of mutants for genetic analysis, one can use spontaneous mutants only [21], which is practicable when positive selection is possible, or one can use a low dose of mutagen and aim at 80% survival. Only in special situations, when we want to have a mutant with a certain phenotype, can higher doses of mutagen be used in order to have a greater chance of getting that type of mutant.

4. ISOLATION OF MUTANTS

Not all types of mutant are equally suitable for genetic research as illustrated by Pontecorvo and co-workers [32]. It is important that they have clear expression; leaky mutants are very difficult to use. As the selection might be a main stumbling block, mutants are often typed in view of this problem as:

1. Morphological mutants, which can usually be recognized on the basis of their appearance. A dissecting microscope might be used to detect the mutants, and often the cells must be plated in rather low density.
2. Resistance mutants, which can be selected on plates with the corresponding inhibitor. It is often easy to screen large numbers of cells on a plate because only resistant mutants will grow. In the case of dense platings the plate might be covered with a confluent film of poor growing colonies, and higher drug concentrations might be needed.
3. Auxotrophic mutants, which are very useful for physiological studies. Because so many mutations can result in a metabolic defect, these mutants are also very valuable for genetic studies. They are difficult to isolate, however, and special isolation techniques are required.
4. Substrate utilizers can be useful, as mutants that can use specific substrates can be selected directly. Nonutilizers have the same problems as auxotrophic mutants.
5. Revertants from auxotrophic toward prototrophic are easy to select. They are very suitable to study mutation induction. Not all revertants are real back-mutations. Many of them are suppressor mutants or mutations that open other ways to achieve that phenotype.

A. Selection of Auxotrophic Mutants

Because of the low frequency of mutants among the survivors, enrichment of special selection techniques are needed to select auxotrophic or other deficiency mutants. Several procedures have been described for the selection of auxotrophic mutants of fungi.

Marking of Presumed Mutants

When a treated cell suspension is plated on a complete medium some colonies are often seen to grow slowly. A primitive way to detect auxotroph mutants is to mark all early-growning colonies and see what grows upon additional incubation. Not all unmarked colonies are auxotrophs, but it works when low doses of mutagen are used. A great disadvantage is that many fast-growing mutants (most useful) escape. Sometimes mutants can be made recognizable by a vital stain.

Replica Plating

The suspension is plated on nonselective medium, and mutants are detected on replica plates made on a selective medium, on minimal medium, or on indicator plates. The mutants can be rescued from the master plate. The master plate must have separate colonies, which limits the number of colonies that can be screened. For various systems the number of colonies per plate may vary from 50 to 500. This can sometimes be improved by reducing the colony size by adding Triton X-100 or another substance to the medium. The replica plating technique with transfer on velvet was originally developed for *E. coli* by Lederberg and Lederberg [2], but it is also applicable in yeast and several other fungi. Variations in which the velvet is replaced by cellophane, filter paper, or another type of sheet can adjust the method for specific fungi. Mostly replicas are made upside-down, but sometimes the colonies are allowed to grow through the carrier sheet.

Filtration Enrichment

Filtration enrichment is a classical method based on selective removal of prototrophs of hyphal fungi and was originally developed for *Ophiostoma multiannulatum* by Fries [33] and subsequently adapted to *N. crassa* by Woodward et al. [34] and Catcheside [35]. The filtration enrichment method has proved to be very efficient for the isolation of *A. nidulans* mutants unable to use specific carbon sources [36–38]. The treated spores are incubated in a liquid medium, and the suspension is filtered through a plug of glass wool after each of several periods of incubation. It may be important to refresh the medium in order to avoid enrichment with growth substances.

Inhibition of Prototrophs

A certain specific type of mutants can sometimes be isolated by plating on a medium with an inhibitor in combination with the specific growth substance. The basis for this method is that the mutants that are blocked in that metabolic step have no problem with the presence of the inhibitor whereas the growth of prototrophic cells is inhibited on such media. The use of 5-fluoro-orotic acid to isolate mutants blocked in the pyrimidine biosynthesis is a good example of the approach [39]. In several cases the mutants show resistance to the inhibitor and, at the same time, deficiency for a metabolic pathway. Such mutants, for example, chlorate-resistant mutants in *A. nidulans* and *A. niger*, provide a very efficient two-way selection system [40,41].

Selective Killing of Prototrophs

Various methods based on the selective killing of prototrophs have been used for the enrichment of fungal mutants—e.g., by antibiotics [42–44]. The biotin starvation method [45] has often been used for *A. nidulans*, and comparable methods based on starvation of specific double mutants have been used in other fungi. Such methods tend to be species-specific or even strain-specific.

Cell-wall–degrading enzymes have been used for the isolation of auxotrophic mutants of yeasts [46–49]. The lytic enzyme preparation Novozyme 234 proved to be very efficient for the enrichment of auxotrophic mutants of *A. niger* [50]. The method can also be applied to other fungi after adaption to the specific situation. In general, only metabolically active conidiospores with very young hyphae (germ tubes) can bc eliminated by the Novozyme treatment. Upon chilling for a few hours, many hyphae become insensitive to lysis by Novozyme 234.

For many selection methods a population of synchronously germinating conidiospores or, at least, a homogeneous growing cell population, is a prerequisite, as the sensitivity to the treatment often exists for only a short period. As mentioned earlier, even a very low dose of mutagen can cause physiological damage that results in inhibition of germination or growth [50]. So, it is advisable to introduce a propagation step after mutagenic treatment. There are also reasons to include subculturing as a segregation step. When a gene has become nonfunctional by a mutation in a multinucleate cell, a diploid cell, or a haploid cell in G2 phase, the cell will not have the mutant phenotype because functional alleles are still present. In addition, for some mutations it takes some time before full phenotypic expression is realized. The consequences of a propagation step are that the relative frequency of the mutants can change and that some mutants might be present as clones. In order to ensure that the mutants that are isolated are different, only one mutant of a specific phenotype should be taken from a certain batch. In view of this the original suspension can be split up in several small fractions. Induced mutants from parallel suspensions can be expected to descend from different events.

5. CHARACTERIZATION OF MUTANTS

After isolation mutants are reisolated from monospore cultures and characterized for deficiency by supplementation with individual growth factors. Phenotypically similar mutants are combined in complementation tests to see if they concern the same gene of different genes. Then, in the case of new genes, the mutants are analyzed to locate the gene in the genome.

A. Growth Tests

Auxotrophic mutants are mostly isolated on the basis of nongrowth on minimal medium. The next step is to identify the end product deficiency or, in other words, the metabolic pathway that is blocked. An efficient way is to use combinations of growth factors. When it is known which metabolic pathway is blocked, it is sometimes possible to restore growth with intermediates, but intermediates are often unstable substances or they are also involved in other metabolic pathways. The most conclusive evidence is obtained from enzyme assays. In some cases an intermediate before the block will accumulate, but not necessarily the intermediate just before the blockage. Nevertheless, cross-feeding experiments with different mutants of the same pathway can give valuable information on the steps that are blocked.

B. Complementation Test

Even without information on the metabolic pathway mutants can be characterized genetically. Mutants of different genes can usually complement each other. Mutants of one complementation group are supposed to be allelic. It is possible, however, that mutants blocked in the same step of a biosynthetic pathway belong to different complementation groups. Two obvious reasons are intragenic complementation or the enzyme can be composed of two different peptides determined by different genes. It is also possible that an additional intermediate step is involved or that, for example, one of the genes controls a cofactor.

Complementation tests in heterokaryons are much more reliable than cross-feeding tests. In a heterokaryon both types of nuclei are present in the same cell. In certain situations when a heterokaryon test does not give conclusive evidence, complementation tests can be done with heterozygous diploids that have been isolated from a heterokaryon (see Chapter 4). Heterokaryosis can be forced by using strains with different auxotrophic markers. In view of this it is advisable to use various strains with different auxotrophic markers for the isolation of new mutants.

The complementation test has been called a cis–trans test. When two allelic mutations are in trans position (i.e., on different homologous chromosomes), no complementation is found, but when both allelic mutations are in cis position (i.e., one chromosome with two mutations and the homologue with no mutant allele), wild-type growth is observed. Complementation is discussed by Fincham [51] and Ratner and Rodin [52].

C. Nomenclature of Genes and Mutants

Phenotypically similar mutants get the same symbol and a unique reference number (*argl, artg2,* etc.). It is a good practice to use three-letter symbols as proposed by Demerec et al. [53]. Mutants belonging to the same complementation group (same gene) get the same capital character. Gene symbols are written in *italics.* The phenotypes are usually written beginning with a capital (Arg mutants, Lys mutants). The unique number (isolation number) of a mutant is used as allele number. When different groups are working on the same organism, good contacts and agreements are necessary to avoid confusion due to the time between experiments and publication. The same holds for research on related organisms.

6. DIRECTED MUTAGENESIS

The induction and isolation of mutants that have been discussed up to this point are the result of a random process. What we get depends mainly on the efficiency of the isolation procedure, and many valuable mutants escape our detection. If we know exactly what we want, there are now sometimes other possibilities with the use of cloned genes. The molecular genetic aspects are discussed in Chapters 5, 7, and 8.

A. Insertion Mutagenesis

It is possible to inactivate a gene by insertion of a piece of DNA, as in the case of a transposon (see Chapter 5). Gene disruption may be achieved by nonhomologous integration of transforming DNA, but one can also aim at mutants of a certain gene. When a related gene (which may be from another organism) has already been cloned, a copy of it can be made inactive in vitro. A plasmid with this inactive gene is used to transform a strain that has the wild-type gene. In most cases the plasmid also has another functional gene that is used for selection of transformants, or else cotransformation with two different plasmids is done. When a cell has taken up DNA, as the transformants for the selected gene have done, there is a chance that in some cases a plasmid has been inserted in the target gene because of the homology between the plasmid and the target gene. Transformants isolated on the basis of the selected gene are tested to see if they are deficient for the target gene function. These insertion mutants can be used for genetic and physiological studies, but their use has some limitations because they are not point mutations. Sometimes this is called gene replacement, which can be correct only if the mutant site is exchanged for the corresponding part of the target gene by homologous recombination. This approach has, for example, been used to isolate *trpC* mutants of *A. niger* with the aid of an inactive *A. nidulans trpC* gene [54].

B. Site-Directed Mutagenesis

When a gene has been cloned it is possible to introduce base substitutions surrounding a certain restriction site in vitro and to replace the corresponding gene by the constructed mutant allele. It is, however, also possible to create a mutation at a specific site if the base sequence of that part of the gene is known. The gene is cloned in a single-stranded phage such as M13, and short synthetic nucleotides are used as primers for the in vitro synthesis of the complementary strand of the vector. At the site chosen for change, an incorrect nucleotide is incorporated in the primer. Hybridization will proceed in the presence of a one-base-pair mismatch when done at low temperature. The in vitro synthesized vector is subsequently multiplied in *E. coli* and can be used to transform the fungal strain.

7. EXPERIMENTS

A. Mutagenesis in *A. nidulans* (Fig. 3)

Aim

To study the relation between survival and mutant yield.

Procedure

We use the *metG1* system in *A. nidulans* [31]. A suspension of conidiospores of a *metG1* strain of *A. nidulans* is irradiated with UV light and samples are taken at several short intervals. The samples are plated on CM for survival count and plated on MM to count *Met*$^+$ revertants. The number of the cells in the sample is counted to correct for inhomogeneous sampling. (Note: When it is not possible to do accurate cell counts it is better to plate the desired dilutions first and to irradiate the plates for the desired time. A similar dilution scheme can be followed as described below.)

Literature

Bos, C. J. (1987). *Curr. Genet. 12*:471–474.
Haynes, R. H., Eckardt, F. (1976). *Can. J. Genet. Cytol. 21*:277–302.
Lilly, L. J. (1965). *Mutat. Res. 2*:192–195.
Munson, R. J., Goodhead, D. T. (1977). *Mutat. Res. 42*:145–160.

Materials

The complete medium (CM) and minimal medium (MM) are essential according to Pontecorvo and co-workers [32]. For details see References 39, 56.

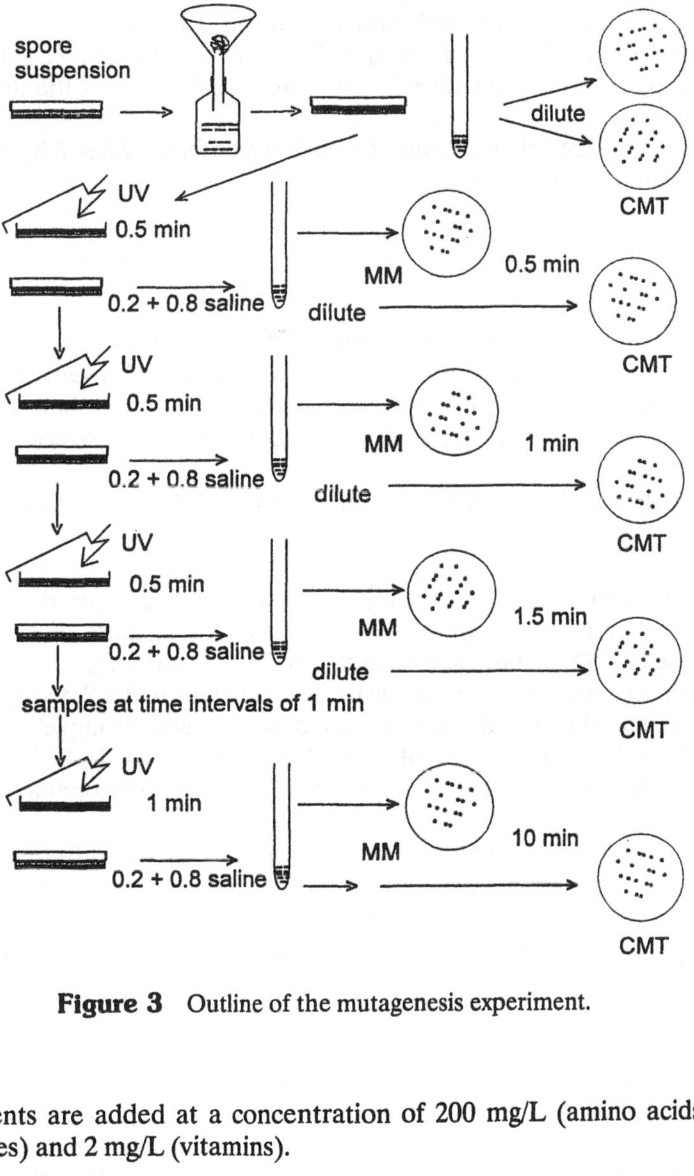

Figure 3 Outline of the mutagenesis experiment.

Supplements are added at a concentration of 200 mg/L (amino acids and nucleotides) and 2 mg/L (vitamins).

- A 3-day-old culture of the *metG1* strain WG282. The genotype of WG282 is *pabaA1, yA2, AcrA1, metG1, lacA1, choA1* (i.e., deficiency for *p*-aminobenzoic acid, choline, methionine; lactose nonutilizing; acriflavin resistant; yellow conidiospores).

- A sterile glass Petri dish (grease-free).
- 12 tubes with 0.8 mL saline on ice (for the samples).
- 24 plates CMT (= CM with 0.01% Triton X-100 to reduce colony size).
- 12 plates SM-methionine (= MM+cho+pab, to meet the deficiencies of WG282).
- Particle-free saline (filtered through a membrane filter 0.2 μm) for the Coulter counter.

Experiment

1. Spore suspension

   - Prepare 14 mL spore suspension in saline-Tween in a 30-mL screw-cap bottle. Collect the spores from the plate and avoid release of spores in the room. Shake vigorously to break the conidiospore chains.
   - Filter through a small cotton wool plug in a funnel to remove mycelium debris.
   - Dilute to $2–4 \times 10^7$ spores/mL and transfer 12 mL into the glass Petri dish.

2. Treatment

   - The irradiation is done by placing the covered Petri dish with the suspension under a prewarmed UV tube at a distance of 30 cm in the case of a 30-W tube (at a dose rate of 20 erg/mm^2/sec).
   - The treatment starts when the lid is removed from the Petri dish. The lid is replaced to end the first treatment period, and a sample of 0.2 mL is taken and added to the tube with 0.8 mL saline (= 2×10^{-1}). The samples are stored on ice. Then the next dose is given by opening the Petri dish.

     This works as follows: Mix the suspension in the Petri dish and take two samples before irradiation, and one after the following irradiation periods: 0.5, 1, 1.5, 2, 3, 4, 5, 6, 8, 10 min. The first treatment is 0.5 min, then an additional treatment of 0.5 min gives a total of 1.0 min, and so on.

3. Platings

   - Spread 0.2 mL from each sample on MM to count the number of revertants.
   - Dilute the samples by adding 0.2 mL to 1.8 mL saline and dilute further according to the scheme below. Transfer and spread 0.1 mL onto each of two CMT plates for survival count. The colonies can be counted after 2 days of incubation at 37°C.

	Concentrations plated in duplicate on CMT						
Dose UV (min)	10^{-1}	10^{-2}	10^{-3}	10^{-4}	S	Log S	Revertants
0 sample duplicate				x			
0.5				x			
1			x				
1.5			x				
2			x				
3		x					
4		x					
5		x					
6	x						
8	x						
10	x						

4. Total count

   Samples of 0.2 mL are added to 9.8 mL clean saline (without Tween) if a Coulter counter is used. The Coulter counter probably determines the number of spores in a volume of 0.5 mL. With these counts we can make a correction for the total amount of spores in the samples.

Results

1. Write the survival counts in the table and calculate the fraction that has survived the treatment (S). Write in the next column the logarithm of this value (log S). Use the next column for the number of revertants that are found.
2. Make a graph of the survival with the log S on the Y axis and the dose of UV (time) on the X axis.
3. Add at the right of the graph a second Y axis with a linear scale and put the numbers of revertants in the graph.
4. Draw a conclusion from the results.

B. Isolation of Auxotrophic Mutants of *A. niger* (Fig. 4)

Aim

Mutagenesis of *A. niger* and enrichment of auxotrophic mutants and preliminary classification of the mutants in growth test.

Performance

Mutants are induced at low mutagen dose, and the treated cells are subcultured. Then the relative frequency of auxotrophic mutants is enhanced by filtration enrichment. The surviving cells are sown on CM, and replicas are made on MM to identify auxotrophic mutants. These mutants are collected and tested on mixtures of growth factors.

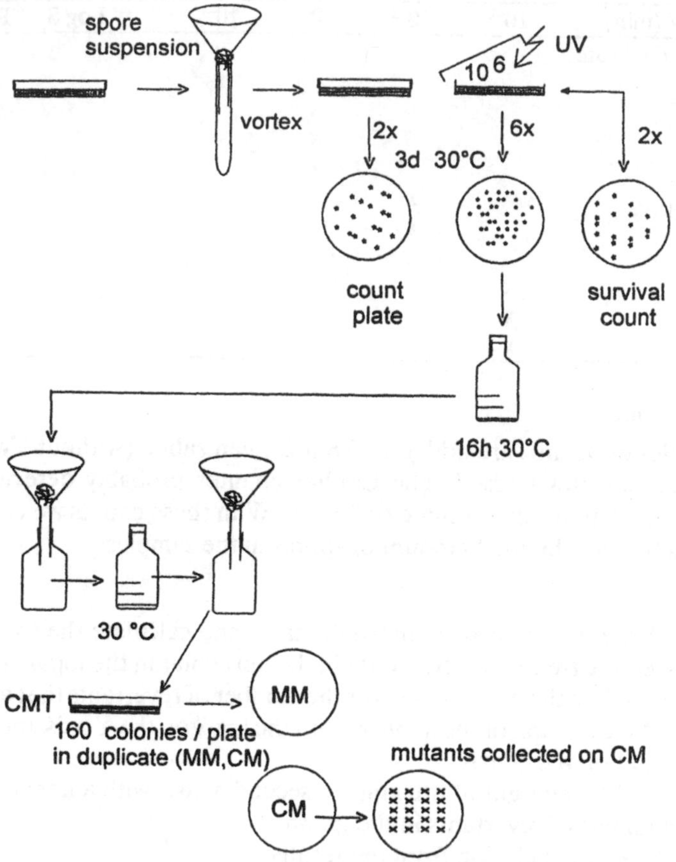

Figure 4 Isolation of auxotrophic mutants.

A dense suspension of conidiospores is needed for this experiment. This suspension can also be used for experiment C; in fact, the two experiments can be done together. In view of later use of the mutants it is desirable to use two different strains—e.g., N502 (*olvA1*, *bioB2*), or N522 (*fwnA1*, *metB11*).

Requirements

Day 1

- Suspension of conidiospores
- 8 dilution tubes with 1.8 mL saline

- 8 plates CMT (+0.05% Triton X-100 for survival counts)
- Two 100-mL flasks with CM(atu) culture medium (CM(atu) = CM + arginine, tryptophan, uridine)
- 6 mL molten CM(atu) medium (keep at 45°C)

Day 3
- 100-mL flask with 30 mL liquid SM (= MM + met + bio), pH 2.0)
- Funnel with glass wool plug

Day 4
- Sterile 100-mL flask, funnel with glass wool plug

Day 5
- 2 plates CM; saline
- Funnel with glass wool plug; sterile 100-mL flask; 15 mL centrifuge tubes

Day 6
- 8 plates CMT(atu)
- Sterile filter paper

Day 8
- 8 plates MM + met + bio (to replicate filter-grown colonies)
- Wood-block 9 cm, forceps, 70% alcohol

Day 9
- 1 plate CM(atu) to collect mutants of all groups (glass Petri dish)

Day 11
- Plates to test auxotrophic mutants for groups of growth factors

Experimental Procedure

Day 1

a. Mutagenesis
   - Take a sample of 0.2 mL of the spore suspension and dilute till 10^{-4}.
   - Plate 0.1 mL of 10^{-3} and 10^{-4}, both in duplicate, on CMT. Incubate at 30°C. From these plates you get the viable count of the suspension.
   - Take 2 mL spore suspension apart (for use in experiment C).
   - Bring 10 mL suspension in a glass Petri dish and put this in the cabinet with UV lamp. Irradiation 45 seconds at a dose of 20 erg/mm^2/sec by taking away the cover of the dish for the desired time.
   - Transfer the suspension in a sterile flask using a 10-mL pipet.
   - Take a sample of 0.2 mL and dilute till 10^{-4}.
   - Plate 0.1 mL of 10^{-3} and 10^{-4}, both in duplicate, on CMT. Incubate at 30°C. From these plates and the viable count you can calculate the percent survival.

b. Isolation of auxotrophic mutants
  - In duplicate: add 3 mL of the irradiated suspension (prewarmed at 30°C) to 3 mL molten CM(atu) (in 45°C water bath) and pour this mixture onto a CM(atu) medium layer in a 100-mL flask. Incubate 3 days at 30°C.

Day 2

- Count the colonies on the CMT plates and calculate the percentage survival.

Day 3

- Make a spore suspension of the cultures in the 100-mL flasks (combined). Count the spores and add 10^8 spores to the 30 mL liquid SM in a 100-mL flask.
- Incubate 24 h in a reciprocal shaker at 30°C (200 rpm).

Day 4

- Filter the suspension through a funnel with glass wool plug and in a sterile 100-mL flask and incubate this for another 24 h.

Day 5

- Filter again through glass wool plug in a sterile flask.
- Transfer in each of two centrifuge tubes 10 mL of the suspension and spin the spores down for 5 min at 3000 rpm.
- Resuspend both pellets each in 1 mL saline and pool them in one tube.
- Prepare a dilution 10^{-1} and plate the undiluted and the 10^{-1} suspension on CM. Incubate 1 day at 30°C. Save the suspensions in the refrigerator.

Day 6

- Count the colonies on the plate of day 5. Calculate how much suspension you have to plate to get ± 90 colonies on a plate.
- Put sterile filter paper on top of 8 plates CM(atu) + Triton X-100.
- Place on top of the filter paper a quantity of the suspension that will give rise to ± 90 colonies (this should be at least 0.2 mL because of the absorbtion into the filter paper). Incubate 2 days at 30°C.

Day 8

- Make replicates of the filter paper grown colonies on MM + met + bio to find out whether you have auxotrophic mutants among these colonies. This should be done in the chemical hood to prevent scattering of spores. Transfer the filter paper on top of a wooden block using a sterile forceps with the colonies upwards. Put the MM plate on top of the filter paper, press slightly, remove the MM plate, and put back the filter paper in the CM(atu) plate. Mark the corresponding plates with a number. Incubate the MM plates 1 day at 30°C and store the CM(atu) plate in the refrigerator.

Day 9

- Score the MM plates for nongrowing colonies and retrieve these on the corresponding CM(atu) plate. Pick up with a needle a spore sample of these colonies and inoculate them (in square position) onto a CM(atu) plate (one or two plates to collect all mutants of all groups). Incubate 2 days at 30°C.

Day 11

- Replicate the master plate onto test plates to determine auxotrophic requirement (amino acids, vitamins, and nucleosides). Incubate test plates 2 days at 30°C.

Day 12

- Score test plates provisionally and incubate 1 day more.

Day 13

- Score test plates definitely. Discuss the results.

C. Isolation of Chlorate-Resistant Mutants of *A. niger* (Fig. 5)

Aim

To show the positive selection of chlorate-resistant mutants, which in fact is a way to isolate mutants blocked in the nitrogen metabolism. Because of the possibility of positive selection we can compare the frequency of spontaneous mutants with that after a low dose of UV. Growth test on different N-sources can be done to classify the mutants. For this purpose some reference strains are available. In addition, a complementation test can be done to see which *crn* mutants are allelic.

Growth characteristics of some N-source mutants [40]:

Phenotype	NO_3^-	NO_2^-	Hypoxanthin	Urea
Nia	—	+	+	+
Nir	—	—	+	+
Cnx	—	+	—	+
Crn	+	+	+	+

Requirements

Day 1

- Spore suspensions of N522 (even groups) or N502 (odd groups) as prepared in exp. B
- 8 dilution tubes with 1.8 mL saline; 8 plates CMT (same as in exp. B)

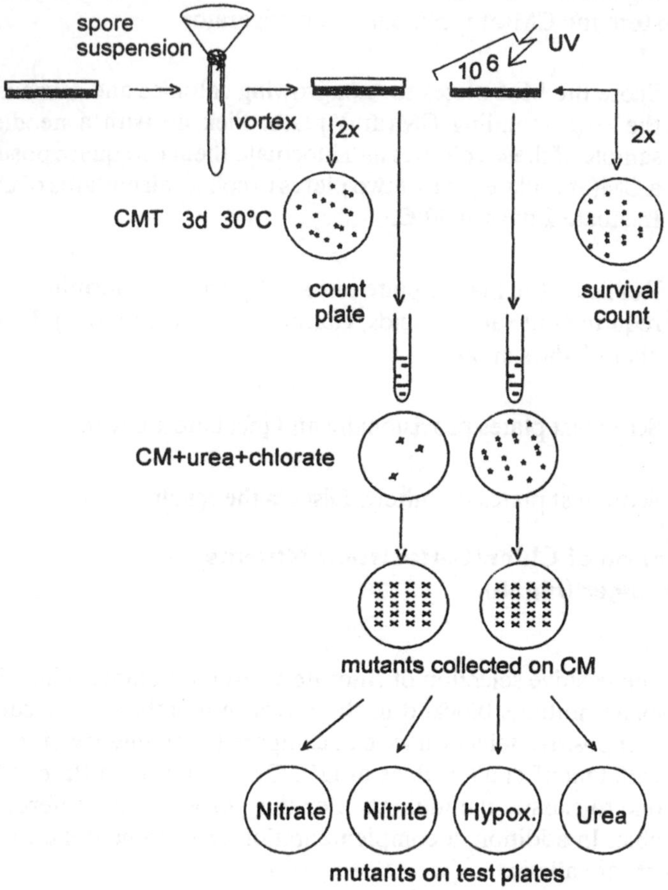

Figure 5 Isolation of chlorate-resistant mutants.

- 4 plates CM+ClO_3^-
- 12 mL selection medium (= top agar CM + urea + chlorate), also 45°C

Day 3

- 1 plate CM+ClO_3^- (isolation plate, glass Petri dish)

Day 5

- Test plates: SM + 10 mM NO_3^-; SM + 1 mM NO_2^-; SM + 1 mM hypoxanthine and SM + 5 mM urea

- Plate with water agar to cool needles of the replicator
- Parental strains as reference

Day 7

- 2 plates CM+ClO_3^-, 2 plates SM(NO_3^-), 2 plates SM(urea)

Day 8

- 10 tubes with 2.5 mL liquid CM + urea, to make mycelial mats of combinations of mutants

Day 10

- 3 plates MM + NO_3^- and 3 plates MM + urea + ClO_3^-

Experimental Procedure

See for mutagenesis the procedure described in experiment B. Samples of the nonirradiated and the irradiated suspension are plated both in duplicate on chlorate medium (CM+urea+ClO_3^-).

Day 1

- Add 0.5 mL suspension of both the nonirradiated spores and the irradiated spores to 3 mL molten CM+ClO_3^- medium in a short tube (45°C), mix the content by rolling very briefly between the hands, and pour the contents on a plate with similar solid medium. Incubate 3 days at 30°C.

Day 2

- Count the colonies on the CMT plates and calculate the survival.

Day 3

- Count the ClO_3^- resistant colonies and calculate the frequency of resistant mutants (resp. spontaneous mutants, UV-induced mutants).
- Isolate (and purify) 20 chlorate-resistant colonies by inoculating from one spore head onto a new plate CM+ClO_3^- (in square according to the design of the replicator with at place "0" the parental strain).

Day 5

- Make replicates using the needle replicator onto the SM test plates with NO_3^-, NO_2^-, hypoxanthin, and urea as N-source. Incubate 2 days at 30°C.

Day 6

- Record provisionally the growth on the different N-sources and incubate for 1 day more.

Day 7

- Record the growth definitely and classify the mutants.
- Purify two *cnx* mutants; take with a wet needle a few spores of one spore head and transfer to 0.5 mL saline in a tube; vortex well; plate

0.1 mL onto CM+urea+ClO_3^-; SM(NO_3^-), and SM(urea). Incubate 2 days at 30°C.

Day 8

- Do a complementation test of your selected mutants with some mutants of other groups. Make of both mutants a spore suspension each in 1–2 mL saline, by taking spores of single cell colonies of the CM+ClO_3^- plates.
- Prepare for each of your mutants 5 tubes with liquid CM + urea and 0.1 mL spore suspension of your mutant. Add to each of the 5 tubes 0.1 mL spore suspension of a different *cnx* mutant obtained from a group working with the other parental strain. Record what you do! Incubate 1 day at 30°C and store the master plate in the refrigerator.

Day 10

- Wash mycelial mats grown in the tubes and transfer 4 small pieces from each combination to plates MM(NO_3^-) as well as to MM(urea)+ ClO_3^- in 4 rows of 4 pieces on one plate. Incubate 2 days at 30°C.

Day 12

- Record the complementation test provisionally and incubate 1 day more and draw conclusions.

D. Protocols for Mutagenesis

General protocols are given in this section for some procedures that are used frequently. Methods for mutation induction can also be found in Kafer [55].

Mutagenic Treatment

In fungi mutants can be induced by ultraviolet light irradiation of vegetative spores. We shall concentrate on this mutagen because it is effective, cheap, and safe to handle. It is necessary to protect the eyes for UV light.

During mutagenic treatment and for 1 or 2 days afterwards the cells should not be exposed to normal light if the fungus has an active photorepair mechanism. The experiment can be done under low-intensity red light or on ice under dimmed light. The mutagenic treatment can be done with a suspension of spores (e.g., 10 mL in a Petri dish), which should be agitated gently during treatment. The UV source can be a 30-W UV tube at a dose rate of 100 $J/m^2/min$. The intensity can be measured with a short-wave UV meter. For light-colored spores about 50 J/m^2 and for dark spores about 150 J/m^2 might be a suitable dose. It is worthwhile to make a survival curve for the strain being used. It is difficult to take representative samples because many spores rest on the bottom despite attempts to agitate the suspension gently. Reliable results are obtained when the total spores in the samples are counted afterwards with, for example, a Coulter counter.

The suspension of treated spores can be subdivided into several aliquots which are subcultured separately on CM. Spore suspensions from these parallel cultures are each subjected to an enrichment procedure or directly to selection if positive selection is possible. Colonies that appear after enrichment can be tested by transfer onto supplemented MM and CM, respectively, using wooden toothpicks (e.g., 160 colonies per plate). Colonies failing to grow on supplemented MM are tested for auxotrophy on test plates with combinations of amino acids or nucleotides or a vitamin mixture.

Filtration Enrichment

About 10^7 conidiospores are added to liquid MM (supplemented for the deficiencies of the original strain) in a bottle or conical flask (e.g., 50 mL in a 200-mL flask) and incubated on a reciprocal shaker for in total 24 hours (sometimes 36 or 48). After intervals of about 8–12 h, the germinated conidia are removed by filtering through a glass funnel with a small plug of glass wool and the medium is refreshed. Finally, the conidia are collected and plated for viability count, and later the suspension is plated on CM for rescue.

Enrichment by Novozyme

The basis of this enrichment procedure is that germinating conidiospores will be killed by lytic enzymes, whereas mutant spores that do not germinate will survive. Novozyme 234 from Novo Industries Denmark is very effective, but batches may differ strongly in activity. The technique was originally developed for the isolation of auxotrophic diploid recombinants [56]. Essential aspects are 1.) a propagation step between mutation induction and enrichment to avoid retarded germination due to the mutagenic treatment; 2.) the germination time before treatment with Novozyme was 12 h (10 h for diploids); 3.) lytic treatment was given immediately after the germination period without chilling.

REFERENCES

1. Luria, S. E., M. Delbrück (1943). Mutations of bacteria from virus sensitivity to virus resistance. *Genetics 28*:491.
2. Lederberg, J., E. M. Lederberg (1952). Replica plating and indirect selection of bacterial mutants. *J. Bacteriol. 63*:399.
3. Stadler, D., H. Macleod, D. Dillon (1991). Spontaneous mutation at the *mtr* locus of *Neurospora*: the spectrum of mutant types. *Genetics 129*:39.
4. Bos, C. J., et al. (1987). Induction and isolation of mutants in fungi at low mutagen doses. *Curr. Genet. 12*:471.

5. Dillon, D., and D. Stadler (1994). Spontaneous mutation at the *mtr* locus in *Neurospora*: the molecular spectrum in wild-type and mutator strain. *Genetics 138*:61.
6. Russell, W., L. Russell, and E. Kelly (1958). Radiation dose rate and mutation frequency. *Science 128*:1546.
7. Russell, W., E. Kelly (1982). Mutation frequencies in male mice and the estimation of genetic hazards of radiation in men. *Proc. Natl. Acad. Sci. 79*:542.
8. Cairns, J., J. Overbaugh, S. Miller (1988). The origin of mutants. *Nature 335*:142.
9. Alper, T., N. E. Giles, M. M. Elkind (1960). The sigmoid survival curve in radiobiology. *Nature 1986*:1062–1063.
10. Bos, C. J., P. Stam, J. H. van der Veen (1988). Interpretation of UV-survival curves of *Aspergillus* conidiospores. *Mutat. Res. 197*:67–75.
11. Atwood, K. C., A. Norman (1949). On the interpretation of multi-hit survival curves. *Proc. Natl. Acad. Sci. Wash. 35*:696–709.
12. Giles, N. H. (1951). Studies on the mechanism of reversion in biochemical mutants of *Neurospora crassa. Cold Spring Harbor Symp. Quant. Biol. 15*:283–313.
13. Norman, A. (1954). The nuclear role in the ultraviolet inactivation of *Neurospora* conidia. *J. Cell. Comp. Physiol. 44*:1–10.
14. Chang, I., R. W. Tuveson (1967). Ultraviolet-sensitive mutants in *Neurospora crassa. Genetics 56*:801–810.
15. Griffin, D. H. (1984). *Fungal Physiology*. John Wiley & Sons, New York.
16. Bridges, B. A., R. J. Munson (1968). Genetic radiation damage and its repair in *Escherichia coli.* In: Ebert, M., A. Howard (eds.). *Current Topics in Radiation Research*. North Holland Press, Amsterdam, *4*:95–188.
17. Haynes, R. H. (1966). The interpretation of microbial inactivation and recovery phenomena. *Radiat. Res. Suppl. 6*:1–29.
18. Haynes, R. H., and F. Eckardt (1979). Analysis of dose-response patterns in mutation research. *Can. J. Genet. Cytol. 21*:277–302.
19. Davis, P. J., R. S. Tippins, J. M. Parry (1978). Cell-cycle variations in the induction of lethality and mitotic recombination after treatment with UV and nitrous acid in yeast, *Saccharomyces cerevisiae. Mutat. Res. 51*:327–346.
20. Kafer, E. (1977). Meiotic and mitotic recombination in *Aspergillus nidulans* and its chromosomal aberrations. *Adv. Genet. 19*:33–131.
21. Upshall, A., B. Giddings, S. C. Teow, I. D. Mortimore (1979). Novel methods of genetic analysis in fungi. In: Sebek, O. K., A. Laskin (eds.). *Proceedings of the Third International Congress on the Genetics of Industrial Micro-organisms*, pp. 197–204.
22. Ball, C. (1983). Genetics of beta-lactam producing fungi. In: Demain, A. L., N. A. Solomon (eds.). *Antibiotics Containing the Beta-Lactam Structure I*. Springer-Verlag, Berlin,
23. Kolmark, H. G., B. J. Kilbey (1968). Kinetic studies of mutation induction by epoxides in *Neurospora crassa. Mol. Gen. Genet. 101*:89–98.
24. Kilbey, B. J., T. Brycky, A. Nasim (1978). Initiation of UV mutagenesis in *Saccharomyces cerevisiae. Nature 274*:889–891.

25. Witkin, E. M. (1956). Time, temperature, and protein synthesis: a study of ultraviolet-induced mutation in bacteria. *Cold Spring Harbor Symp. Quant. Biol.* *21*:123–140.
26. Munson, R. J., D. T. Goodhead (1977). The relation between induced mutation frequency and cell survival—a theoretical approach and an examination of target hypothesis for eukaryotes. *Mutat. Res.* *42*:145–160.
27. Lawrence, C. W., R. B. Christensen (1978). Ultraviolet-induced reversion of *cyc1* alleles in radiation sensitive strains of yeast. I. *rev1* Mutant strains. *J. Mol. Biol.* *122*:1–21.
28. Burnett, J. H. (1975). *Mycogenetics.* John Wiley, London, p. 34.
29. Fincham, J. R. S., P. R. Day, A. Radford (1979). *Fungal Genetics*. Blackwell, Oxford, p. 269.
30. Alderson, T., M. J. Hartley (1969). Specificity for spontaneous and induced forward mutation at several loci in *Aspergillus nidulans. Mutat. Res.* *8*:255–264.
31. Lilly, L. J. (1965). An investigation of the suitability of the suppressors of met1 in *Aspergillus nidulans* for the study of induced and spontaneous mutations. *Mutat. Res.* *2*:192–195.
32. Pontecorvo, G., J. A. Roper, L. M. Hemmons, K. D. MacDonald, A. W. J. Bufton (1953). The genetics of *Aspergillus nidulans. Adv. Genet.* *5*:141.
33. Fries, N. (1947). Experiments with different methods of isolating physiological mutations of filamentous fungi. *Nature* *159*:199.
34. Woodward, V. W., J. R. de Zeeuw, A. M. Srb (1954). The separation and isolation of particular biochemical mutants of *Neurospora* by differential germination of conidia, followed by filtration and selective plating. *Proc. Natl. Acad. Sci.* *40*:192–200.
35. Catcheside, D. G. (1954). Isolation of nutritional mutants of *Neurospora crassa* by filtration enrichment. *J. Gen. Microbiol.* *11*:34.
36. Armitt, S., W. McCullough, C. F. Roberts (1963). Analysis of acetate non-utilizing (*acu*) mutants in *Aspergillus nidulans. J. Gen. Microbiol.* *93*:263.
37. Bos, C. J., M. Slakhorst, J. Visser, C. F. Roberts (1981). A third unlinked gene controlling the pyruvate dehydrogenase complex in *Aspergillus nidulans. J. Bacteriol.* *148*:594.
38. Uitzetter, J. H. A. A., C. J. Bos, J. Visser (1986). Isolation and characterization of *Aspergillus nidulans* mutants in carbon metabolism after D-galaturonate enrichment.
39. Bos, C. J., A. J. M. Debets, G. Kobus, S. M. Slakhorst, K. Swart (1989). Adenine and pyrimidine genes of *Aspergillus niger* and evidence for a seventh linkage group. *Curr. Genet.* *16*:307.
40. Cove, D. J. (1976). Chlorate toxicity in *Aspergillus nidulans*: the selection and characterisation of chlorate resistant mutants. *Heredity* *36*:191.
41. Debets, A. J. M., K. Swart, C. J. Box (1990). Genetic analysis of *Aspergillus niger*: isolation of chlorate resistance mutants, their use in mitotic mapping and evidence for an eighth linkage group. *Mol. Gen. Genet.* *221*:453.

42. Snow, R. (1966). Enrichment method for auxotrophic mutants using the antibiotic nystatin. *Nature 211*:206–207.
43. Ditchburn, P., K. D. MacDonald (1971). Two differential effects of nystatin on growth of auxotrophic and prototrophic strains of *Aspergillus nidulans. J. Gen. Microbiol. 67*:299–306.
44. Bal, J., E. Balbin, N. J. Pieniazek (1974). Method for isolating auxotrophic mutants in Aspergillus nidulans using N-glycolsyl polifungin. *J. Gen. Microbiol. 84*:111–116.
45. Pontecorvo, G., J. A. Roper, E. Forbes (1953). Genetic recombination without sexual reproduction in *Aspergillus nidulans. J. Genet. 52*:198–210.
46. Piedra, D., L. Herrera (1976). Selection of auxotrophic mutants in *Saccharomyces cerevisiae* by snail enzyme digestion method. *Fiola Microbiol. 21*:337–341.
47. Ferenczy, L., M. Sipiczky, M. Szegedi (1975). Enrichment of fungal mutants by selective cell wall lysis. *Nature (Lond.) 253*:46.
48. Sipiczki, M., L. Ferenczy (1978). Enzymic methods for the enrichment of fungal mutants. I. Enrichment of *Schizosaccharomyces pombe* mutants. *Mutat. Res. 50*: 163–173.
49. Delgado, J. M., L. S. Herrera, C. Perez, R. Lopez (1979). Optimization on the selection of auxotrophic mutants in *Candida utilis* by snail enzyme treatment. *Can. J. Microbiol. 25*:486.
50. Bos, C. J., A. J. M. Debets, H. Nachtegaal, S. M. Slakhorst, K. Swart (1992). The Isolation of auxotrophic mutants of *Aspergillus niger* by filtration enrichment and lytic enzymes. *Curr. Genet. 21*:117.
51. Fincham, J. R. S. (1966). *Genetic Complementation*. Benjamin, New York, p. 143.
52. Ratner, V. A., S. N. Rodin (1976). Theoretical aspects of genetic complementation. *Prog. Theoret. Biol. 4*:1.
53. Demerec, M., E. A. Adelberg, A. J. Clark, P. E. Hartman (1966). A proposal for a uniform nomenclature in bacterial genetics. *Genetics 54*:61.
54. Goosen, T., F. van Engelenburg, F. Debets, K. Swart, K. Bos, H. van den Broek (1989). Tryptophan auxotropic mutants in *Aspergillus niger*: inactivation of the trpC gene by cotransformation mutagenesis. *Mol. Gen. Genet. 219*:282.
55. Kafer, E. (1994). Genetics. In: May, G., W. Timberlake (eds.). Aspergillus *Handbook*. Springer-Verlag, Berlin.
56. Debets, A. J. M., K. Swart, C. J. Bos (1989). Mitotic mapping in linkage group V of Aspergillus niger based on selection of auxotrophic recombinants by Novozyme enrichment. *Can. J. Microbiol. 35:982.*

3

Meiotic Recombination*

Karl Esser
Ruhr-Universität, Bochum, Germany

1. INTRODUCTION

Genetic material possesses the capacity for self-duplication. The total amount of genetic information as well as its order within the genome remains the same from nuclear generation to nuclear generation. With each mitotic division copies are transmitted to the daughter cells. The identity of the genetic information is thus assured for each cell of a multicellular organism. Nevertheless, such an inflexible transmission of hereditary material would prevent

*Some decades ago we described fungal recombination in a comprehensive manner in the textbook *Genetik der Pilze* [1], where the pertinent, still valid literature is compiled. Due to the increasing importance of molecular genetics, this area of classical genetics now holds less interest for geneticists. As a result, there are very few new data, and they are not leading to a new breakthrough. Thus the file "Research in Classical Genetics" has to be considered closed. Naturally, classical genetics is needed as a basis for molecular genetics in both fundamental research and biotechnology. Therefore, adopted from the English translation in this chapter a short presentation on the principles and the application of classic recombination genetics is given. For more details and for the references one is referred to *Genetics of Fungi* [2]. If not otherwise indicated, figures and tables originate from that book.

evolution. This problem is overcome by two fundamental properties of the genetic material, namely recombination and mutation.

Recombination is the reassortment of the genome during nuclear divisions. It occurs regularly in meiocytes during reduction division (meiotic recombination). New combinations also occur in vegetative cells in rare cases; this is called somatic recombination. It takes place most commonly as a recombination in diploid somatic cells during mitosis (mitotic recombination; see Chapter 4).

In both meiotic and mitotic recombination entire chromosomes as well as parts of chromosomes may recombine. In the former case we speak of *interchromosomal*; in the latter, of *intrachromosomal* recombination. The various modes of these events may be seen from the scheme of Figure 1. Following are some key facts to keep in mind.

- Meiotic recombination consists of the exchange between nonsister chromatids and of the reassortment of chromosomes during the prophase and anaphase, respectively, of the first meiotic division.
- Meiotic recombination may concern entire chromosomes as well as parts of chromosomes—interchromosomal or intrachromosomal recombination, respectively.
- Since in interchromosomal recombination the assortment of different chromosomes through the poles of the spindle proceeds according to chance, one always obtains recombination values of 0.5. This can be easily shown in using markers that are located on different chromosomes.
- Intrachromosomal recombination concerns only markers that are located on the same chromosome.
- As a consequence of the two meiotic divisions alleles of homologous genes segregate during the first or the second meiotic division (pre- or post-reduction). The homologous centromeres are always prereduced.

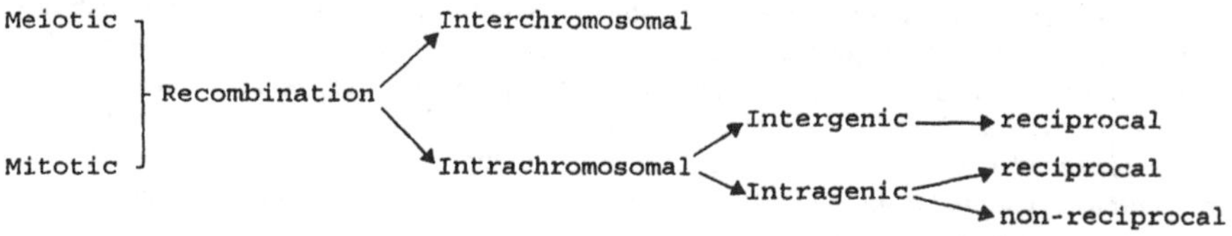

Figure 1 Schematic survey of the different modes of recombination; details in the text.

- The frequency of intrachromosomal recombination between two loci depends on the distance between them and can be used to establish genetic maps.
- The two meiotic divisions result in a tetrad. Linear ordered tetrads provide information directly on the frequency of pre- and postreduction.
- In the case of unordered tetrads the frequency of recombinants provides indirect information on the frequency of recombination.
- Random distribution of crossover may be hampered by interference. *Chromosome interference* refers to the distribution of crossovers in a tetrad while *chromatide interference* is concerned with the participation of individual strains in multiple crossing over.
- In the absence of interference the maximum frequency of recombinants approaches 0.5 as in the case of loci of different chromosomes.
- In presence of interference, specific mapping functions are needed to obtain so-called corrected chromosome maps.

2. INTERCHROMOSOMAL RECOMBINATION

The prerequisites for the mechanism of interchromosomal recombination are 1.) the union of two different haploid genomes through karyogamy, and 2.) the reduction of the diploid to the haploid chromosome complement in meiosis or in a series of irregular mitotic divisions.

A. Mechanism of Chromosomal Distribution

The segregation of the chromosomes during meiosis is such that each of the two daughter nuclei receives one chromosome of each bivalent (Fig. 2). In this way reciprocal genomes arise because two homologous chromosomes are prevented from going to the same pole. Thus the assortment of different chromosomes to the poles of the spindle proceeds according to chance. Therefore, parental combinations of chromosomes occur only rarely. A combination of maternal and paternal chromosomes is the most common result. Because each chromosome of a bivalent ordinarily has the same chance of going to one or the other of the poles, a recombination for every two nonhomologous chromosomes occurs in half the cases (Fig. 2). If each chromosome carries a different marker gene, the recombination frequency is 50%.

B. Preduction and Postreduction

From the arrangement of genetically marked spores in the asci of certain fungi, it was concluded that one and the same allelic pair may be reduced at certain times in the first meiotic division and at others in meiosis II. In the first case

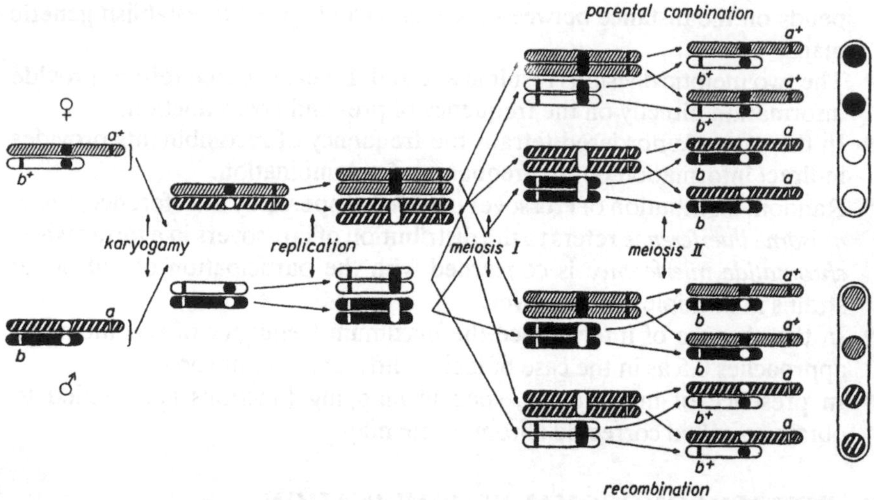

Figure 2 Distribution of two pairs of chromosomes at meiosis. Homologous chromosomes are indicated by units of equal size. Chromosomes of the male parent are black, or white with heavy lines; those of the female parent white, or white with thin lines. Centromeres are represented by black or white circles or elipsoids, while the marker genes are shown by narrow, vertical black or white bands. The segregation of the chromosomes and markers in the four-spored ascus is shown at the right: black = parental combination of the female parent: a^+b^+; white = parental combination of the male parent: ab; lined = recombinant: a^+b; stippled = recombinant: ab^+.

the gene is *pre*reduced, and in the latter *post*reduced. The segregation pattern of Figure 3 is based on the following general rules:

1. Homologous centromeres always segregate during the first meiotic division; i.e., they are always prereduced. The adjacent chromosomal segments up to the first points of crossover are prereduced with the centromeres.
2. Postreduction of an allelic pair results from a crossover between the centromere and a particular genetic locus.
3. Crossing over occurs in the four-strand stage; only two of the four strands take part in a single crossover.

C. Mitotic Recombination

In addition to the reassortment of entire chromosomes, which regularly occurs at meiosis, an interchromosomal recombination in somatic cells may take place. This involves an occasional irregular chromosome distribution in diploid

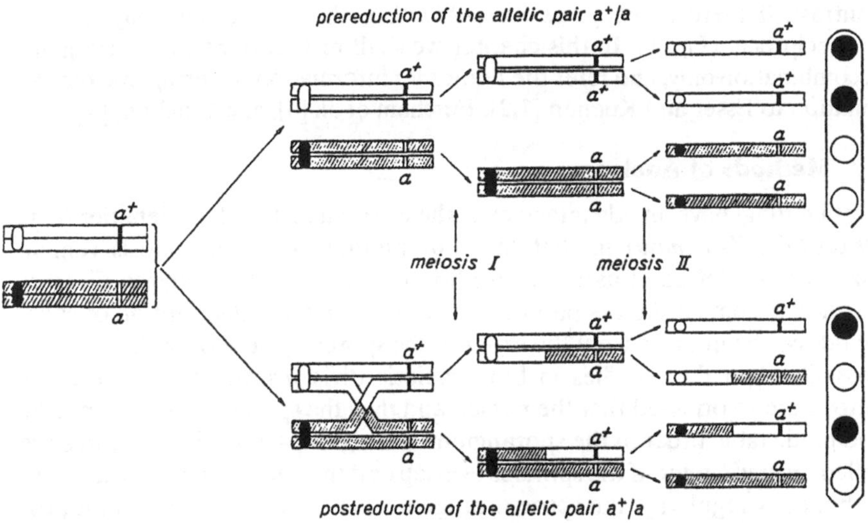

Figure 3 Segregation of the allelic pair a^+/a in pre- and postreduction—i.e., the separation of alleles in the first or in the second meiotic division. *Above*: Segregation of the four chromatids without a crossover having occurred between the centromere and the genetic marker (compare with Fig. 2). *Below*: Segregation of the four chromatids two of which are new combinations resulting from a crossover between the centromere and marker. Unlined and lined units represent homologous chromosomes of the two parents. Centromeres and marker genes are shown as in Figure 2. Open and solid circles (right) represent spores with the a and a^+ markers, respectively, in a four-spored ascus.

mitoses which, in a few cases, may lead to balanced haploid nuclei (mitotic recombination). These may carry combinations of genes that were previously located in different nuclei (see Chapter 4).

To sum up, interchromosomal recombination is caused by random separation of homologous chromosomes during meiosis (and in rare cases during mitosis), leading to a recombination frequency of 50%.

3. INTRACHROMOSOMAL RECOMBINATION

Recombination within a chromosome is brought about by mutual exchange of chromosomal parts by breakage and fusion as a result of a crossing over. If the exchange is achieved between different genes, the phenomenon is called *intergenic recombination*, which always leads to reciprocal segregation. In contrast, if there is *intragenic recombination*, the segregation may also be nonreciprocal (Fig. 1). In this chapter we shall restrict our text to intergenic recombination only, and refer the reader for information on intragenic recombination to Esser and Kuenen [1,2], Fincham et al. [3], and Fincham [4].

A. Methods of Analyses

Certain fungi have an advantage over the classical materials of genetics (e.g., *Drosophila, Zea mays*) in that the four products of each meiosis remain together as a tetrad of asco- or basidiospores and can be analyzed directly. Further, in some instances particular exchanges and the distribution of chromatids can be inferred from the order of the spores in the sporangium. This is possible when the spindles in both meiotic divisions and the postmeiotic mitosis are so oriented that the nuclei, and thus the spores, assume a specific and predictable order in the sporangium. Such groups of nuclei or spores are called *ordered tetrads*. If the spindles overlap or if the nuclei slip by one another without any regularity, the spore arrangements in the sporangia cannot provide any information about the events occurring in meiosis. In such a case the tetrads are said to be *unordered*.

Ordered Tetrads

The total information that can be obtained through the analysis of ordered tetrads is greater than that obtained by analysis of unordered tetrads. Table 1 comprises the experimental possibilities that are discussed in this chapter and which are available for use by the geneticist. The additional information obtained in the former case results primarily from being able to distinguish between the pre- and postreduction of an allelic pair. Since postreduction is a consequence of crossing over between a gene and its centromere, the centromere serves as an additional marker in the analysis of ordered tetrads. Representatives of the genera *Neurospora*, *Sordaria*, and *Podospora* (Sordariaceae) are suitable for the analysis of ordered tetrads. The segregation pattern of mutants that differ from the wild type in spore characters (e.g., color, size) can be determined directly by viewing the spores in the asci (Fig. 4).

Figure 5 shows how the eight genetically marked spores come to be arranged linearly in the ascus, using *Sordaria macrospora* as an example. The determining factors are 1.) orientation of the spindles; 2.) distribution of the nuclei; and 3.) mode of reduction of the marker genes.

A single factor cross such as $a^+ \times a$ in *Sordaria macrospora* and certain other euascomycetes results in six distinguishable tetrad types (Fig. 3). Compare with Table 2.

Table 1 Experimental data required for obtaining evidence of the various genetic events discussed in this chapter.

Event	Minimum number of markers required for the investigation	Tetrads and Ordered tetrads (o)	Genotypes necessary for the investigation: Unordered tetrads (u)	Genotypes necessary for the investigation: Single strands (s)
1. Prereduction of centromere	2 (o)	tetrad types (A_1–D_2)		
2. Recombination in the four-strand stage	2 (o, u)	tetratypes (A_3, B, C)	tetratypes (T)	
3. Distribution of homologous chromosomes in				
a) meiosis I	1 (o)	prereduction types		
b) meiosis II	1 (o)	postreduction types		
4. Assortment of nonhomologous choromosomes in meiosis	2 (o, u)	tetrad types (A_1–D_2)	tetrad types (P, R, T)	
5. Reciprocal recombination	1 (o, u)	pre- or postreduction types	tetrad types	
6. Linkage				
a) between gene markers	2 (o, u, s)	tetrad types (A_1–D_2)	parental and nonparental ditypes (P, R)	parental and recombination types (P_1, P_2, R_1, R_2)
b) between gene marker and centromere	1 (o)	pre- and postreduction types		
7. Localization and mapping of				
a) gene markers	2 (o, u, s)	tetratypes (A_3, B, C)	tetrad types (P, R, T)	parental and recombination types (P_1, P_2, R_1, R_2)
b) centromeres	1 (o)	postreduction types	tetratypes (T)	recombination types (R_1, R_2)

Abbreviations in the second column: o = ordered tetrads, u = unordered tetrads, s = single strands. For explanation of the symbols used in columns 3–5, see text.

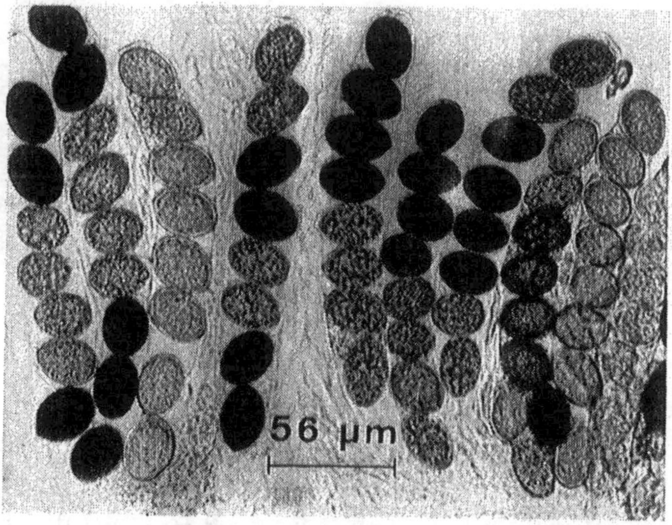

Figure 4 Asci from a perithecium of *Sordaria macrospora*, derived from a cross of a black-spored (r^+) with pink-spored (r) strain. All asci show a 4:4 segregation for black and pink spores.

Two types (1, 2) result from the separation of the two alleles at meiosis I (prereduction type). The remaining four types (3–6) are the results of second-division segregation (postreduction type). Types 3 and 4 represent asymmetric, and types 5 and 6 symmetric, segregations. The prereduction types occur in a 1:1 ratio, and the postreduction types in a 1:1:1:1 ratio with a random distribution of the centromeres in the two meiotic divisions.

In considering that in a single-factor cross there are six different ascus types, in a two-factor cross $6 \times 6 = 36$ ascus types may be observed. This may be easy visualized if one writes in a diagram in the abscissa "a" the six ascus types for the first factor (a/a^+) and in the ordinate the six ascus types for the second factor (b/b^+). In combining the different ascus types for the segregation of a/a^+ and b/b^+, one obtains the 36 different ascus types (Table 2).

To use this scheme for genetic analysis is too complicated. If top and bottom of the ascus types are not considered, because this information is not necessary for a genetic analysis, a diminution of these 36 types to seven types is possible, as indicated in Figure 6. To simplify the ascus structure, as shown in Figure 5, the situation in *Sordaria macrospora* in the four nuclear stages before the postmeiotic mitosis is shown, since this division leads to pairs of identical nuclei.

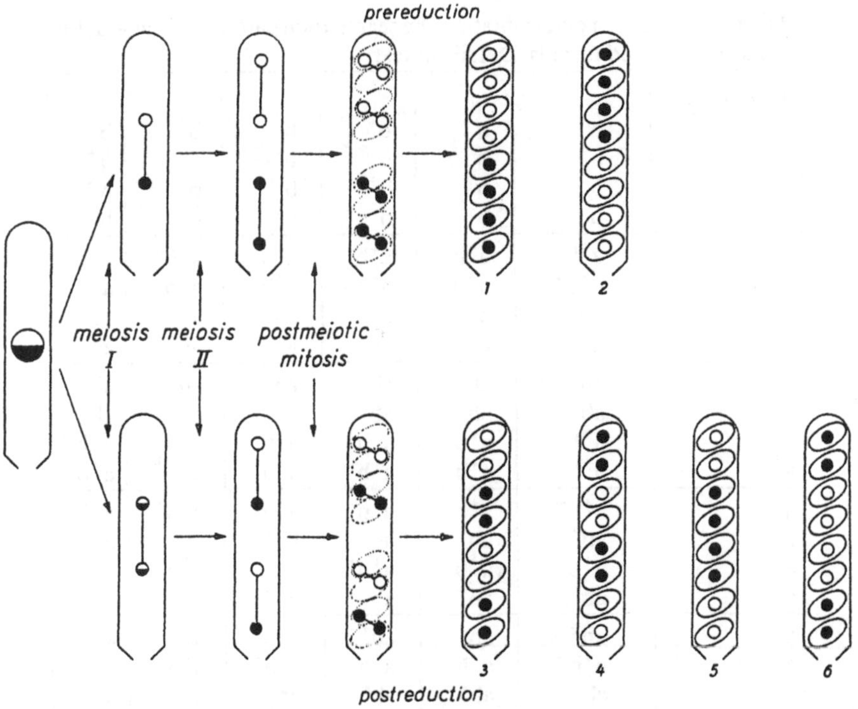

Figure 5 Development of the linear spore arrangement in an ascus of an eight-spored ascomycete (e.g., *Sordaria macrospora*). The asci are derived from a unifactorial cross. Two prereduction types (1, 2) and four postreduction types (3–6) may be distinguished (compare Table 3). Open circle and semicircle = allele of one parent; solid circle and solid semicircle = allele of the other parent (see explanation in text).

As Figure 6 reveals, the four groups *A*, *B*, *C*, and *D*, which are determined by the time of reduction of the two genes, can be subdivided in part. The following tetrad types can be distinguished: 1.) parental ditypes (*P*) with exclusively parental combinations (A_1 and D_1); 2.) nonparental or recombination ditypes (*R*) with exclusively nonparental combinations (A_2 and D_2); and 3.) tetratypes (*T*) whose four nuclei are all genetically different (A_3, *B*, and *C*).

In the following the tetrad types will be designed by capital letters—e.g., A_1, D_2, *P*, *T* (see Fig. 6). These symbols are also used in indicating the numbers of different tetrad types—e.g., A_1 = 243 (see Table 6). Lowercase letters are used to designate the frequencies of the tetrad types—e.g., a_1, d_2, *p*, *t* (see Tables 5, 6). By multiplying these values by 100 the percentage of each type

Table 2 Mode of determination the segregation pattern of a two-factor cross with ordered tetrads. Details in text.

Tetrad types b^+b / a^+a	b^+ b^+ b b	b b b^+ b^+	b^+ b b^+ b	b b^+ b b^+	b^+ b b b^+	b b^+ b^+ b
a^+ a^+ a a	a^+b^+ a^+b^+ ab ab	a^+b a^+b ab^+ ab^+	a^+b^+ a^+b ab^+ ab	a^+b a^+b^+ ab ab^+	a^+b^+ a^+b ab ab^+	a^+b a^+b^+ ab^+ ab
a a a^+ a^+	ab^+ ab^+ a^+b a^+b	ab ab a^+b^+ a^+b^+	ab^+ ab a^+b^+ a^+b	ab ab^+ a^+b a^+b^+	ab^+ ab a^+b a^+b^+	ab ab^+ a^+b^+ a^+b
a^+ a a^+ a	a^+b^+ ab^+ a^+b ab	a^+b ab a^+b^+ ab^+	a^+b^+ ab a^+b^+ ab	a^+b ab^+ a^+b ab^+	a^+b^+ ab a^+b ab^+	a^+b ab^+ a^+b^+ ab
a a^+ a a^+	ab^+ a^+b^+ ab a^+b	ab a^+b ab^+ a^+b^+	ab^+ a^+b ab^+ a^+b	ab a^+b^+ ab a^+b^+	ab^+ a^+b ab a^+b^+	ab a^+b^+ ab^+ a^+b
a^+ a a a^+	a^+b^+ ab^+ ab a^+b	a^+b ab ab^+ a^+b^+	a^+b^+ ab ab^+ a^+b	a^+b ab^+ ab a^+b^+	a^+b^+ ab ab a^+b^+	a^+b ab^+ ab^+ a^+b
a a^+ a^+ a	ab^+ a^+b^+ a^+b ab	ab a^+b a^+b^+ ab^+	ab^+ a^+b a^+b^+ ab	ab a^+b^+ a^+b ab^+	ab^+ a^+b a^+b ab^+	ab a^+b^+ a^+b^+ ab

may be obtained—e.g., $100 \times a_I$ = percentage of A_I types. The recombination value gives the percentage of recombination and is 100 times greater than the corresponding recombination frequency (*rho*).

If the individuals crossed differ in more than two factors, the number of tetrad types that can be distinguished increases exponentially. Ordered tetrads have the advantage of allowing a larger number of types to be identified than unordered tetrads or single spores selected at random (Table 3).

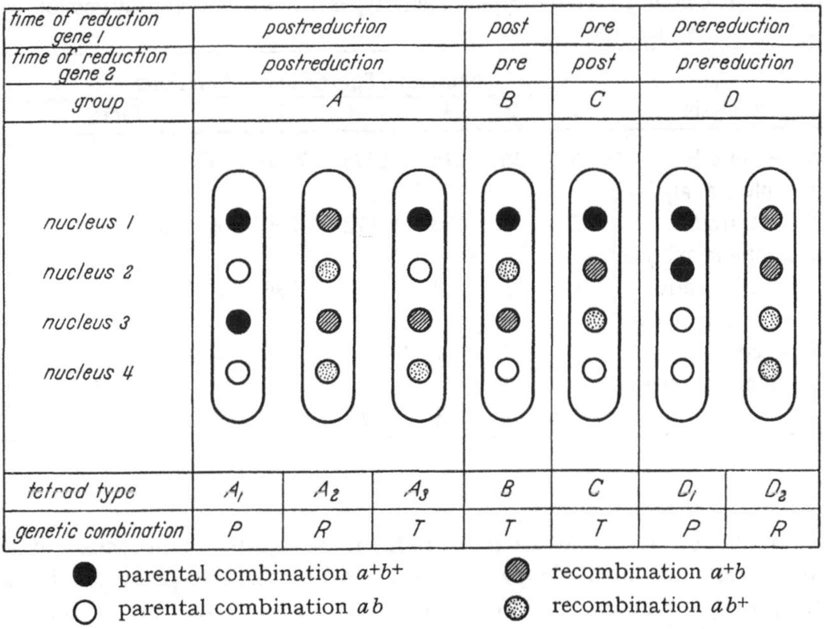
time of reduction gene 1
postreduction
post
pre
prereduction
time of reduction gene 2
postreduction
pre
post
prereduction
group
A
B
C
D
nucleus 1
nucleus 2
nucleus 3
nucleus 4
tetrad type
A_1
A_2
A_3
B
C
D_1
D_2
genetic combination
P
R
T
T
T
P
R
parental combination a^+b^+
parental combination ab
recombination a^+b
recombination ab^+

Figure 6 The seven types of ordered tetrads that are theoretically possible in the two-factor cross $a^+b^+ \times ab$ (compare with Table 2). The nuclei, which arise through meiosis I and II, are designated by circles. In eight-spored ascomycetes each of the meiotic products divides by mitosis into two identical daughter nuclei. Further explanations in text.

The tetrad types described for uni- or multifactorial crosses are characteristic of fungi that exhibit the same mode of nuclear division in the ascus as *Sordaria macrospora*. These include some eight-spored Euascomycetes, such as *Neurospora crass, N. sitophila, Sordaria fimicola*, and *Bombardia lunata*.

The formation of spores occurs in a different way in *Podospora anserina* (Fig. 7) [5], *Neurospora tetrasperma*, and *Gelasinospora tetrasperma.* The asci of these fungi contain only four dikaryotic spores. The distribution of nuclei in *Podospora anserina* is controlled by the position of the spindles. The spindles are oriented longitudinally in the two meiotic divisions, but in the postmeiotic mitosis they lie obliquely to the long axis of the ascus. Because of the arrangement of nuclei that results from these spindle orientations, a pair of nonsister nuclei from the postmeiotic mitosis is included in each spore.

Table 3 The number of tetrad types or genotypes that are theoretically possible with 1, 2, 3, 4, 5 . . . *n* markers.

	Number of possible combinations					
Type of analysis	1	2	3	4	5	n factors
Ordered tetrads (complete analysis)	6	36	216	1296	7776	6^n
Ordered tetrads (incomplete analysis)	2	7	32	172	4860	$\frac{6^n + 5 \times 2^n}{8}$
Unordered tetrads (complete analysis)	1	3	12	60	336	$\frac{6^{n-1} + 3 \times 2^{n-1}}{4}$
Unordered tetrads (incomplete analysis)	1	3	11	48	236	$\frac{6^n + 3 \times 4^n + 15 \times 2^n}{48}$
Single strands	2	4	8	16	32	2^n

With prereduction of an allelic pair the two spores in an ascus half are homokaryotic for one allele (types 1 and 2 in Fig. 5). With postreduction all four spores are heterokaryotic. In contrast to the situation in *Sordaria macrospora*, distinguishing the four postreductional types (types 3–6 in Fig. 5) is not possible. In two-factor crosses only five, instead of seven (compare with Fig. 6), tetrad types can be recognized directly. Nevertheless, it may be possible to detect the missing tetrad types through further analysis of dikaryotic spores of ascus type *A*. Thus an analysis of ordered tetrads is possible.

Unordered Tetrads

These differ from ordered tetrads in that the spore arrangement does not allow any conclusions regarding the mode of reduction of marker genes. Thus the centromere cannot be used as a marker for genetic studies. Unordered tetrads occur in certain eight-spored euascomycetes, in many four-spored yeasts, and in all basidiomycetes.

Single factor crosses ($a^+ \times a$) reveal only whether segregation is reciprocal ($4a^+$:$4a$) or nonreciprocal (e.g., $6a^+$:$2a$; Fig. 4). Pre- and postreduction types cannot be distinguished. Three tetrad types occur in two-factor crosses: the parental ditype (*P*), the nonparental or recombination ditype (*R*), and the tetratype (*T*) (Fig. 6, bottom row).

If parents differ at three loci, the number of tetrad types is 12, provided complete analysis is possible. If only the tetrad types for each two loci, and not the genotypes of the single spores, are considered (incomplete analysis), only 11 types are distinguishable (Table 4). In the latter case the *TTT* types fall into a single group. The number of distinguishable tetrad types increases exponen-

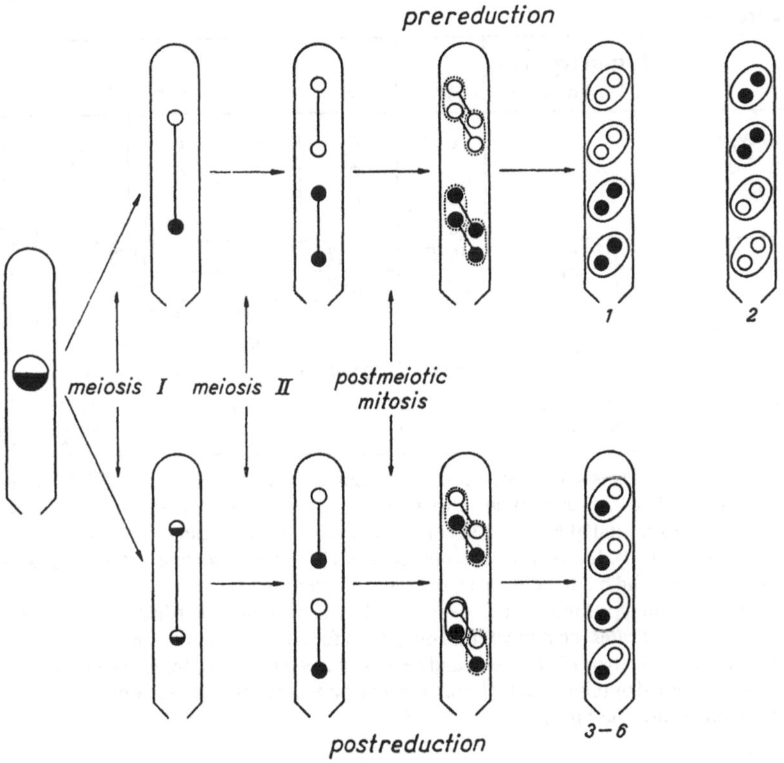

Figure 7 Ascospore formation in *Podospora anserina* showing the segregation of a single pair of alleles. Symbols are the same as in Figure 5. Further explanation in the text.

tially with an increase in the number of allelic differences. Because pre- and postreduction types cannot be distinguished, the position of the centromere cannot be determined directly in unordered tetrads. From this follows that, instead of the centromere, which is used as a marker in experiments with ordered tetrads, an additional marker is required for most genetic experiments. Indirect methods have been developed to determine the position of the centromere, which, however, are rather difficult to apply for standard procedures.

Single Strands

When tetrad analysis cannot be performed, the isolation study of single spores remains as a possibility. Because only one of the four meiotic products is

Table 4 Distribution of three marker genes which are theoretically possible in unordered tetrads.

	Tetrad types for a^+/a and c^+/c	*P*	*R*	*T*			
Tetrad types for a^+/a and b^+/b	a^+ a^+ a a	c^+ c c	c c^+ c^+ c^+	c^+ c c^+ c	c c c c^+	c^+ c^+ c c^+	c c c^+ c^+ c
P	b^+ b^+ b b	*PPP*	*PRR*	*PTT*			
R	b b b^+ b^+	*RRP*	*RPR*	*RTT*			
T	b^+ b b^+ b b b^+ b b^+	*TTP*	*TTR*	*TPT* *TRT* *TRT* *TPT*		*TTT*	
	b^+ b b b^+ b b^+ b^+ b			*TTT*		*TPT* *TRT* *TRT* *TPT*	

Distribution of markers is shown for the three-factor cross a+b+c+ × abc. With the order a+a+aa constant (line 2, column 2) there are six sequences possible for b+/b (column 2) and for c+/c (line 2), giving a total of 6 × 6 or 36 combinations in all. These can be reduced to 11 or 12 different types (see text). In the triple combinations of *P*, *R*, and *T*, the first letter designates the tetrad type for a+/a and b+/b (column 1), the second letter the type for b+/b and c+/c, and the third letter the tetrad type for a+/a and c+/c (line 1). For example, the triple combination *RRP* (line 4, column 3) indicates the nonparental ditype (*R*) for a+/a and b+/b—i.e., a+b, a+b, ab+, ab+; the nonparental ditype (*P*) for a+/a and c+/c—i.e., a+c+, a+c+, ac, ac. The four products of meiosis represented by this tetrad therefore possess the genotypes a+bc+, a+bc+, ab+c, ab+c. (Adapted from Whitehouse [6].)

identified in such a case, this type of study has been called single-strand analysis. Such an approach is necessary when the spores of a tetrad do not remain together, as in the case of lower fungi such as *Phycomyces blakesleanus*. However, single-strand analysis is often used in organisms having ordered or unordered tetrads, if a large number of spores can be obtained quickly and one wishes to avoid the tedious isolation of single asci or basidia.

At least two marker genes are required for these investigations. In two-factor crosses two parental types (P1 and P2) and two nonparental or recombination types (R1 and R2)—i.e., a total of four genotypes—are obtained. With each additional factor, the number of different genotypes is doubled (Table 2).

In summary:

1. A large number of fungi have been particularly useful for genetic investigation because the four products of each meiosis remain together as a tetrad and may be analyzed directly. Analysis of ordered tetrads is possible,

however, only if the mechanism of nuclear division leads to a regular linear arrangement of spores in the ascus.

2. The amount of information that can be obtained is greatest from ordered tetrads and least from single spores. The same is true with respect to the number of distinguishable tetrad types and genotypes in crosses involving a particular number of markers (Table 3).

3. Ordered, in contrast to unordered, tetrads indicate the mode of reduction of particular allelic pairs and their relation to the centromere. Isolation and analysis of unordered tetrads as well as single spores, on the other hand, require much less time and thus allow a more rapid solution of certain types of genetic problems.

4. The applicability of the three types of analysis to genetic investigations is indicated in Table 1.

Table 5 Summary of criteria for linkage and independent assortment of two allelic pairs.

Unordered tetrads			Unordered tetrads	Single strands	
Tetrad distribution		Map distance	Tetrad distribution	Recombination frequency	Inferences
$d_1 > d_2$	$a_1 > a_2$	$w = y - x$	$p > r$ $r{:}t < 1{:}4$ (L-distribution)	$e < 0.5$	Markers linked and on the same side of the centromere
	$a_1 = a_2$	$w = x + y$			Markers linked and on different sides of the centromere
$d_1 = d_2$	$a_1 = a_2$	$w = y - x$ or $w = x + y$	$p = r$ $r{:}t \leq 1{:}4$ (N-, T-distribution)	$e = 0.5$ $e \leq x + y$	Linkage or nonlinkage of markers cannot be determined
$d_1 = d_2$	$a_1 = a_2$	$w \neq y - x$ and $w \neq x + y$	$p = r$ $r{:}t > 1{:}4$ (F-distrubtion)	$e = 0.5$ $e > x + y$	Markers are not linked

Abbreviations and further explanations in text. Examples for the application of these criteria in Table 6.

Table 6 Examples of the application of the linkage criteria described in Table 5. In each cross three markers are involved—two genes and the centromere.

Object and cross	No. of asci analyzed	Tetrad types (ordered tetrads)							Tetrad types (unordered tetrads)			Recombination value (%) (100 *w*)	½ Post-reduction value (%) for	
		A1	*A2*	*A3*	*B*	*C*	*D1*	*D2*	*P*	*T*	*R*		Gene 1 (100 *x*)	Gene 2 (100 *y*)
Podospora anserina $t_1 \times i$	615	67	1	6	2	399	136	4	203	407	5	33.9	6.2	38.5
Podospora anserina *m* × *un*	1036	20	24	47	76	692	166	11	186	815	35	42.7	8.1	37.8
Podospora asnerina $t_1 \times m$	400	2	3	6	30	57	153	149	155	93	152	49.6	5.1	8.5

Recombination values and postreduction frequencies can be calculated as indicated in Figure 6 from the segregation pattern A_1–D_2, independently whether there is linkage or not. ½ Postreduction frequency corresponds to the uncorrected distance gene–centromere. Evidence for linkage according to the criteria of Table 5. Two examples for statistical evaluation are given in Table 7.

5. In this context it must be emphasized that the most notable advantage of tetrad analysis is that very few tetrads are necessary to obtain precise information. As compared to random analysis sometimes four or five tetrads are required to establish linkage. Direct information about chromosomal or chromatid interference is only available from tetrad segregation patterns.

B. Linkage

Linkage between genetic markers refers to the predominance of parental combinations among the offspring of crosses; i.e., the genetic markers do not recombine at random. The genes that are linked with one another constitute a *linkage group*. Unlinked genes belong to different linkage groups. Cytogenetic investigations have proved that the number of linkage groups corresponds to the number of chromosomes in the haploid complement. Existence of linkage may be determined experimentally by analyzing the meiotic products of a cross. The criteria commonly used to determine linkage or independent assortment of two genes are summarized in Table 5. Examples for linkage tests based on tetrads analysis are given in Table 6 and will be discussed later.

Evidence for linkage or independent assortment from analysis of						
Ordered tetrads			Unordered tetrads		Single strands	
D_1:D_2	A_1:A_2	100w:100(x+y) [+] 100w:100(y–x) [–]	P:R	R:T	e	Interpretations
136:4 $d_1 > d_2$	67.1 $a_1 > a_2$	33.9:44.7 [+] 33.9:32.3 [–] $w \approx y - x$	203:5 $p > r$	5:407 r:t < 1:4	0.339 $e < 0.5$	Markers located on the same side of centromere
166:11 $d_1 > d_2$	20:24 $a_1 \approx a_2$	42.7:45.9 [+] 42.7:29.7 [–] $w \approx x + y$	186:35 $p > r$	35:815 r:t < 1:4	0.427 $e < 0.5$	Markers located on opposite sides of the centromere
153:149 $d_1 \approx d_2$	2:3 $a_1 \approx a_2$	49.6:13.6 [+] 49.6: 3.4 [–] $w \neq x + y$ $w \neq y - x$	155:152 $p \approx r$	152:93 r:t > 1:4	0.496 $e \approx 0.5$ $e > 0.136$	Markers are unlinked

Ordered Tetrads

Analysis of Crosses A maximum of seven tetrad types are expected from the two-factor cross, $a^+b^+ \times ab$, if the meiotic products are ordered (Fig. 6). This number is independent of the linkage or independent assortment of the markers, and it does not depend upon the amount of crossing over between the markers, if the genes are linked. On the other hand, the small number (seven) of tetrad types corresponds to a significantly greater, theoretically unlimited number of crossing-over configurations. Figure 8 diagrams those which account most simply for the seven types.

Figure 8 shows:

1. Only the D_1 type results from the absence of crossover, regardless of whether the genes are on the same side (I) (Fig. 8, left) or on different sides (II) (Fig. 8, right) of the centromere.
2. A single crossover accounts for Type C in both cases, type A in case I and type B in case II.
3. The following types result from a double crossover: type D_2 (a four-strand double in a single region, case I and II); types B and A_3 (two- or three-strand double in case I); types A_1, A_2, and A_3 (two-, four-, or three-strand

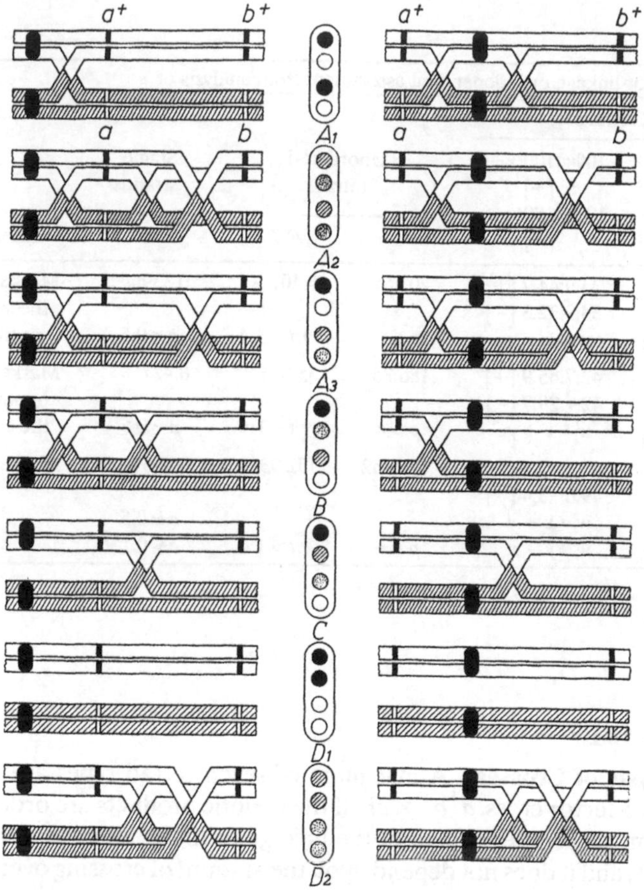

Figure 8 Interpretation of the seven tetrad types in terms of crossovers. Only the simplest crossover configurations are shown. *Left*: Both loci are located on the same side of the centromere—i.e., in the same chromosome arm. *Right*: The two loci are located on opposite sides of the centromere—i.e., in different arms of the chromosome. *Center*: The genotypes are labeled as in Figure 6. The symbols for the chromosomes, centromeres, and marker genes are the same as in Figure 3.

double in case II). The ratios $a_1{:}a_2{:}a_3$ in case II and $a_3{:}b$ in case I give information on the distribution of chromatids in double crossing over. The random expectations are $a_1{:}a_2{:}a_3 = 1{:}1{:}2$ and $a_3{:}b = 1.1$. A deviation from these ratios indicates chromatid interference.

4. A triple crossover accounts only for type A_2 of case I.

In calculating crossover frequencies one must remember that many possible crossover configurations are not shown in Figure 8, even if no more than two crossovers per region are considered. For example, on the basis of random distribution of chromatids from a double crossover (two-strand:three-strand:four-strand = 1:2:1) if $d_2 = 0$, a few D_1 types (as a result of two-strand doubles), and double the number of C types (as a result of three-strand doubles) would be expected with the frequencies d_2 and $2d_2$.

Evidence for Linkage The stronger the linkage between two markers, the closer the frequency of parental types approaches 1 and the recombinant frequency approaches 0. The percentage of crossover also approaches 0 because recombination can result only when at least one crossover takes place between the two genes. Therefore, when genes are linked, D_2 types are always fewer than D_1 types ($d_1 > d_2$). Further, for the case shown on the left in Figure 8, A_2 types are expected at a lower frequency than A_1 types ($a_1 > a_2$). Each inequality is a sufficient condition for establishing linkage (Tables 1 and 5). Thus the following generalization holds: A significant deviation of the observed distribution of tetrads from the random is sufficient to establish linkage.

The frequency of crossing over between two markers is a measure of the distance between them. Accordingly in the case of linkage the distance w between two genes is equal either to the sum or the difference of the distances x and y between the genes and the centromere ($w = x + y$) or either to the sum or the difference of the distances, x and y, between the genes and the centromere ($w = x + y$ or $w = y - x$ when $x \leq y$) depending on whether the genes are located on different sides or on the same side of the centromere (Fig. 8, right and left; Tables 1 and 5).

Criteria to use tetrad data for the determination of the mode of intragenic segregation patterns of markers (linkage or nonlinkage) are summarized in Table 5. Some examples are given in Table 6.

Unordered Tetrads

Analysis of Crosses An analysis of unordered tetrads reveals a maximum of three types: the parental ditype (P); the nonparental ditype—i.e., recombination ditype (R); and the tetratype (T) (Fig. 6). Certain crossover combinations fail to yield recombination of marker genes as in the case of ordered tetrads.

Evidence for Linkage Parental and nonparental ditypes occur in equal frequency from a cross involving unlinked markers ($p = r$). A statistically significant deviation from such a distribution is a sufficient condition for proof of linkage, provided the number of parental ditypes predominates ($p > r$) (see Table 5).

As in the case of ordered tetrads, information about linkage and the location of the genes can be obtained by comparing the distances between each two of the three genes. However, three-factor crosses are necessary to establish the order of loci with certainty.

Single Strands

Analysis of Crosses A two-factor cross yields four different genotypes either in independent assortment or in partial linkage. These may be detected even if the products of meiosis cannot be analyzed by tetrads. If recombinants occur, at least one crossover must have taken place in the marked region. No further conclusions can be drawn beyond this correlation between crossing over and recombination with a two-factor difference.

Evidence for Linkage With random assortment of two pairs of alleles, the four possible genotypes occur in equal frequency. The frequency of recombinants (recombination frequency *rho*) equals 0.5 in the absence of linkage. By definition the value *rho* < 0.5 holds for linked genes. This is sufficient condition for linkage (see Table 5).

C. Chromosome Maps

Chromosome maps or linkage maps are graphical linear representations of chromosomes in which the markers of a linkage group are arranged according to the relative distances between them. The frequency of crossover between two linked markers serves as a measure of the map distance between them. The construction of accurate linkage maps depends primarily on the precision with which the number of crossovers can be determined from the recombination data.

If two genes are located in close proximity where only one crossover is possible, there is no problem in calculating from the experimentally obtained exchange rate. However, if chromosome markers are located in larger distances, the recombination values obtained from experiments are not in accordance with the number of crossovers. This is brought about by double or multiple crossovers between the markers concerned. In addition, in most organisms the formation of crossovers is hindered by interference.

Interference is defined as the nonrandom distribution of crossovers among the four chromatids of the tetrad. Two types of interference are recognized:

1. Chromatid interference occurs if a crossover either decreases or increases the probability that a second crossover will involve the same strand. In the first case the interference is positive, while in the second, it is negative. In both cases chromatid interference leads to a nonrandom distribution of two-, three-, and four-strand double crossovers (Fig. 8).

2. Chromosome interference occurs if, following a crossover, there is a less or greater than random chance that additional crossovers will occur in the vicinity of the first. Here also the former is termed positive, the latter negative, interference.

Thus in both cases the frequencies of multiple crossovers deviate from a random expectation. *Chromosome interference* refers to the distribution of crossovers in a tetrad, while *chromatid interference* is concerned with the participation of individual strands in multiple crossing over.

The functional relationship between crossing over and the experimentally confirmed exchange of markers can be understood under simplified assumptions with the aid of mathematical interference models. The mathematical formulation of this relationship leads to mapping functions which allow a localization of markers and also permit a determination of the degree of chromosome interference.

Single-Strand Analysis

Since single-strand or random analysis distinguishes only between parental and recombinant types, empirical and theoretical mapping functions have been derived from which the distance between two genes can be determined from the frequency of recombination [5].

Tetrad Analysis

Mapping functions are also used for localizing markers in organisms in which the meiotic products remain together as tetrads. They allow the calculation of crossing over frequencies from postreduction and tetra-type frequencies. The postreduction values give the relative location of the centromere.

An interference model for both ordered and unordered tetrads was developed by Kuenen [7] (Fig. 9). It allows the derivation of mapping functions for different degrees of interference and the graphical and arithmetical determination of the degree of chromosome interference from tetrad distribution. If ordered tetrads are available, it is necessary to determine the value of the maximum crossover frequency. If only unordered tetrads are available, the tetra-type frequency is a function of the crossover frequency.

The application of this model for the determination of the map distances between centromere and marker (ordered tetrads) or between two markers of a chromosome will be explained using two fungi which differ extremely in their degree of interference.

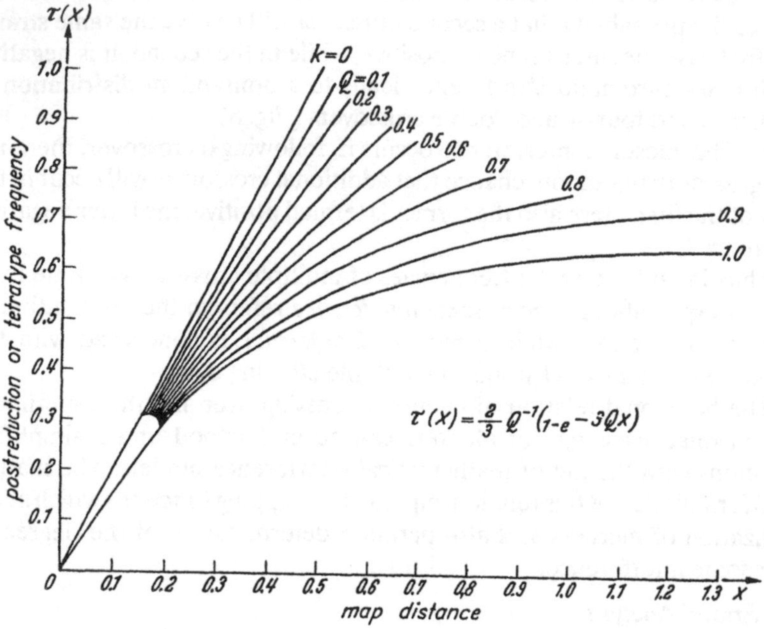

Figure 9 Graphical representation of tetrad mapping functions for 11 different degrees of interference based on the interference value Q. Complete interference $k=0$; positive interference $Q=0.1$ to $Q=0.9$; no interference (random distribution of crossovers) $Q=1.0$. For application see text.

In *Podospora anserina* the maximum of postreduction frequency is almost 1.0 (0.98 for the mating-type genes). As may be seen from Figure 9, there is a linear correlation between postreduction frequency and map distance. In other words, half the value of the postreduction frequency is equal to the map distance.

In contrast, in *Sordaria macrospora*, where the maximum of postreduction frequency does not exceed the value 0.66, there is almost no interference, leading to a Q value close to 1.0. It may be seen from Figure 9 that there is no linear correlation between postreduction frequencies and values from map distance. It further becomes evident from Figure 9 that, if postreduction frequencies exceed the value of 0.5, a graphical determination of the map distance is no longer possible. It can, however, be achieved in a mathematical way, using the equation for the mapping functions of Figure 9.

Table 7 Statistical evaluation (χ^2 test)for linkage for segregation patterns of ordered tetrads from two-factor crosses as listed in Table 6. In each cross three markers are involved—two genes and the centromere.

Segregation	Groups							
t1 × m	A1	A2	A3	B	C	D1	D2	Sum
Obtained (obt)	2	3	6	30	57	153	149	400
Expected % (exp %)	0.43	0.43	0.86	8.50	15.30	37.30	37.30	
Expected (exp)	1.7	1.7	3.4	34	61.2	149.2	149.2	
Deviation (d)	0.3	1.3	2.6	4.0	4.2	3.8	0.2	
Deviation 2/ Expected d2/exp)	0.05	0.99	1.99	0.47	0.29	0.09	0.00	

Sum d2/exp = χ_2 = 3.88 — Degrees of freedom (F) = 6

Probability (*P*) = 0.7–0.5 — Both genes are unlinked

Segregation	Groups							
t1 × i	A1	A2	A3	B	C	D1	D2	Sum
Obtained (obt)	67	1	6	2	399	136	4	615
Expected % (exp %)	1.2	1.2	2.4	7.6	33.7	26.9	26.9	
Expected (exp)	7.4	7.4	14.8	46.7	207.3	165.7	165.7	
Deviation (d)	59.6	6.4	8.8	44.7	191.7	29.7	161.7	
Deviation 2/ Expected (d2/exp)	480.0	40.9	12.9	999.7	92.1	6.5	6536.0	

Sum d2/exp = χ^2 = 8168.1 — Degrees of freedom = 6

Probability (P) = far below 0.001 — Both genes are linked

The principle of these statistics consists in determining the significance between the segregation pattern obtained from tetrad data (obt) and the segregation pattern that is expected on the basis of random assortment of the two genes (exp).

In using the postreduction values (see Table 6) the frequencies (exp%) of the segregation pattern can be calculated as follows: group A = post × post, both genes; group B = post gene 1 × pre gene 2. Group C = pre gene 1 × post gene 2; Group D = pre × pre, both genes. The distribution within group A is 1:1:2 for A1:A2:A3 and in group D 1:1, as explained above (p. 61).

The absolute values (exp) are calculated from the frequencies (exp %), in order to be able to apply the χ^2 test. As generally agreed, probability (*P*) values below 0.1 show that the obt values are not in accordance with the exp values, and free assortment of the marker genes is therefore not statistically proved.

As an example for the calculation of map distances, data from a cross with Podospora anserina may serve ($t_l \times t$; see Table 6).

1. Determination of interference value Q. It can be shown that the two markers t_l and i lie on the same side of the centromere on the basis of the linkage criteria presented in Table 6 (compare with Table 7). In this instance the interference value is:

$$Q = \frac{a_3 + b}{(a_1 + a_2 + a_3 + b)(a - 3 + b + c)}$$

The small letters in this equation represents the frequencies of corresponding tetrad types A_1, A_2, A_3, B, C (compare Fig. 6). Thus

$$Q = \frac{(0.010 + 0.003)}{(0.109 + 0.002 + 0.010 + 0.003)(0.010 + 0.003 + 0.649)}$$

$$Q = \frac{0.013}{0.124 \cdot 0.662} = 0.16$$

2. Determination of the map distance between the centromere and the markers t_l and i. In the mapping function shown above t represents the postreduction frequency for t_l and i. Table 7 shows the values 0.124 (t_l) and 0.770 (i). Thus the map distance between the centromere and t_l is:

$$\begin{aligned} x_1 &= \frac{1}{3} \cdot 0.16^{-1}\, ln\, (1 - \frac{3}{2} \cdot 0.16 \cdot 0.124) \\ &= -2.083\, ln\, 0.97024 = 0.0628 \end{aligned}$$

and the map distance between the centromere and i:

$$\begin{aligned} x_2 &= \frac{1}{3} \cdot 0.16-1\, ln\, (1 - \frac{3}{2} \cdot 0.16 \cdot 0.770) \\ &= -2.083\, ln\, 0.8152 = 0.4254 \end{aligned}$$

3. Determination of the map distance between t_l and i. In this case t represents the frequency of the tetratypes. The value $t = 0.662$ is obtained from $T = 407$ (Table 7). Therefore, the map distance is:

$$\begin{aligned} x_3 &= \frac{1}{3} \cdot 0.16^{-1}\, ln\, (1 - \frac{3}{2} \cdot 0.16 \cdot 0.662) \\ &= -2.083\, ln\, 0.84112 = 0.3631 \end{aligned}$$

The calculated map distances are additive. Since the centromere lies beyond the region marked by t_l and i, the relationship $x_2 - x_l = x_3$ holds, or in specific values: 0.4254 – 0.0628 = 0.0628–0.3631. It is readily apparent in

Figure 9 that the points P_1 (0.0628/0.124), P_2 (0.4254/0.770), and P_3 (0.3631/0.662) lie between the two curves $Q = 0.1$ and $Q = 0.2$; i.e., for the ordinate values (postreduction and tetra-type frequencies) the calculated values on the abscissa (map distances) are found on the curve for $Q = 0.16$.

To sum up: 1.) Genetic markers can be arranged linearly into linkage maps on the basis of recombination frequencies. The map distance between two markers is defined as the average number of crossovers between these markers in 100 single strands. 2.) Map distances may be calculated with the aid of mapping functions directly from recombination frequencies—i.e., crossover frequencies. Such mapping functions have been derived using data from single-strand analysis as well as data from tetrad analysis.

4. CONCLUSIONS

This paper was written to inform geneticists and mycologists about the possibilities to exploit in an optimal manner data from sexual crosses of fungi for both fundamental and applied genetics. It should become obvious that tetrad analysis, and especially the evaluation of ordered tetrads, gives more precise information with respect to segregation patterns and gene linkage than random analysis, because the evaluation of only a few tetrads is required to be able to make precise statements. Furthermore, comprehensive information about the cytological events is obtained by genetic analysis, which allows a better understanding of the recombinational events and thus enhances the possibilities for concerted breeding in the mushroom industry and biotechnology. Because of the limited space available in this book, I had to refrain from dealing with intragenic recombination (Fig. 1), which for breeding procedures is mostly not used.

5. EXPERIMENTS

A. Evaluation of Ordered Tetrads in *Sordaria macrospora*

Aim

The segregation pattern in ordered tetrads in both a one- and a two-factor cross can be easily demonstrated in using marker genes concerning the spore color.

One-Factor Cross

Strains The mutant *sterilis* contains the gene s_1, which is responsible for the blockage of the normal life cycle before spore formation. The perithecia contain no asci, which is easily proved by squeezing them open with forceps.

The mutant *rosea* contains the gene *r*, which causes the production of pink spores in crosses.

Each mutant contains the corresponding wild gene of the other, so that the complete genetic formula is given by: *sterilis* = s_1r^+ and *rosea* = s_1^+r.

Procedure

Inoculate a Petri dish containing cornmeal agar with mycelial inocula of s_1 and r, at a distance of 4–6 cm apart. Grow at room temperature or in an incubator at 25°C; light is not necessary.

Evaluation

The first fruit bodies arise after 5–6 days. Open the perithecia and establish that the fruit bodies on the s_1 mycelium are empty and contain only a gelatinous matrix. On the other hand, perithecia are found in the contact zones, which show a segregation pattern of the pink and black spores. Fruit bodies found on the r mycelium contain pink spores only. Transfer the contents of four or five fertile perithecia to a water drop on a microscope slide, using a preparation pen. After applying a cover slip, carefully apply pressure, absorb excess water and, ring the cover slip with nail varnish. In ideal preparations it can be seen that the asci, arising from a gelatinous center (ascogonial cell with ascogenous hyphae), are arranged radially and that paraphyses are absent. The segregation pattern of the colored spores may be observed under low to medium magnification (Fig. 4).

For this experiment it is necessary only to remove perithecia that contain 20 or 30 mature asci. These can be distinguished from the many immature perithecia whose asci contain eight white or yellow spores while still in the Petri dish, or in the slide preparations. In mature perithecia, the asci rupture very easily, and the large numbers of individual spores make interpretation difficult.

Before we interpret the segregation of the spore colors, it must be borne in mind that we are only examining the segregation of the allele pair r^+/r. The segregation of the pair s_1^+/s_1, used for marking purposes only, can be ignored. To examime the segregation of this allele pair, the spores must first be germinated.

In principle, two ascus types occur: 1.) the spore colors are distributed equally in the two halves of the asci (4:4 segreation); 2.) in each half of an ascus there are two black and two pink spores (2:2:2:2 segregation) (Fig. 5). The type 2 asci result from the fact that at least one crossing over has taken place during prophase of meiosis I, between the centromere and the gene locus responsible for spore coloration, so that the marker r^+/r becomes postreduced (second-division segregation). In type 1 asci prereduction has occurred (first-division segregation). The distance between the centromere and the gene can be estimated using the proportion of type 2 asci. The greater the number of postreduced asci, the greater the number of crossovers, and the greater the distance from the centromere to the gene.

B. Two-Factor Cross

Strains

Mutant *rosea* (characteristics given under the one-factor cross). Mutant *lutea* contains the gene *lu*, which blocks the production of the black pigment in the spores, which thus remain yellow. As *lu* is capable of self-fertilization, the production of self-fertilized perithecia containing exclusively yellow spores can be avoided by marking with *sterilis*, or with another sterilizing gene. The gene *spadix* (*spd*) is especially suitable for this purpose, the *spd* strain being fully

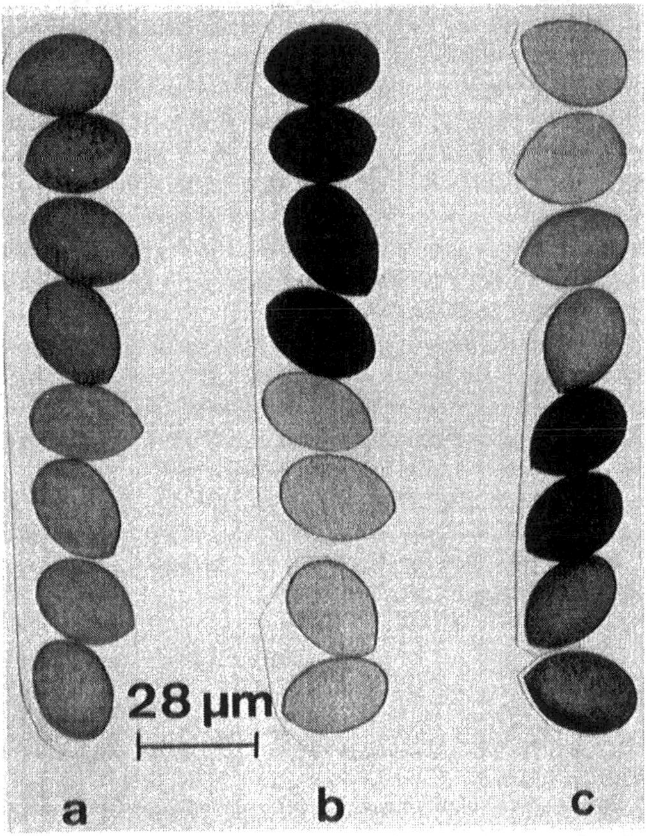

Figure 10 *Sordaria macrospora*. Ascus types from a two-factor cross of the color spore mutants pink and yellow. (a) Parental type, yellow and pink spores; (b) recombinant type, black and white spores; (c) tetratype, all four spore colors. For explanations see text. (Adapted from Esser [8].)

sterile. The complete genetic formula of the two mutants is thus: *rosea* = $r\ lu^+$ and *lutea* = $r^+ lu$.

Procedure

In the same way as in the one-factor cross, Petri dishes should be inoculated with blocks of *rosea* and *lutea* mycelium.

Evaluation

Fruit bodies produced on the *rosea* side contain pink spores only. The perithecia produced in the contact zone contain spores expressing the spore color genes, whereas those forming on the *lutea* side (if unmarked with a sterilization factor) contain exclusively yellow spores.

In the slide preparations, made as described above, three types of asci can be identified: 1.) *parental type*, containing pink ($r\ lu^+$) and yellow ($r^+ lu$) spores only; 2.) *recombinant type*, containing black and white spores only, arising from the new combination of the genes to give wild spores ($r^+ lu^+$) and white spores ($r\ lu$), showing both defects; 3.) *tetratype*, containing all four spore colors. With this relatively simple cross it is possible to demonstrate the basic principles of Mendel's laws of segregation and independent assortment of genes from a single perithecium (Fig. 10).

Segregation data obtained from this cross may be used for a number of different calculations, such as the distance of the two genes from the centromere, the demonstration of whether the genes are linked or segregate independently, and the existence of interference. Details to how evaluate data for various purposes may be found earlier in this chapter.

Strains may be obtained from 1.) Prof. Dr. U. Kück, Lehrstuhl für Allgemeine Botanik, Ruhr-Universität, D-44780 Buchum, Germany, or 2.) Prof. Dr. B. Hock, Lehrstuhl für Botanik, TU München (Weihenstephan), D-83350 Freising, Germany.

REFERENCES

1. Esser, K., R. Kuenen (1967). *Genetik der Pilze*. Springer, Berlin, Heidelberg, New York. Ergänzter Neudruck.
2. Esser, K., R. Kuenen (1967). *Genetics of Fungi*. Translated by Erich Steiner. Springer, Berlin, Hieldelberg, New York.
3. Fincham, J. R. S., P. R. Day, A. Radford (1979). *Fungal Genetics. Botanical Monographs*, Vol. 4. Blackwell, Oxford, Edinburgh, 4th Ed.
4. Fincham, J. R. S. (1983). *Genetics*. Wright PSG, Bristol, London, Boston.
5. Esser, K. (1974). *Podospora anserina*. In: King, R. C. (ed.). *Handbook of Genetics*, Vol. I:531–551. New York.

6. Whitehouse, H. L. K. (1942). Crossing over in *Neurospora. N. Phytol. 41*:23–62.
7. Kuenen, R. (1962). Ein Modell zur Analyse der crossover-Interferenz. *Z. Verebl. 93*:35–65.
8. Esser, K. (1982). *Cryptogames. Cyanobacteria, Algae, Fungi, Lichens*. University Press, Cambridge.

4

Somatic Recombination

Cees J. Bos
Wageningen Agricultural University, Wageningen, The Netherlands

1. INTRODUCTION

In this chapter we study somatic recombination and how we can use it for genetic analysis and strain construction. We shall use the term *somatic recombination* for all (re-)combination of genetic information of somatic cells (also for protoplast fusion, heterokaryosis), and the term *mitotic recombination* only with respect to the recombination processes that occur during mitosis (mitotic crossing over, nondisjunction, gene conversion).

Six key concepts in somatic recombination are these:

1. The parasexual cycle may be an important alternative to sex. A complete parasexual cycle consists of fusion of different haploid somatic nuclei (karyogamy), mitotic recombination in the diploid cell and haploidization.

2. Somatic recombination occurs when genetic material (nuclei, mitochondria) is exchanged.

3. Mitotic recombination refers to recombination processes in a diploid cell during mitosis and covers mitotic crossing over, nondisjunction, and gene conversion.

4. Mitotic crossing-over affects intrachromosomal recombination and results in homozygosis of all loci distal of the point of exchange. It can be used to determine the linear arrangement of genes in a linkage group.

5. Nondisjunction during mitosis leads to aneuploids and, consequently, to diploids homozygous for one chromosome, or to haploidization.

6. Haploidization follows nondisjunction in aneuploids that lack a chromosome. It results in the reassortment of whole chromosomes and can be used efficiently for determining linkage groups.

2. THE PARASEXUAL CYCLE

In eukaryotes the occurrence of sex or sexual reproduction includes recombination of genetic material during meiosis. Recombination is, however, not exclusively restricted to meiosis. In prokaryotes there are various alternatives to sex [1] that result in recombination. The intermediate state is mostly a partial zygote (merozygote) in contrast to meiotic recombination, where the zygote consists of two complete sets of chromosomes. In principle transformation of fungi (Chapter 9) leads in the first instance also to a merozygote (with only a very small part of donor DNA). An alternative to sex called the parasexual cycle [2] proceeds by a holozygote, and it is based on mitotic recombination. It consists of a sequence of heterokaryosis, karyogamy, and haploidization in somatic cells. In some fungi the parasexual cycle is fully functional and can be studied very well. Mitotic recombination occurs also in other eukaryotes, but when vegetative reproduction is absent or incidental, there is no parasexual cycle. The parasexual cycle can be studied in organisms where sexual and parasexual recombination occur side by side. In *Aspergillus nidulans* Pontecorvo and co-workers were able to compare the two systems [3]. At that time it was also found in *A. niger* [4] and in some other fungi, but it has not been found in *Neurospora crassa*, the fungus that is best known genetically. In this fungus mitotic recombination has been found in colonies derived from diploid ascospores [5]. The parasexual cycle has been demonstrated in many other imperfect fungi and in some ascomycetes (e.g., in *A. flavus* [6,7], *A. fumigatus* [8], *A. sojae* and *A. oryzae* [9], *A. parasiticus* [10], *Verticillium albo-atrium* [11,12], *Fusarium oxysporum* [13,14], *Penicillium chrysogenum* [15,16], and *Cochiobolus sativus* [17]. It has also been exploited for genetic analysis of fungi of which no sexual reproduction is known.

In imperfect fungi of industrial interest the parasexual cycle provides possibilities for breeding production strains, and in general it enables genetic analyses. In plant pathogenic fungi these processes confer genetic flexibility on the populations; in imperfect fungi of industrial interest somatic recombination provides possibilities for breeding production strains. In general somatic recombination enables genetic analyses of imperfect fungi.

The parasexual processes that may lead to somatic recombination consist of a number of steps that will be discussed in the following paragraphs. Since the review on fungal flexibility of Hansen [18], several reviews on heterokaryosis and parasexuality in fungi have been published [19–22].

3. HETEROKYAROSIS AND PROTOPLAST FUSION

Between two hyphae of the same strain or of compatible strains hyphal fusions (anastomoses) may occur, so that nuclei of different hyphae can migrate. If fusion is between hyphae of genotypically different mycelia, this gives rise to heterokaryons. When hyphal fusions occur, the nuclei can also migrate further in the mycelium, because the septae have a pore. In the conidiospores, however, the two parental types will reappear. When the cells that produce a chain of conidiospores are uninucleate, all spores of a chain will be of the same genotype. When the conidiophore contains nuclei of only one genotype, all conidiospores of a conidial head will be similar.

Heterokaryosis is the first step in somatic recombination and the simplest form of somatic recombination. Heterokaryons can also arise through mutation in a homokaryon, but under laboratory conditions the main origin is the combination of existing genotypical different nuclei.

Heterokaryons can be selected for by combining on minimal medium (MM) two strains with different auxotrophic markers—i.e., strains that are blocked in essential metabolic pathways. So, in principle heterokaryons will grow on MM. If, in addition, the strains differ in color, these heterokaryons show a dense mixture of the two conidial colors. This process is illustrated in Figure 1. A heterokaryon can only be maintained if effective selection for the two complementary auxotrophies takes place, for on complete medium (CM) or supplemented medium (SM) one finds segregation (sorting out) of the two parental types. However, also on MM sorting out of nuclei occurs and, depending on the types of deficiency in the parents, some homokaryotic hyphae may grow among the otherwise heterokaryotic mycelium by cross-feeding. In general a heterokaryon will only be partly truly heterokaryotic. We have cut hyphal tips from the border of a very well balanced heterokaryon of *A. nidulans* and transferred them in turn to MM and to CM. Only 10% of the viable tips proved to be heterokaryotic. Usually heterokaryons are propagated by transfer of a piece of mycelium. In fungi with multinucleate conidiospores (e.g., *Botritys* sp.), the heterokaryotic condition can be maintained upon vegetative propagation by conidiospores.

Köhler [23] was probably the first to study anastomosis in imperfect fungi (in *Botrytis, Fusarium*). Since then anastomoses have been observed in many imperfect fungi. In Fusarium anastomoses occur only at a certain distance of the border of a colony [24]. Hoffmann [25] found in *F. oxysporum* that about two-thirds of the hyphal fusion was between older parallel situated hyphae and

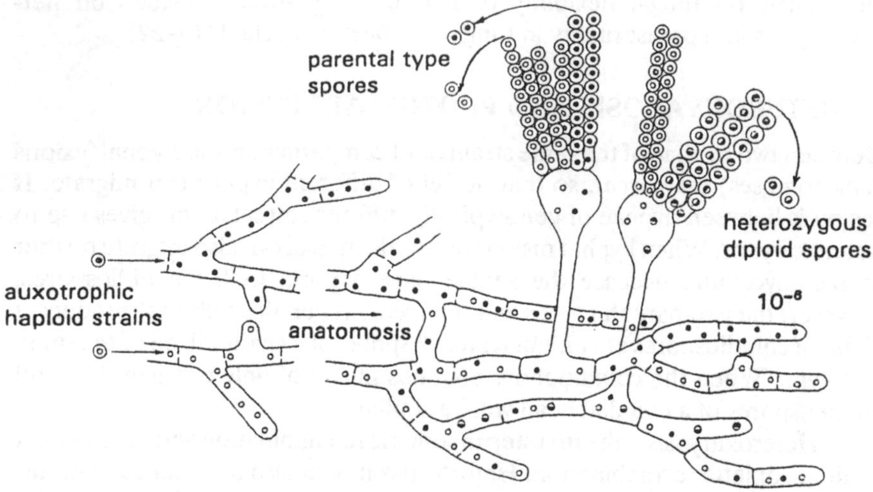

Figure 1 Heterokaryosis in *A. nidulans*. Balanced heterokaryons produce conidia with parental genotypes and at low frequency, also heterozygous diploid conidia. The haploid strains should have different auxotrophic requirements and different colors.

less than 4% between near the tip of the hyphae. In *Helminthosporium* fusion has been observed just behind hyphal tips [26]. In *Verticillium*, on the other hand, fusion between germ tubes of conidia has also been observed [27]. Anastomosis is a complex process. In moist chamber cultures with *Colletotrichum lindemuthianum* hyphae, we have observed that hyphae that grow parallel may attract each other so that side tubes arise at a certain place (Fig. 2). When the hyphae have moved a little the side tubes curve and try to find each other again. Heterokaryosis sometimes seems to be possible between species. Hansen and Smith [28] found interspecific fusion between *Botrytis allii* and *B. ricini*. Uchida et al. [29] between *A. oryzae* and *A. sojae*, Hastie [30] between *V. albo-atrum* and *V. dahliae*.

In several fungi heterokaryon incompatibility has been found. Jinks and co-workers [31] found several incompatibility groups in *A. nidulans* and proved that eight different het-genes proved to be responsible for incompatibility reactions. Incompatibility may have different causes. Sometimes anastomosis does not occur. Sometimes anastomosis proceeds but is followed by an incompatibility reaction, resulting in cell death. The phenomenon is studied in *N. crassa* where anastomosis was not followed by plasmogamy [32,33]. In *A. nidulans* some vegetative incompatible strains produced heterokaryons after prolonged incubation [34]. Papa found 22 incompatibility groups in *A. flavus* [35].

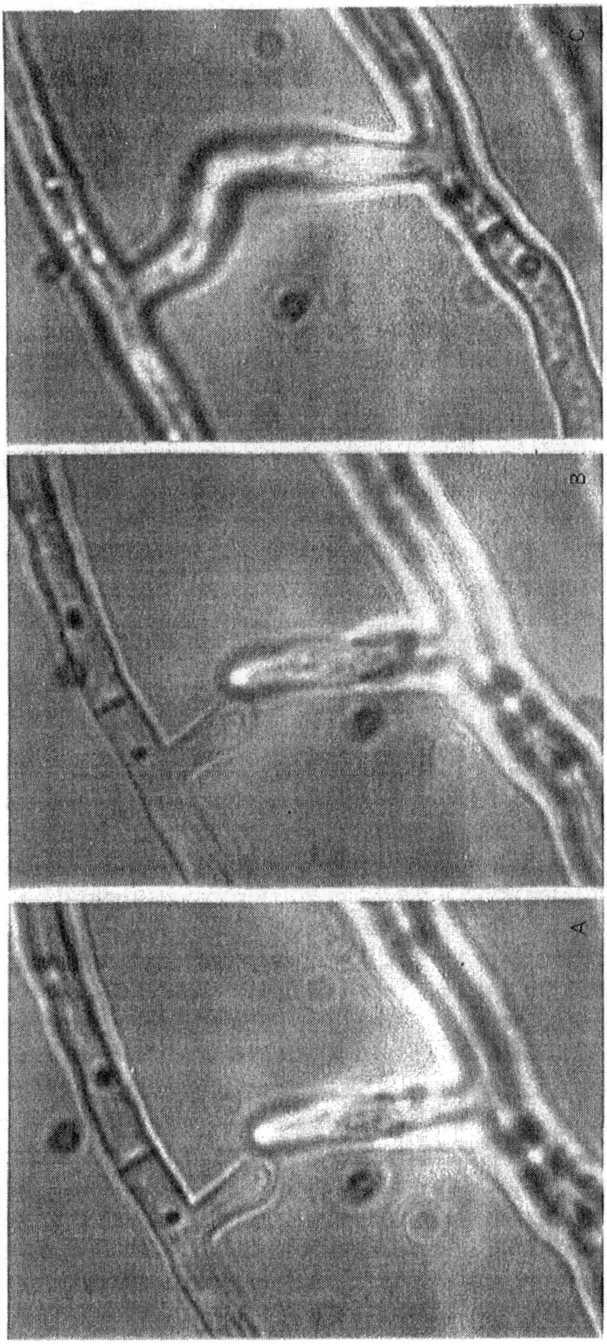

Figure 2 The process of anatomosis in *Colletotrichum lindemuthianum*. Anastomosis in a moist chamber in a drop of liquid complete medium at 22°C. Photos taken at intervals of 1 h.

Heterokaryons are useful for complementation studies. Two phenotypically similar mutants usually complement each other in a heterokaryon when different genes are involved. Because in some cases intragenic complementation can take place, only when two mutants do not complement are they considered as alleles of the same gene. However, some types of nonallelic mutants only complement in diploid nuclei and not in a heterokaryon [36].

Since about 1975 protoplast fusion techniques have been developed for many fungi. Usually young hyphae (germ tubes) are treated with lytic enzymes in an osmotically stabilized medium, and protoplasts can be isolated. Fusion of protoplast can be achieved in stabilizing medium containing PEG (polyethylene glycol). Fusion products are isolated on stabilized minimal medium. In a protoplast fusion experiment including protoplasts of three different *A. nidulans* strains it has been shown that only few protoplasts fuse under such conditions [37]. Protoplast fusion can bypass anastomosis [35], and in some cases fusion of protoplasts from related species may result in heterokaryons [38–40]. In general no or only few heterokaryotic fusion products arise in the case of unrelated strains. Our attempts to fuse *A. nidulans* and *A. niger* protoplasts were not successful; only poorly growing fusion products, which probably grow due to cross-feeding, have been found. Protoplast fusion between unrelated species may be useful for exchange of mitochondria or virus or plasmid transmission [41,42].

4. DIPLOIDS

In *A. nidulans* the frequency of diploid conidia is $10^{-5} - 10^{-6}$. Under laboratory conditions we can force two strains to make heterokaryons (by anastomosis or protoplast fusion) when we start with two strains that differ in auxotrophic requirements. In a heterokaryon two types of nuclei are present, and fusion of two genetically different nuclei results in a heterozygous diploid nucleus. In somatic diploid nuclei two recombination processes may occur: mitotic crossing over (between nonsister chromatids of homologous chromosomes) and haploidization. Mitotic crossing over results in recombinant chromatids. Haploidization results from mitotic nondisjunction of sister chromatids which leads to aneuploid nuclei (2n-1, 2n-1-1, etc.) and by successive losses of chromosomes ultimately to haploid nuclei. During haploidization genes on the same chromosome segregate simultaneously as linkage groups. Mitotic crossing over results in recombination of genes within a linkage group. Both processes are schematically shown in Figure 3.

Occasionally two nuclei in a hypha can fuse by restitution of all chromosomes in one nucleus during mitosis. Somatic karyogamy can only be observed when different nuclei in a heterokaryon fuse to give heterozygous diploid

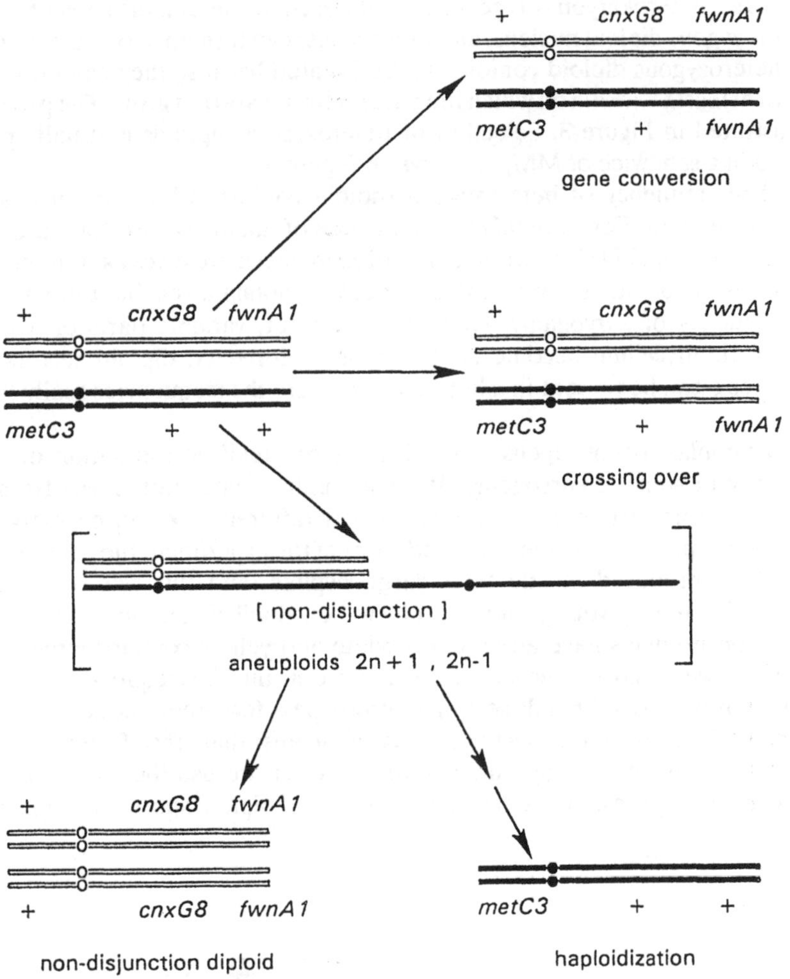

Figure 3 Isolation of heterozygous diploid colonies. Conidiospores collected from a heterokaryon are plated in a layer of minimal medium (MM bottom) and covered with another layer of MM. After 4–6 days of incubation at 37°C (*A. nidulans*), diploid colonies (and also some heterokaryons) appear.

nuclei. In a heterokaryon two genetically different nuclei can fuse, resulting in a heterozygous diploid nucleus, and these nuclei can turn up in conidiospores. The heterozygous diploid conidia can be isolated because they can grow on MM or at least on media on which the parental strains do not grow. The process is illustrated in Figure 3. Selection of heterozygous diploids is usually performed in a sandwich of MM, as shown in Figure 4.

The frequency of heterozygous diploid conidia is both species- and strain-dependent. For *A. nidulans* frequencies of about 10^{-6} are found. Clutterbuck and Roper [43] mentioned that diploid nuclei are twice as frequent in hyphae as in conidia. However, it is difficult to obtain a reliable estimate of the frequency of karyogamy. Due to sorting out, variable parts of a heterokaryotic mycelium become homokaryotic. Moreover, diploid sectors of varying size can be present in a heterokaryon. Are the frequencies really that low?

Protoplast fusion opens a way for a more realistic estimation of the frequency of somatic karyogamy. We used uninucleate protoplasts from a yellow and from a white *A. nidulans* strain with different auxotrophic markers in a fusion experiment. Under the conditions of the experiment the generating protoplasts produced mostly three large hyphae, and this was often still recognizable in very young colonies selected on stabilizing minimal medium. The fusion products have already few white and yellow conidiospores, and sometimes we found a green sector. With those strains the frequency of green sectors was 0.35%. With a dissecting microscope a few green spots were also found in heterokaryotic sectors. This indicates that the frequency of karyogamy may be strongly underestimated when we use the frequency of diploid conidiospores. In practice the frequency of diploid spores is important

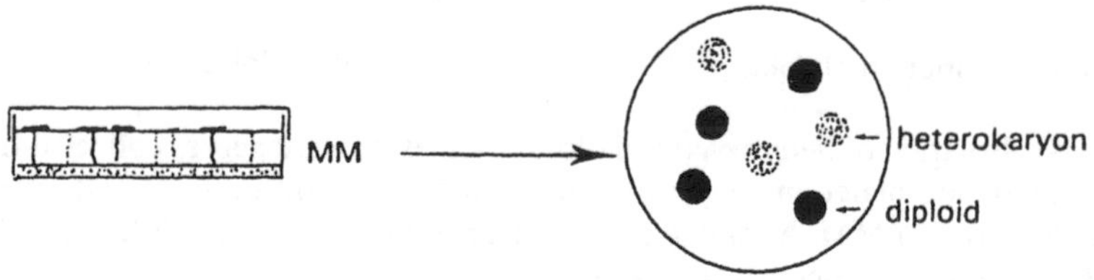

Figure 4 Gene conversion and recombination during mitosis. One pair of homologous chromosomes is drawn to illustrate the consequences of the recombination events that can occur during mitosis. (a) Gene conversion concerns only one gene; (b) mitotic crossing over leads to homozygosity distal to the point of exchange; (c) nondisjunction of chromatids results in aneuploids which in turn will lead to a diploid or a haploid.

for the isolation diploid strains that are needed for genetic analysis. In general, we find on the isolation plates also heterokaryotic colonies. That might be due to small pieces of mycelium left in the conidial suspension. But in some fungi the frequency of heterokaryotic colonies will be high in spite of the effort to filter the suspension. In *A. niger* we often found 20 times more heterokaryotic than diploid colonies. An obvious explanation is that a fraction of the conidiospores has two nuclei and that some are heterokaryotic. In this case the colonies are much more homogeneous than when they started from heterokaryotic mycelium fragments.

Naturally occurring diploid strains have been found in plant pathogenic fungi [44] as well as in biotechnology by important fungi [45]. In some fungi (e.g., *Cladosporium fulvum* [46]) diploids have been found while no balanced heterokaryons could be established.

Although heterozygous diploids can arise by a mutation in a homozygous diploid strain, the existence of a heterokaryon is essential for genetic analysis. A heterokaryon can be forced by mixing two complementing auxotrophic strains and growing them on a minimal medium. As most of the nuclei are uninucleate, a heterokaryon will produce conidiospores of both parental genotypes. Among the conidiospores collected from a heterokaryon the frequency of the heterozygous diploids is often between 10^{-6} and 10^{-5}. They can be selected on basis of the growth on minimal medium. In the case of *A. niger* many more heterokaryons were found on the MM plate than diploid colonies (up to 20 times as many), but these can be distinguished on basis of their morphology (diploid colonies are more compact and have a homogeneous color). Additional proofs that a colony is diploid are volume of the spores and sensitivity to benomyl. Upshall [47] used the frequency of chlorate-resistant mutants as one of the ploidy-determining tests.

Heterozygous diploid strains are isolated by plating of conidiospores from a heterokaryon in 5 mL MM with an additional toplayer of 20 mL MM. The frequency of the diploids is often between 10^{-6} and 10^{-5}. Although these plates will select for heterozygous diploid colonies, varying numbers of heterokaryotic colonies may arise as well.

There can be some complication for the identification of diploids. Some strains have conidiospores with more than one nucleus. When the sterigmata are uninucleate, this is no problem; otherwise the spores might be heterokaryotic and might grow on MM. In *A. flavus*, *A. parasiticus*, and *A. sojae* the color of the spores on a heterokaryon is rather the same as that of heterozygous diploid conidiospores. Consequently it is often difficult to detect diploid colonies. In *A. nidulans* the spore color is determined exclusively by the genotype of the conidiospore. In A. niger the spore color is determined by the genotype of the conidiophore, so that on a heterokaryon, homokaryotic conidiophores show the parental colors and heterokaryotic conidiophores produce the wild-type (black) color due to complementation.

5. HAPLOIDIZATION

Diploids can usually be propagated by transfer of conidiospores. Curing mitosis irregular distribution of choromosomes is more frequent than in higher organs. The main feature of mitosis is that sister chromatids are separated and go to different daughter cells (see Chapter 1). Nondisjunction of sister chromatids results in a 2n+1 nucleus and a 2n–1 nucleus. Aneuploids seem to suffer more from additional chromosome loss. A 2n+1 nucleus results in a diploid identical to the original heterozygous diploid nucleus or with a similar probability to a diploid with two identical chromosomes: a nondisjunction homozygote (see Fig. 3). A 2n–1 nucleus only can lose other chromosomes, and at the end various haploid nuclei will occur. Which chromosome of a pair of homologues will be lost is usually a matter of change. The haploid can have one of the chromosome combinations that can result from free recombination of chromosomes. So, markers on the same chromosome will in principle stay together, whereas free recombination is found for markers on different chromosomes in principle, because mitotic recombination might also have occurred. We shall discuss that below. In *A. nidulans* aneuploid nuclei arise at a frequency of about 2×10^{-2} [48,49].

We have seen that a heterokaryon sorts out producing homokaryotic hyphae, and in this way aneuploid hyphae will also arise. Aneuploids grow usually less vigorously than diploid or haploid hyphae so that sectors can arise. In addition there are few substances that disturb the mitotic spindle, which is responsible for the partition of the sister chromatids over the daughter nuclei. Roper [4] found that *A. nidulans* diploids grow very badly on complete medium with *p*-fluorophenylalanine (*fpa*) and that the poorly growing colonies produce faster-growing sectors that often are haploid. Chloralhydrate stimulates also haploidization, and in many cases benomyl [49] is the most effective nondisjunction agent.

Haploidization is a very reliable way to determine linkage groups. The assignment of a gene to a linkage group can also be very fast if master strains with markers for each linkage group are available. When too many auxotrophic markers are involved in an experiment, slow-growing colonies might be found in the progeny, and then it is easier to use two different test strains with all chromosomes marked in total. For some markers complications in genetic analysis can be found. Certain types of auxotrophic segregants (e.g., adenineless) may be scarce in haploidization experiments, other markers may be difficult to score or may interfere with each other. For *A. nidulans* as well as for *A. niger*, sets of master strains are available.

Markers that are terminal on a chromosome can show some mitotic crossing over, but usually much lower than expected for free recombination (50%) between markers in different linkage groups. So, mitotic crossing over in general will not disturb the results of the haploidization experiments, and the assignment of a gene to a linkage group can mostly be done without doubt. In fungi where the frequency of mitotic recombination is high, propagation of diploids should be avoided. In that case the isolation plates should be stored, and single conidial heads from the original plates should be used. For genetic analyses in *A. niger* we use six small parallel suspensions of single conidial heads from different diploid colonies on the isolation plate. The results will show if, in one of them, mitotic crossing over has disturbed the linkage. If this is the case, that particular suspension will be skipped. The other five showing comparable results can be used for the genetic analysis.

Haploidization is often done by transfer of a master plate with 20–25 diploid colonies with a replicator on complete medium containing *fpa* or benomyl. Such point inoculations usually show colonies with several sectors, but only one of them should be used. It is often necessary to purify the segregants by transfer of a small sample (a spore head) to complete medium.

When a suspension of diploid spores is plated, it is often easier to purify the segregants. When the segregants have been purified, they can be tested for growth requirements. It is advisable to score per contrast and to compare always with the growth on a plate of fully supplemented medium. The test plates differ only for one single growth factor.

Few diploid colonies might be present among the presumed haploids. Absence of any auxotrophic marker makes a colony suspicious when the original diploid contained markers on all or many linkage groups. It might be the original heterozygous diploid. On the other hand, an auxotrophic marker might be present in the case of a nondisjunction diploid, and diploids homozygous as a result of mitotic crossing-over can also be found. Visual observation can give an indication which colonies are suspected not to be haploid. Colonies without or with only a single deficiency might be diploid. The safest decision can be made when the size of the spores or the DNA content is determined. The Coulter counter is very suitable for this purpose especially in combination with a channelizer that visualizes the size distribution. Diploids have twice the volume of haploid conidia.

6. MITOTIC CROSSING OVER

Mitotic crossing over was first observed by Stern [50] in somatic cells of *Drosophila* from so-called twin spots. This phenomenon has been studied extensively in *A. nidulans*, which is very suitable for this purpose. The frequency of mitotic crossing over is high (approximately 10^{-3} per diploid nucleus), collections of several hundreds of genetic markers are available, and heterozygous diploids can be easily selected for. The phenomenon has been observed in several other fungal species (see e.g. reviews by Bradley [19] and Caten [22]), where the frequency of mitotic crossing over might be somewhat higher or lower. Mitotic crossing over provides a tool for the determination of the relative distances of genes with respect to the centromere (mitotic mapping).

Essentially, mitotic crossing over may take place either in the G1 phase between homologous chromosomes or after DNA replication in the G2 phase—i.e., between nonsister chromatids of homologous chromosomes. From experimental data is is concluded that mitotic crossing over occurs predominantly or exclusively in the G2 phase [48,49,51]. Both complementary crossing-over products can be isolated by haploidization of the diploid crossover [52,53]. As mitotic crossing over is fairly rare, in general only one crossover will take place between locus and centromere. In a heterozygote, mitotic crossing over between nonsister chromatids can result in homozygosity of all markers distal of the point of exchange. Homozygous daughter cells result in only one of the two possible anaphase assortments of the chromatids (1+3 and 2+4 in Fig. 5). One of them leads to a phenotype similar to that of the heterozygous diploid. The other can be identified. Strains carrying several linked recessive markers as well as the selective marker in cis position must be constructed by recombination. It is often difficult to find a suitable marker that is selective and also situated terminally on the chromosome. Kafer [54] used strains with reciprocal translocations to overcome this difficulty in *A. nidulans*.

In general, good selection markers terminal on the chromosome arm are essential. In the case of meiotic genetic analysis all ascospores of a cleistothecium descend form the cross, whereas mitotic crossing over occurs incidentally and so mitotic genetic analysis depends strongly on the availability of selection markers. On the basis of different types of markers a number of selection systems have been developed for *A. nidulans* [3]. Some can also be used in other fungi, and others are only for a particular situation. Although the parasexual cycle of *A. niger* has been known about as long as that of *A. nidulans*, the genetic analysis of *A. niger* has been hampered for several decades by the absence of suitable selection markers, but now markers are available for all linkage groups [5,55,56].

Although the frequencies of crossing over and the relative frequencies in various regions differ for mitotic compared with meiotic analysis, the linear arrangement of the genes in a chromosome is the same, and on that basis data from a mitotic and a meiotic approach can be combined. When both meiotic and mitotic genetic analyses are possible, as in *A. nidulans*, the meiotic genetic

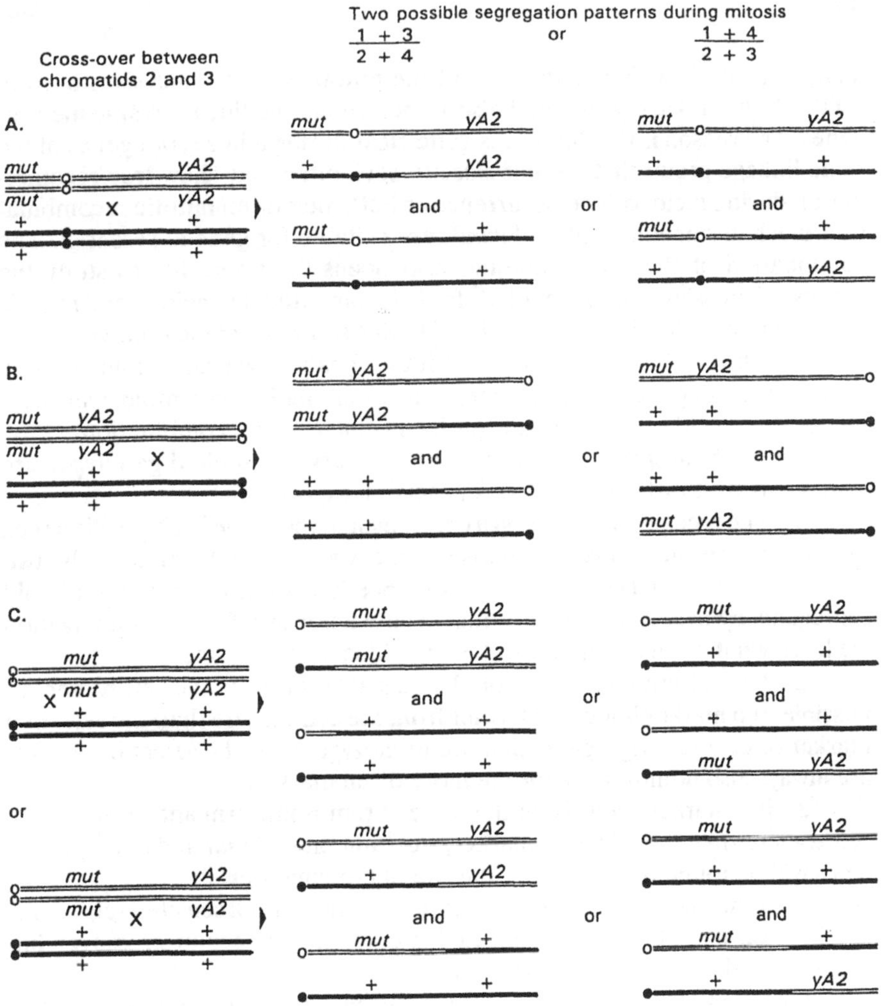

Figure 5 Localization of markers in respect of the centromere. This example is a heterozygous *A. nidulans* diploid, and *yA2* is used as selective marker. Yellow sectors are isolated and the diploid segregants (homozygous for *yA2*) are tested for another marker (*mut*) which may be any mutation in the same linkage group. Note that only crossovers between *yA2* and centromere will give rise to yellow diploids. After mitosis two segregation patterns occur: a daughter nucleus with the chromatids 1+3 and the other with 2+4 or the chromatids segregate as 1+4 and 2+3. Only one crossover situation (crossover between chromatids 2 and 3) has been drawn. In this case the conidiospores with the chromatids 1+3 can be isolated. When all yellow diploid segregants have the mut-phenotype, the arrangement is: *mut——yA2——centromere* (situation B), if none has the mutant phenotype the centromere is probably in the middle (A), and in situation C one can expect both types (ratio depending from the relative distances). The experiment can be done with more than two markers, which gives information on the arrangement of the markers in relation to each other.

analysis is used for fine mapping and the mitotic genetic analysis is used to establish the linkage group and also to determine the linear arrangement of genes. The reason is that linkage is quite clear during mitosis, but genes of the same linkage group that are not close to each other may show free recombination during meiosis. The occurrence of both meiotic en mitotic recombination mechanisms makes *A. nidulans* very suitable for genetic research. The combination of the two techniques also opens the possibility to study the process of crossing over in detail. Kafer [57] compared the meiotic and mitotic genetic maps of the chromosomes I and II. She showed that the relative genetic distances differ. It is not unlikely that future analyses will show that crossing over strongly depends on specific regions. Combination with molecular techniques can elucidate more details of the mechanism of recombination.

Figure 5 illustrates which types of heterozygous diploid genotypes are suitable to start with. We shall look at several aspects:

1. The recessive markers on a chromosome should be in cis position (i.e., on the same chromosome); otherwise, homozygous recombinants for the two recessive genes do not arise. It is still possible to analyze other diploid segregants by haploidization, but that is very laborious. From haploidization we learn what has happened with the individual chromosomes.
2. The selection marker should be distal, as far from the centromere as possible. If a marker is located farther from the centromere than the selection marker (e.g., *yA2*), segregants that are homozygous for the selection marker are always also homozygous for the more distal marker.
3. If a marker (*mut*) is on the same chromosome arm and closer to the centromere than the selection marker, a certain ratio of *mut* and mut^+ phenotypes will be found depending on the site of crossing over.
4. If the marker is on the other chromosome arm, all selected segregants are heterozygous for the nonselected marker, if no second crossover has occurred in the second chromosome arm.

Selection of homozygous diploid segregants can be achieved in different ways:

1. The classical selection markers are color markers—e.g., yellow and white color markers in *A. nidulans* [3]. Segregants are isolated by visual inspection of plates under a dissecting microscope.
2. Suppressors of auxotrophic mutations have also been used in *A. nidulans* from the beginning [58].
3. The vast majority of the available mutants are auxotrophs, and auxotrophic segregants are difficult to isolate. Moreover, there is a risk that auxotrophic segregants suffer from retarded growth on selection medium when segregants are isolated as sectors from point inoculations. Although positive selection of auxotrophic segregants is often impossible, filtration enrichment

has been used to isolate homozygous segregants of *A. nidulans* deficient in carbon metabolism [59] and also for auxotrophic recombinants [60]. Enrichment by lytic enzymes was even more successful and allowed quantitative estimates of the frequency of mitotic crossing over in *A. niger* [59].

4. Resistance to antimetabolites or fungicides can be used, even in the case of a semidominant marker (*acrA*1 in *A. nidulans*). Markers that enable two-way selection deserve special attention. The chlorate resistance (of nitrate nonutilizing mutants) has been used in *A. nidulans* [60] and in *A. niger* [51]. Selenate resistance has been also used in *A. nidulans* to isolate sulphate nonutilizing mutants [61]. Fluoro-orotic acid resistance of pyrimidine deficient mutants has been applied [62].

5. Selection markers can also be introduced by transformation. The *A. nidulans amdS* gene can be expressed in other Aspergilli [63] and has been used as a marker for genetic analysis. A diploid is constructed from a strains with an *amdS* insert and a normal test strain. Mutants with an heterologous *AmdS* insert have been used extensively in genetic analyses of *A. niger* by Debets et al. [64]. The insertion of the plasmid could take place at different places in the genome, and in these experiments it happened that the inserts were often at a terminal end of a chromosome arm, thus extending the genetic map remarkably. Fluoroacetate-resistant (FA^r) diploid segregants can be isolated from such diploids that are hemozygous for the heterologous *amdS* sequence (i.e., the plasmid insert is present on only one of the two homologous chromosomes in the diploid).

Markers suitable for positive selection can produce quantitative information on genetic distances when the necessary precautions are taken. The genetic maps for eight linkage groups of *A. niger* are based on only mitotic crossing over [66]. Mitotic linkage group analysis of (compatible) strains of different origin might elucidate if chromosomal rearrangements are present. Mitotic genetic analysis is also very useful in analyzing transformants to localize inserts. A fast way is to start with a diploid as recipient for transformation [64]. The diploid was heterozygous for several markers in such a way that all chromosomes were marked. Transformants could easily be analyzed by haploidization.

7. STRATEGY FOR MITOTIC GENETIC ANALYSIS

The construction of a heterozygous diploid followed by haploidization is a fast and safe method to construct special genotypes, to introduce an additional marker, or to combine properties of two strains. It can be very useful in the breeding of strains. For some fungi master strains with markers for various linkage groups are available (*A. nidulans* [65], *A. niger* [66]), which can be used in applied projects [67,68]. For strain construction it may often be sufficient to do only recombination between linkage groups. The possibilities for use with strains of different origin depends on heterokaryon compatibility. The *A. niger* strains Lhoas [56] proved to be incompatible with the master strains constructed by Bos and co-workers [60,66], but others proved to be compatible [68,69]. It may be fruitful to study first if groups of strains are compatible. A list of available markers can be found in *Genetic Maps* [70].

8. EXPERIMENTS

A. Linkage Group Analysis in *Aspergillus Niger*

Aim

Linkage group analysis is based on haploidization of a heterozygous diploid that has markers in different linkage groups. Nondisjunction is stimulated by growth on medium with benomyl. As mentioned before we prefer plating of diploid conidiospores on CM + benomyl in stead of transfer with a replicator. We use four different suspensions, and we isolate only one segregant from a colony.

We want to localize the *cysA* gene, and we use the following diploid constructed form tester N722 and the mutant strain N475:

N722	*fwn*A1	+	*his*D4	*bio*A1	*lys*A7	*leu*A1	*pab*A1	*cnx*C5	+
N475	+	*olv*A1	+	+	+	+	+	+	*cis*A4
		I	II		III	IV	VI	VII	?

The diploid is isolated from a heterokaryon of both strains on MM. A conidial suspension obtained from the heterokaryon is plated in a thin layer and covered with a thick top layer. Diploid as well as heterokaryotic colonies will appear on this isolation plate. We make four suspensions by taking very small samples (on conidial head) of six different colonies. The suspensions are plated separately.

Requirements

Day 1: 4 plates CM + benomyl
Day 2: 4 plates CM
Day 3: 4 plates CM
Day 4: 4 times the test series, 1 plate WA 3% for cooling (see Fig. 6).

Experimental Procedure

Day 1: Plate the four suspensions each at a density of about 100 conidia per plate on a plate CM + benomyl.
Incubate at 30°C for 4 days.

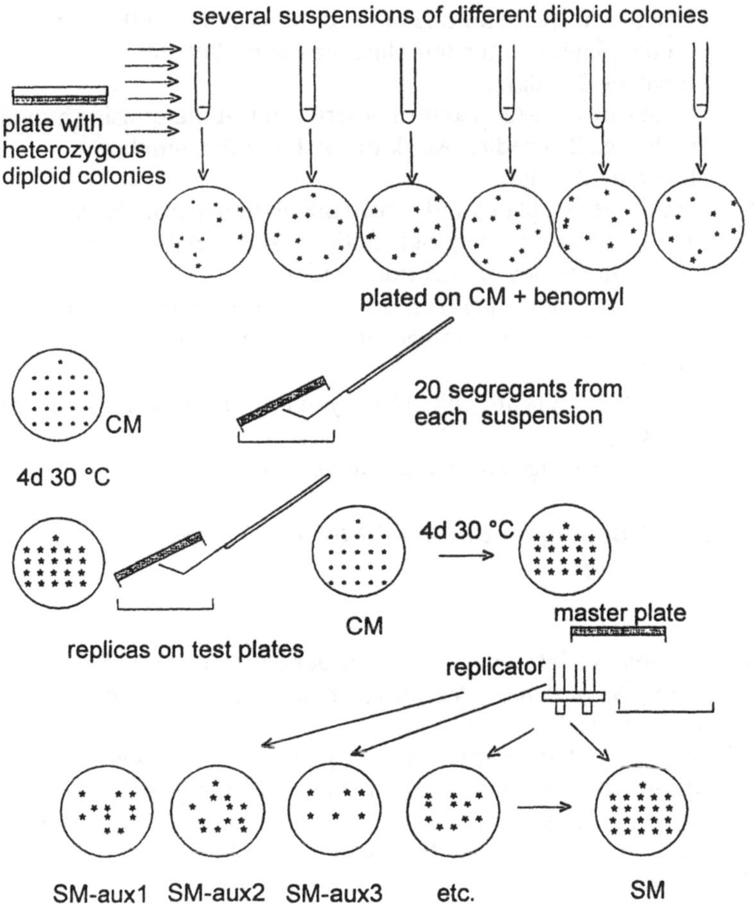

Figure 6 Outline of linkage analysis.

Day 2: Several colonies have formed sectors on the benomyl plates. The haploid sectors can be recognized on basis of the color. Transfer one conidial head from only one haploid sector of each colony onto a plate CM, 21 different ones from one suspension.
Incubate 2–3 days.

Day 3: Purify the isolated segregants (20 of each suspension) by taking only one spore head.

Use the "0" place for one of the parents and note the strain number on the plate. Use the template for the replicator.
Incubate 2–3 days.

Day 4: Replicate the segregants on a series of test plates using a replicator with, e.g., 21 needles. Mark the plates with a number.
Incubate 2–3 days.

Day 5: Score the test plates and record the phenotypes in the scoring table.
Try to determine the ploidy of the colonies on basis of the markers.
Check the ploidy of uncertain colonies.
Make a list of the genotypes (you can use a computer program).
Determine the percentage of the recombinants for the different markers.
Which markers are linked? Do you find intrachromosomal recombination?
In which linkage group is the unknown marker?

B. Genetic Mapping on Basis of Mitotic Crossing Over in *A. niger*

Aim

In this experiment we determine the gene order and location of the centromere in linkage group VI of *A. niger*. Various techniques for the selection of mitotic recombinants are used.

The markers involved are: *pabA*1, *pyrB*4, *cnxA*1. The *cnxA* locus is terminal on the chromosome arm and therefore very suitable as selection marker. As all markers are recessive, they should be in cis position. The chlorate-resistant segregants can be isolated, and they become analyzed for the other markers.

The diploid used is:

N761	*fwnA*1	+	*pabA*1	*pyrB*4	*cnxA*1
N640	+	*metB*10	+	+	+
	I	V		VI	

Requirements

Day 1: The diploid on MM, 4 plates $CMClO_3$
A suspension of conidiospores from a diploid colony

Day 2: 3 plates CM + ClO_3, bottle with saline or sterile water

Day 3: 3 plates CM

Day 4: 3 plates CM + ClO_3, 3 plates CM + oli
3 plates SM (= MM + ureum + uridine + pab)
3 plates SM-pab, 3 plates SM-uri, 1 plate WA 3% for cooling.

Experimental Procedure

Day 1: Plate a suspension of diploid conidiospores on four plates CM + ClO_3 at a density of about 1000 conidiospores per plate. From the literature we expect about 2% *cnxA* recombinants.
Incubate at 30°C for 3 days.

Day 2: Transfer one spore head from the chlorate-resistant colony onto a new plate CM + ClO_3 (3 plates with 21 colonies per plate).
Incubate 2–3 days.

Day 3: Purify the isolated segregants by inoculating one spore head on CM now 3 × 20, inoculate the parent strains now on the "0" place.
Incubate 2–3 days.

Day 4: Replicate the segregants on the test series using the needle replicator. Mark the replicas of a master plate so that it is known which belong together.
Incubate 2–3 days.

Day 5: Score the test series and record the phenotypes in the table.
Try to determine the ploidy of the colonies.
Determine the frequency of chlorate-resistant diploid recombinants and conclude the linear arrangement of the markers with respect to the centromere.

REFERENCES

1. Haldane, J. B. S. (1955). Some alternatives to sex. *New Biol. 19*:7.
2. Pontecorvo, G. (1954). Mitotic recombination as the genetic system of filamentous fungi. *Caryologia Suppl. 6*:192.
3. Pontecorvo, G., J. A. Roper, L. M. Hemmons, K. D. MacDonald, A. W. J. Bufton (1953). The genetics of *Aspergillus nidulans. Adv. Genet. 5*:141.
4. Roper, J. A. (1952). Production of heterozygous diploids in filamentous fungi. *Experimentia 8*:14.
5. Smith, D. A. (1974). Unstable diploids of *Neurospora* and a model for their somatic behaviour. *Genetics 76*:1.
6. Papa, K. E. (1973). The parasexual cycle in *Aspergillus flavus. Mycologia 65*:1201.
7. Papa, K. E. (1976). Linkage groups in *Aspergillus flavus. Mycologia 68*:159.
8. Berg, C. M., E. D. Garber (1962). A genetic analysis of colour mutants of *Aspergillus fumigatus. Genetics 47*:1139.

9. Ishitani, C., Y. Ikeda, K. Sakaguchi (1956). Hereditary variation and genetic recombination in Koji-molds (*Aspergillus oryzea* and *A. sojae*). VI. Genetic recombination in heterozygous diploids. *J. Gen. Application Microbiol.* (*Jpn.*) *2*:401.
10. Papa, K. E. (1978). The parasexual cycle in *Aspergillus parasiticus. Mycologia 70*:766.
11. Hastie, A. C. (1964). The parasexual cycle in *Verticillium albo-atrum. J. Gen. Microbiol. 27*:373.
12. Hastie, A. C. (1967). Mitotic recombination in conidiospores of *Verticillium albo-atrum. Nature 214*:249.
13. Buxton, E. W. (1956). Heterokaryosis and parasexual recombination in pathogenic strains of *Fusarium oxysporum. J. Gen. Microbiol. 15*:133.
14. Garber, E. D., E. G. Wyttenbach, T. S. Dhillon (1961). Heterokaryosis involving formae of *Fusarium oxysporum. Am. J. Bot. 48*:325.
15. Pontecorvo, G., G. Sermonti (1954). Parasexual recombination in *Penicillium chrysogenum. J. Gen. Microbiol. 11*:94.
16. MacDonald, K. D., J. M. Hutkinson, W. A. Gillet (1963). Formation and segregation of heterozygous diploids between a wild type strain and derivatives of high penicillin yield in *Penicillium chrysogenum. J. Gen. Microbiol. 33*:383.
17. Tinline, R. D. (1962). Cochiobolus sativus. V. Heterokaryosis and parasexuality. *Can. J. Bot. 40*:425.
18. Hansen, H. N. (1938). The dual phenomenon in imperfect fungi. *Mycologia 30*:442.
19. Bradley, S. G. (1962). Parasexual phenomena in microorganisms. *Annu. Rev. Microbiol. 16*:35.
20. Parmeter, J. R., Jr., W. C. Snyder, R. E. Reichle (1963). Heterokaryosis and variability in plant-pathogenic fungi. *Annu. Rev. Phytopathol. 1*:51.
21. Tinline, R. D., B. H. Macneill (1969). Parasexuality in plant pathogenic fungi. *Annu. Rev. Phytopathol. 7*:147.
22. Caten, C. E. (1980). Parasexual processes in fungi. In: K. Gull, S. G. Oliver (eds.). *The Fungal Nucleus*, Cambridge Univ. Press, Cambridge, p. 191.
23. Köhler, E. (1930). Zur Kentnisse der vegetativen Anastomosen der Pilze. II. *Mitt Planta 10*:3.
24. Dickinson, S. (1932). The nature of saltation in *Fusarium* and *Helminthosporium. Univ. Minn. Agr. Exp. Sta. Bull. 88*:1.
25. Hoffmann, G. M. (1967). Untersuchungen über die heterokaryosebildung und den Parasexualcyclus bei *Fusarium oxysporum*. III. Paarungsversuche mit auxotrophen Mutanten van *Fusarium oxysporum* f. *callistephi. Arch. Mikrobiol. 56*:40.
26. Hrushovetz, S. B. (1956). Cytological studies of *Helminthosporium sativum. Can. J. Bot. 34*:321.
27. Schreiber, L. R., R. J. Green (1966). Aanastomosis in *Verticillium albo-atrum. Phytopathology 56*:1110.
28. Hansen, H. N., R. E. Smith (1932). The mechanism of variation in imperfect fungi: *Botrytis cinera. Phytopathology 22*:953.
29. Uchida, K., C. Ishitani, Y. Ikeda, K. Sakaguchi (1958). An attempt to produce interspecific hybrids between Aspergillus oryzea and A. sojae. J. Gen. Microbiol. 4:31.

30. Hastie, A. C. (1973). Hybridization of *Verticillium albo-atrum* and *Verticillium dahliae*. *Trans. Br. Mucol. Soc. 60*:511.
31. Jinks, J. L., C. E. Caten, G. Simchen, J. H. Croft (1966). Heterokaryon incompatibility and variation in wild populations of *Aspergillus nidulans*. *Heredity 21*:227.
32. Garnjobst, L. (1955). Further analysis of genetic control of heterokaryosis in *Neurospora crassa*. *Am. J. Bot. 42*:444.
33. Garnjobst, L., J. F. Wilson (1956). Heterocaryosis and protoplasmic incompatibility in *Neurospora crassa*. *Proc. Natl. Acad. Sci. 42*:613.
34. Dales, R. B. G., J. H. Croft (1977). Protoplast fusion and the isolation of heterokaryons and diploids from vegetatively incompatible strains of *Aspergillus nidulans*. *FEMS Microbiol. Lett. 1*:201.
35. Papa, K. E. (1986). Heterokaryon incompatibility in *Aspergillus flavus*. *Mycologia 78*:98.
36. Roberts, C. F. (1964). Complementation in balanced heterokaryons and heterozygous diploids of *Aspergillus nidulans*. *Genet. Res. 5*:211.
37. Bos, C. J., A. J. M. Debets, A. W. Van Heusden, H. T. A. M. Schepers (1983). Fusion of protoplasts from conidiospores of *Aspergillus nidulans*. *Experientia Suppl. 45*:298.
38. Ferenczy, L., M. Szegedi, F. Kevei (1977). Interspecific protoplast fusion and complementation in Aspergilli. *Experientia 33*:184.
39. Kevei, F., J. F. Peberdy (1977). Interspecific hybridization between *Aspergillus nidulans* and *Aspergillus rigulosus* of fusion of somatic protoplasts. *J. Gen. Microbiol. 102*:255.
40. Mellon, F. M., J. F. Peberdy, K. D. MacDonald (1983). Hybridization of *Penicillium chrysogenum* and *P. baarnense* by protoplast fusion; genetic and biochemical analysis. *Experientia Suppl. 45*:310.
41. Ferenczy, L. (1981). Microbial protoplast fusion. In: Glover, S. W., D. A. Hopwood (eds.). *Genetics as a Tool in Microbiology*. Cambridge Univ. Press, Cambridge, p. 1.
42. Kaiying, C., L. Pingyan (1984). Interspecific somatic hybridization between *Penicillium patulum* and *P. chrysogenum*. II. Viruses transmit through protoplast fusion. *Acta Genet. Sin. 11*:159.
43. Clutterbuck, A. J., J. A. Roper (1966). A direct determination of nuclear distribution in heterokaryons of *Aspergillus nidulans*. *Genet. Res. 7*:185.
44. Caten, C. E., A. W. Day (1977). Diploidy in plant pathogenic fungi. *Annu. Rev. Phytopathol. 15*:295.
45. Yuill, E. (1950). The numbers of nuclei in conidia of Apsergilli. *Trans. Br. Myc. Soc. 33*:324.
46. Van Tuyl, J. M. (1977). Genetics of fungal resistance to systematic fungicides. Thesis. Mededelingen Landbouwhogeschool Wageningen; Wageningen, Netherlands.
47. Upshall, A. (1981). Naturally occurring diploid isolate of *Aspergillus nidulans*. *J. Gen. Microbiol. 122*:7.
48. Käfer, E. (1961). The process of spontaneous recombination in vegetative nuclei of *Apsergillus nidulans*. *Genetics 46*:1581.

49. Kafer, E. (1977). Meiotic and mitotic recombination in *Aspergillus nidulans* and its chromosomal aberrations. *Adv. Genet. 19*:33.
50. Stern, C. (1936). Somatic crossing over and segregation in *Drosophila melanogaster. Genetics 21*:625.
51. Roper, J. A., R. H. Pritchard (1955). The recovery of reciprocal products of mitotic crossing-over. *Nature 175*:639.
52. Pritchard, R. H. (1955). The linear arrangement of a series of alleles of *Aspergillus nidulans. Caryologia 6* (Suppl. 1):1117.
53. Debets, A. J. M., K. Swart, C. J. Bos (1990). Genetic analysis of *Aspergillus niger*: Isolation of chlorate resistance mutants, their use in mitotic mapping and evidence for an eight linkage group. *Mol. Gen. Genet. 221*:453.
54. Kafer, E. (1975). Reciprocal translocations and translocation disomics of *Aspergillus* and their use for genetic mapping. *Genetics 79*:7.
55. Pontecorvo, G., J. A. Roper, E. Forbes (1953). Genetic recombination without sexual reproduction in *Aspergillus nidulans. J. Genet. 52*:198.
56. Lhoas, P. (1967). Genetic analysis by means of the parasexual cycle in *Aspergillus niger. Genet. Res. 10*:45.
57. Kafer, E. (1958). An eight chromosome map of *Aspergillus nidulans. Adv. Genet. 9*:105.
58. Pontecorvo, G., E. Kafer (1958). Genetic analysis based on mitotic recombination. *Adv. Genet. 9*:71.
59. Bos, C. J., S. M. Slakhorst, J. Visser, C. F. Roberts (1981). A third unlinked gene controlling the pyruvate dehydrogenase complex in *Aspergillus nidulans. J. Bacteriol. 148*:594.
60. Bos, C. J., A. J. M. Debets, K. Swart, A. Huybers, G. Kobus, S. M. Slakhorst (1988). Genetic analysis and the construction of master strains for assignment of genes to linkage groups in *Aspergillus niger. Curr. Genet. 14*:437.
61. Debets, A. J. M., K. Swart, C. J. Bos (1989). Mitotic mapping in linkage group V of *Aspergillus niger* based on selection of auxotrophic recombinants by Novozym enrichment. *Can. J. Microbiol. 35*:982.
62. Cove, D. J. (1976). Chlorate toxicity in *Aspergillus nidulans*: the selection and characterisation of chlorate resistant mutants. *Heredity 36*:191.
63. Kelly, J. M., M. J. Hynes (1985). Transformation of *Aspergillus niger* by the *amdS* gene of *Aspergillus nidulans. EMBO J.* 4:475.
64. Debets, A. J. M., K. Swart, C. J. Bos (1990). Genetic analysis of *Aspergillus niger*: isolation of chlorate resistance mutants their use in mitotic mapping and evidence for an eighth linkage group. *Mol. Gen. Genet. 224*:264.
65. Clutterbuck, A. J. (1993). *Aspergillus nidulans.* In: O'Brien, S. J. (ed.). *Genetic Maps.* Cold Spring Harbor Laboratory Press, Cold Spring Harbor, NY, p. 3.71.
66. Bos, C. J., S. M. Slakhorst, A. J. M. Debets, K. Swart (1993). Linkage group analysis in *Aspergillus niger. Appl. Microbiol. Biotechnol. 38*:742.
67. Swart, K., P. J. Van der Vondervoort, C. F. B. Witteveen, J. Visser (1990). Genetic localization of a series of genes affecting glucose oxidase levels in *Aspergillus niger. Curr. Genet. 18*:435.

68. Boschloo, J. G., A. Paffen, T. Koot, W. J. J. Van de Tweel, R. F. M. Van Gorcom, J. H. G. Cordewener, C. J. Bos (1991). Genetic analysis of benzoate metabolism in *Aspergillus niger*. *Appl. Microbiol. Biotechnol.* *34*:225.
69. Valent, G. U., M. R. Calil, R. Bonatelli Jr. (1992). Isolation and genetic analysis of *Aspergillus niger* mutants with reduced extracellular glucoamylase. *Rev. Brasil. Genet.* *15*:19.
70. Bos, C. J., F. Debets, K. Swart (1993). *Aspergillus niger* genetic loci. In: O'Brien, S. J. (ed.). *Genetic Maps*. Cold Spring Harbor Laboratory Press, Cold Spring Harbor, NY, p. 3.87.

5

Molecular Genetic Analysis

Theo Goosen and Fons Debets
Wageningen Agricultural University, Wageningen, The Netherlands

1. INTRODUCTION

Genetic analysis has long been restricted to a few fungi, especially those that could be easily grown on simple media in the laboratory. In such fungi, best exemplified by *Saccharomyces cerevisiae*, *Neurospora crassa*, and *Aspergillus nidulans*, large numbers of mutants could be isolated (see Chapter 2). With these mutants detailed genetic maps [1–3] have been constructed for these organisms, using parasexual analysis (see Chapter 4) as well as results from genetic crossings (see Chapter 3).

In many more fungi, however, such detailed genetic analyses have not been possible. The main reason for this is usually either the impossibility to grow the fungus on a simple, defined medium, as is the case with obligate parasites, or the lack of natural ways to exchange genetic information necessary for mapping, as in those imperfect fungi in which until now no parasexual cycle has been observed. Among these fungi there are quite a few that have an important economic and social impact. In the last decade, considerable progress has been made with the introduction of molecular genetic techniques in fungal research. In this chapter we will first discuss physical karyotyping on basis of the electrophoretic separation of whole chromosomes, and then we will deal with several techniques for the use of short DNA sequences as molecular markers for genetic analysis.

Six concepts are these:

1. Physical karyotyping enables analysis of genome rearrangements and has elucidated an unexpected variability in genomes.

2. Restriction fragment analysis provided new ways for genetic characterization of strains.

3. Restriction fragment length polymorphisms (RFLPs) proved to be good molecular markers for genetic analysis.

4. With probes for repetitive sequences DNA fingerprints can be obtained to characterize individual isolates.

5. Polymerase chain reaction (PCR) techniques made it possible to make probes from very small amounts of DNA, that can be used for RFLP analysis.

6. A further extension is the use of small artificial DNA sequences as probe. This is achieved in the random amplification of polymorphic DNA (RAPD) technique.

As a consequence, many of the aforementioned fungi have attracted new attention. Studies at the molecular genetic level have been initiated in many scientific disciplines, including (phyto)pathology, molecular biology, ecology, and population and evolution genetics. Some of the molecular genetic techniques that are, or can be, applied to such research will be reviewed here.

2. PHYSICAL KARYOTYPING

The characteristics of the genome in terms of the number of chromosomes and the microscopic morphology of each of these are called the karyotype [4]. In plants, the information provided by karyotype analysis is often used as a taxonomical trait in species description and in genetics for studying chromosome number variations and morphological aberrations. In fungi, karyotyping by cytological methods is much more difficult because the chromosomes are relatively small and therefore, the technique has not been applied extensively. Using the better morphology of chromosomes in meiotic cells and in post-meiotic mitosis it is feasible [5], and chromosome counts have been obtained for several fungi [6]. Studying chromosome morphology has not been very successful in fungi. Moreover, in asexual species the approach is not possible. The introduction of the technique of pulsed field gel electrophoresis (PFGE) has allowed the separation of chromosome-sized DNAs [7–9]. Thus "electrophoretic" (also called *physical* or *molecular*) karyotypes can be obtained relatively easily. To date, for many fungal species of all major classes karyotype analysis has been applied, generating valuable information on variation in chromosome size and number within and between species, on genetic linkage of markers as well as applications in gene mapping and cloning and in analyzing transformants and chromosome mutations. Examples of the application of PFGE in analyzing the fungal genome will be presented. At first, some technical aspects of PFGE will be discussed. Some reviews on principles of PFGE systems: [10–13].

A. Principles of PFGE and Technical Aspects

In conventional gel electrophoresis DNA molecules smaller than 50 kb can be separated by size. Larger DNA molecules have a minimal velocity that is not proportional to their length and therefore cannot be separated by size. Since the introduction of the concept that chromosome-size DNA molecules can be separated by using two alternating fields (so-called pulsed field gel electrophoresis), several systems have been developed. Most commonly used is the contour-clamped homogeneous electric field (CHEF) gel electrophoresis. In the CHEF system, the electric field alternates between two orientations at angles of 120° [14]. The duration of the alternating electric fields is called the pulse time. After changing the electric field, DNA molecules first have to reorient before migrating into the new direction. The pulse time can be considered to consist of a reorientation period and a migration time. Size-dependent separation of large DNA molecules in PFGE is based on the principle that reorientation time is size-dependent: small DNA molecules need less time to reorient and thus spend more time migrating than larger molecules. Resolution of DNA molecules in a certain size range requires a specific pulse time. To get maximum resolution of the various chromosome-size DNAs of a fungus, it is often necessary to change the pulse time during the electrophoretic run.

Preparation of Intact Chromosomal DNA

For many fungi protoplast formation is routinely done by using the commercially available Novozym 234 [15,16]. As an example, the protocol we use for *Aspergillus* will be given. Mycelial protoplasts are isolated using standard procedures [17]. Protoplasts are subsequently washed in isotonic medium (1.2 M sorbitol) containing EDTA (50 mM) and resuspended in 0.5–0.8% low-melting-point agarose (e.g., InCert agarose, FMC) in isotonic medium containing 500 mM EDTA and 1–2 mg/mL proteinase K. The final concentration of protoplasts being about 2×10^8/mL. The mixture is subsequently pipetted into a prechilled mold to obtain plugs. Next the embedded protoplasts are lysed in situ, by incubating the plugs in a mixture of 1% N-lauroylsarcosine, 500 mM EDTA and 1 mg/mL proteinase K at 50°C for 48 h. After washing (in 50 mM EDTA) the plugs are stored in 50 mm EDTA at 4°C. Generally, upon storage embedded chromosomal DNA remains intact for several years. For size estimation commercially available standards can be used (*Saccharomyces cerevisiae* and *Schizosaccharomyces pombe*) with chromosomes ranging from several hundred kb to 5.7 Mb.

Electrophoresis Conditions

As an example of CHEF electrophoresis the protocol for the separation of *Aspergillus niger* chromosomal DNAs ranging in size from 3.5 to 6.6 Mb will be given [16,18]. Electrophoresis is performed using the CHEF-DR II (or its successor CHEF-DR III) apparatus of Bio-Rad. Gels are prepared of agarose 0.5% of low electroendosmosis (EEO), like chromosomal-grade agarose (Bio-Rad), Seakem GTC (FMC), or Megarose (Clontec). DNA-agarose plugs are loaded as follows: the plugs are placed next to the comb in an empty mold and fixed with agarose, and then agarose is poured gently into the mold. After cooling, the comb is removed and the holes are filled with agarose. Gels are electrophoresed at 9°C in circulating 0.5 × TAE buffer [19] at 45 V with pulse intervals of 55 min, 47 min, and 40 min each over 48 h. Gels are stained in 0.5 μg/mL ethidium bromide for 1 h and then destained in water for 1 h and photographed under UV illumination (at 302 nm). Gels can be processed to denature the DNA, blotted to appropriate membranes, and used for hybridization analysis using standard processes.

B. Recognition of Linkage Groups in the Electrophoretic Karyotype

The thus obtained banding pattern of choromosomal DNAs is called the primary karyotype. Using the yeast chromosomal DNAs as size standards, chromosome sizes can be estimated. For well-studied fungi like *Neurospora crassa* and *Aspergillus nidulans*, of which genetic linkage groups (LGs) are available, a correlation can be made with specific bands in the physical karyotype by hybridization using LG-specific probes and/or by analyzing translocation strains [15,20]. In this way it can be inferred which bands are singlets (i.e., represent a single chromosome) and which are duplets or even triplets. Subsequently the total genome size can be estimated. A special application of PFGE has been the partitioning of *A. nidulans* cosmid libraries into chromosome-specific subcollections [21]. In the absence of LG-specific probes, chromosome counts can be based on band intensity. Alternatively, LG-specific markers can be introduced.

In the economically important asexual fungus *A. niger* genetic analysis using parasexual processes is possible, and genetic LGs have been obtained and a genetic map has been constructed [22]. On the other hand, several genes have been cloned but have not been assigned to a specific LG, because there are no mutants of the genes with a usable phenotype for genetic analysis. Such cloned genes can be assigned by hybridization to a specific band in the karyotype. However, because of the lack of LG-specific probes in *A. niger*, LGs had to be correlated to the chromosomal DNA bands using an alternative method. By transformation of *A. niger* with the heterologous *amdS* gene, several transformants were obtained. For each of these transformants, the chromosome in which *amdS* was integrated was determined by genetic analysis [23]. In this study the *amdS* insert in each transformant could be assigned to a single site on a chromosome. Seven transformants, each carrying the heterologous *amdS* gene on a different chromosome, were subsequently analyzed by CHEF electrophoresis. Thus, seven of the eight LGs of *A. niger* could be correlated with specific DNA bands in the karyotype upon hybridization with the *amdS* probe [16]. However, several of the chromosomal DNAs of *A. niger* are similar in size and comigrate. Therefore, unambiguous assignment of cloned genes to *A. niger* chromosomes by CHEF analysis could not be done this way. It was observed that one transformant with a relatively large insert showed a significant change in chromosome size and migration position. This principle was further exploited by generating a number of strains with altered electrophoretic karyotypes by the introduction of multiple copies of the *glaA* gene [24]. Thus, a set of tester strains for chromosome assignment has been constructed [18] and can be used as an alternative to or in combination with the available set of tester strains for formal genetic analysis [25].

C. Karyotype Variations

Studying electrophoretic karyotype variability has several applications: 1.) It may be used to detect genomic rearrangements in industrial strains. 2.) Analysis of the variation within and between species may provide insight into the reproductive strategy of the fungus (e.g., chromosome length variations among strains within a population may hinder sexual reproduction) and may be used as a taxonomic tool. 3.) Finally, PFGE allows the detection of supernumerary (B) chromosomes.

1. Commercial strain typification can be facilitated using karyotype variation as detected by PFGE. Genomic rearrangements due to the heavy mutagenesis in an industrial strain improvement program may thus be detected. For example, in commercials trains of *Acremonium chrysogenum* differing in rate of cephalosporin C biosynthesis, but showing no RFLP variation [26], different chromosome patterns were found after using PFGE. More specifically, a number of *A. chrysogenum* strains from a lineage improved in cephalosporin C production were analyzed using PFGE to detect at what points across the lineage chromosome rearrangements had occurred [27]. Genomic rearrangements were also observed in the commercial heterokaryon of *Agaricus bisporus* as indicated by the presence of noncomigrating homologous

chromosomes identified by a number of DNA probes [28]. Similarly, karyotype variations could be detected comparing wild-type and high cellulase-producing mutant strains of *Trichoderma longibrachiatum* (*reesei*) [29].

2. Physical karyotyping may also be used as a taxonomic tool. Karyotypic variability in *Nectria haematococca* was shown to correlate with pathogenic variation [30]. In *Leptosphaeria maculans,* electrophoretic karyotyping differentiates highly virulent from weakly virulent isolates [31]. A high degree of karyotypic variability was observed among individual strains of *Ustilago hordei* representing each of the 14 races but not among strains derived from individual meiotic tetrads, although the segregation of a chromosome length polymorphism in one meiotic tetrad was observed [32]. Chromosomal DNA banding patterns identified by Fekete et al. [33] for eight species of *Fusaria* support current taxonomical classification. Considerable karyotype variation was also found among culture collection strains of the black Aspergilli [34] in members of *Aspergillus* section *Flavi* [35] and in a *Septoria tritici* population [36]. Cooley and Caten [37] found clear karyotype differences both between and within groups of strains of *S. nodorum* adapted to wheat or barley. Moreover, considerable karyotype variation was apparent even among six wheat-adapted strains isolated from the same population. Finally, electrophoretic karyotype differences between mating types of the zygomycete *Absidia glauca* have been detected [38].

3. In plants as well as animals, especially insects, B chromosomes (also called supernumerary chromosomes) accumulate [39]. B chromosomes are dispensable and are about the size of normal chromosomes or somewhat smaller. In others they are considerably smaller. B chromosomes are meiotically unstable. PFGE enabled the detection of small chromosomes in some plant pathogenic fungi characteristic of B chromosomes [40]. In *Nectria haematococca Pda* genes are responsible for resistance to the phytoalexin pisatin, an antimicrobial compound produced by garden pea. Miao et al. [41] found that the *Pda6* gene was mapped by electrophoretic karyotyping to a small, dispensable, and meiotically unstable chromosome only present in virulent strains. In the phytopathogen *Colletotrichum gloeosporioides* a strain-specific dispensable cyclin homolog was encoded on a minichromosome [42]. B chromosomes were also discovered in *Cochliobolus heterostrophus.*

In conclusion, PFGE has enabled the analysis of chromosome size and number in fungi. Since the introduction of the concept it has been widely used in the analysis of fungal karyotypes. In combination with standard genetic analysis techniques it is of great value for gene mapping and cloning, for analyzing genome rearrangements, as tool in taxonomic studies, and for analyzing transformants, studying karyotype variations among populations, and detection and analysis of B chromosomes.

3. RESTRICTION FRAGMENT ANALYSIS

The basis of genetic analysis is to have tools to discriminate biological entities at many different levels. Classically, this has been done using morphological and physiological criteria to distinguish between species. With filamentous fungi discrimination at species level using these traits already can be very difficult or even give erroneous results [43,44]. At even lower levels (e.g., isolates of one species from different populations), these methods are no longer applicable. Molecular markers, however, can be applied at these levels with great reliability.

Molecular markers are not principally different from classical genetic mutations. Both are the result of changes in the DNA of the organism, but where with classical mutants this resulted in a phenotype that can be distinguished at the cellular level, the phenotype of a molecular marker can only be observed by looking directly at the DNA. The first molecular markers that have been used, which are simple to interpret and therefore still are very popular, are restriction fragment length polymorphisms (RFLPs). An RFLP can be the result of the gain or loss of a restriction endonuclease cleavage site at a given chromosomal location (Fig. 1), resulting from point mutations in the DNA. However, deletions and insertions will also lead to alterations in the length of the restriction fragments.

RFLPs can be detected by analyzing restriction digests of chromosomal DNA preparations by Southern blotting. Individual fragments on the blots are visualized by hybridization with labeled probes. These probes are usually generated from cloned genomic DNA fragments or from cDNA. The patterns that result can then be compared and used to evaluate the nature and number of differences between the two samples (Fig. 1).

RFLP analysis thus requires the isolation of substantial amounts of rather extensively purified DNA. Although several good protocols for DNA extraction from fungi have been published [45–48], in many cases obtaining restrictable DNA is quite cumbersome.

RFLP analysis is used in taxonomy and phylogenetics, and single-locus RFLP markers can be applied at species level as demonstrated with *Aspergillus* [49], *Phytophthora* [50], and *Fusarium* [51], as well as at subspecies (forma specialis) level, e.g., within *F. oxysporum* [52] and *Leptosphaeria maculans* [53]. However, in some fungal species only very little sequence variation occurs. Since RFLP detects mutations in restriction sites, which are only very short sequences, this might imply that large numbers of restriction enzymes and probes must be used to find sufficient polymorphisms in such fungi.

The results from RFLP marker analysis can be applied to estimate genetic similarity between samples which then can be used to construct phylogenetic trees [34]. With *L. maculans* [53], it was shown that the tree has

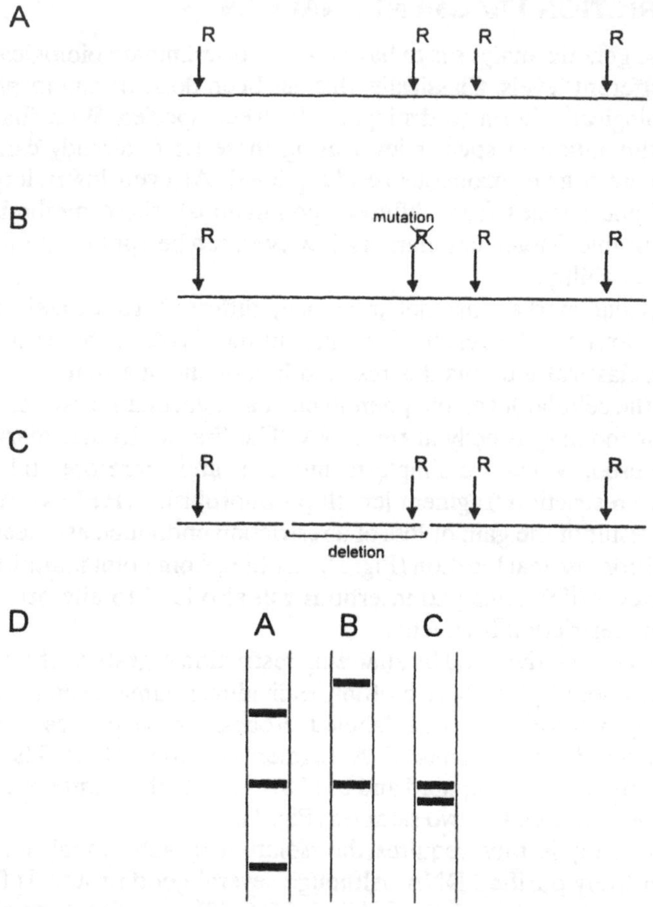

Figure 1 RFLP analysis. Panel A shows a given locus, containing several restriction sites (R). Panels B and C show how point mutations or deletions will alter the restriction pattern of the chromosomal DNA at this locus. Also other rearrangements (insertions, translocations) can bring about such alterations. Panel C shows the resulting patterns in an RFLP analysis, where the different fragments are separated and visualized by Southern blotting and hybridization to a labeled probe.

two main branches—one comprising the aggressive isolates, and the other the nonaggressive ones.

RFLPs can also be suitable markers in population genetic studies. For the wheat pathogen *Septoria tritici* such a study has been performed by McDonald and Martinez [54,55]. First it was shown that RFLPs occur at very high frequency in this fungus [54]. This then allowed the analysis of a natural population in the field [55]. The results showed a high degree of diversity within one location in the field, within one plant, and even within a single leaf. This led to the conclusion that initial infection occurred with a genetically diverse population and that spreading by secondary infections did not play an important role. On top of that it could be shown that individual lesions often contain different haplotypes, which provided evidence for the occurrence of coinfections. A similar analysis with *Erysiphe graminis* f. sp. *hordei*, the causal agent of powdery mildew disease in barley [56], using fingerprinting (see below) showed that for this fungus much less variation occurred. About half of all isolates represented the same clone.

In crosses, RFLP markers behave exactly like other dominant genetic markers. This allows the construction of RFLP maps and even integrated maps with both genetic and RFLP markers. In plants [57], these maps have been shown to be extremely useful in breeding as well as in map-based cloning of specific genes. That a similar approach is feasible in fungi as well has been demonstrated with some phytopathogenic fungi. Using both genomic and cDNA probes, a linkage map with 61 RFLP markers on 13 linkage groups, including eight avirulence loci, could be constructed for the biotrophic oomycete *Bremia lactucae*, which causes lettuce downy mildew [58]. One of the avirulence loci shows a tight coupling to two RFLPs which would allow its cloning. Also, for *E. graminis* f. sp. *hordei*, RFLP markers linked to virulence genes have been identified [59].

In *N. crassa*, which is haploid, a RFLP map [60] was generated by crossing two strains that differ strongly in their genetic background. DNA from progeny derived from ordered asci were used for hybridization. Since this progeny is deposited at the Fungal Genetics Stock Centre, other laboratories can contribute to the map. In the homothallic oomycete *Pythium ultimum*, molecular markers were used to demonstrate that outcrossing can occur [61].

A special application of RFLP analysis, which is particularly well suited for fungi, is the analysis of mitochondrial DNA. Mitochondrial genomes [62] are small (20–180 kb), and mtDNA can be isolated in quantities sufficient to perform restriction analysis with multiple enzymes directly. This can result in quick and sensitive detection of strain differences, which has made RFLP analysis of mitochondrial DNA very popular in taxonomic studies. Moody and Tyler [63] used such methods to discriminate between species within the *Aspergillus flavus* group, and Förster et al. were able to detect differences in *Phytophthora* at both species [64] and subspecies [49] levels, which eventually resulted in taxonomic grouping of 194 *P. megasperma* isolates in nine distinct molecular groups [65]. Also with 35 isolates of *Ophiostoma ulmi* and two races of *O. novo-ulmi* such a molecular relationship has been inferred [66]. With

Fusarium oxysporum, mitochondrial RFLPs could discern not only three subspecies within a group of 25 strains [67,68], but even different vegetative incompatibility groups within one subspecies [69].

Mitochondria in fungi are usually uniparentally inherited [62], but they can be transferred independent of the nuclear genome [70], and mtDNAs do show recombination [71]. Consequently mitochondrial RFLPs might either underestimate or overestimate actual genetic divergence. Therefore combined RFLP analyses of both mitochondrial and nuclear DNA sequences can be useful.

4. DNA FINGERPRINTING

The relative insensitivity of RFLP analysis as described above is partly caused by the fact that the analysis is restricted to the area that is covered by the probe employed. Sensitivity could be enhanced by using probes that detect multiple loci: dispersed repetitive sequences. In many eukaryotic organisms, including several fungi, repetitive sequences are very common. Isolation of such a sequence and using it as a probe on Southern blots of restriction digests of chromosomal DNA leads to patterns with many bands. This technique, which is called fingerprinting, has been applied to the human pathogen *Aspergillus fumigatus* [72]. At least 20 bands were detected, allowing discrimination of individual isolates. Also synthetic simple repeat oligonucleotides can be used as probes, as was demonstrated in typing different strains from the genera *Penicillium*, *Aspergillus*, and *Trichoderma* [73] and in classification of species within the *Trichoderma* aggregate [43].

The most convincing example of the possibilities of fingerprinting, however, is its application to the rice blast fungus *Magnaporthe grisea*. From this fungus several dispersed repetitive sequences have been isolated and characterized [74]. One of these, MGR586, has been used to resolve pathotype diversity in field isolates of rice blast pathogens [75] to establish that rice blast pathogens are a distinct branch of *M. grisea* [76] and to construct a genetic map of the fungus [77–80] which defines eight linkage groups, comprising over 60 molecular markers and several genes.

5. ANALYSIS BY POLYMERASE CHAIN REACTION

A major disadvantage of RFLP analysis is that quite large amounts of chromosomal DNA (10 μg per lane) are needed. In many cases only small samples of the fungi to be analyzed are available, which used to mean that these had first to be propagated. However, now it is possible to extract the DNA of such a small sample and analyze this by amplification of specific sequences using the polymerase chain reaction (PCR).

PCR is based on the ability of DNA polymerases to duplicate a DNA molecule in vitro. This duplication depends on the presence of a single-stranded template and two primers complementary to sequences on either strand to initiate synthesis. Starting with a double-stranded DNA fragment, one cycle of heat denaturation, primer annealing, and synthesis thus results in doubling of the amount of DNA (Fig. 2). By employing DNA polymerase from extreme thermophilic bacteria like *Thermus aquaticus* (*Taq* polymerase), this cycle can be repeated several times, resulting in exponential amplification. Considerable amounts of DNA fragments can thus be obtained, even when the reaction is started with only a few template molecules.

The applications of PCR technology are almost countless. Here we will review the techniques that are applied in fungal genetics. For a review of other PCR applications to fungal research see Foster et al. [81].

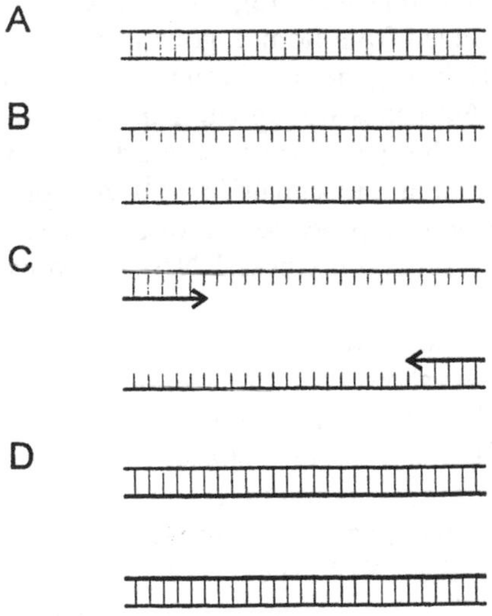

Figure 2 Principle of PCR. Double-stranded DNA (A) is denatured into single strands (B) by heating at 90–98°C. Two short synthetic oligonucleotides (primers), which are complementary to sequences on opposite strands at a moderate distance (a few 100 base pairs up to several kb), are allowed to anneal to the ssDNA at a low temperature (C). The resulting structure is a substrate for DNA polymerases, which will convert them to two double-stranded DNA molecules (D). Repeating these steps results in exponential amplification of the original DNA. The use of thermostable DNA polymerases and of automatic thermo cyclers has made the process simple and very efficient.

In principle, the PCR-RFLP technique is exactly analogous to normal RFLP analysis. With the aid of two primers a specific region of the genome is amplified. The amplified fragment can then be analyzed with restriction endonucleases to detect polymorphic sites. The obvious advantages of PCR-RFLP are its speed and its sensitivity: the whole experiment can be performed within 24 hours, with only 10 ng of chromosomal DNA, compared to several days and 10 μg of DNA for normal RFLP analysis. Moreover, the quality restrictions set to the DNA are less severe, thus allowing very rapid DNA minipreps [47,82]. In fact, for many fungi DNA extraction is not even necessary, because PCR can be directly performed on crushed spores [83] or mycelium. Most applications of this kind of PCR are found in the specific detection of fungi in complex samples—e.g., the wilt fungus *Phoma traceiphila* in lignified branches of lemon [84], and *Candida albicans* in clinical specimens [85].

The disadvantage of PCR-RFLP is that in order to synthesize the necessary primers, the chromosomal region one wants to investigate has to be cloned and sequenced first, unless the region is sufficiently conserved to allow the use of primers, deduced from homologous genes from other organisms. The ribosomal DNA (rDNA) genes are in such a DNA region.

In eukaryotes, rDNA genes are repeated up to several hundreds of times in a clustered manner. The genes are separated by nontranscribed spacer regions (NTS), which contain the signals for rDNA expression. The primary transcript of an rDNA gene is processed to one copy each of 18S, 28S, and 5.8S rRNA. The regions separating these RNAs are called the internal transcribed spacers (ITS).

The nuclear rDNA sequences coding for the small subunit (18S) and large subunit (28S) RNAs show very little evolutionary change and can thus be used to compare distantly related organisms. The internal transcribed spacer region as well as the intergenic spacer of the nuclear rDNA repeat evolve much faster, and sequence differences in these regions occur between species within one genus or even between different populations of one species. Consequently, analyzing the rDNA repeat is very useful for comparisons over a wide range of taxonomic levels but nevertheless has a high resolving power, depending on which part of the ribosomal DNA genes is analyzed.

The nucleotide sequences of the rDNA repeat unit have been determined from a large number of eukaryotes. Compilation of these sequence data can identify stretches of nucleotides within the 18S, 5.8S, and 28S regions that are highly conserved [86]. This allows the design of primers, which can be used in PCR experiments to amplify regions of the rDNA repeat, including the NTS or ITS regions (Fig. 3). Primers may be chosen in such a way that they are specific for a group of genera—e.g., all fungi. Because several hundreds of

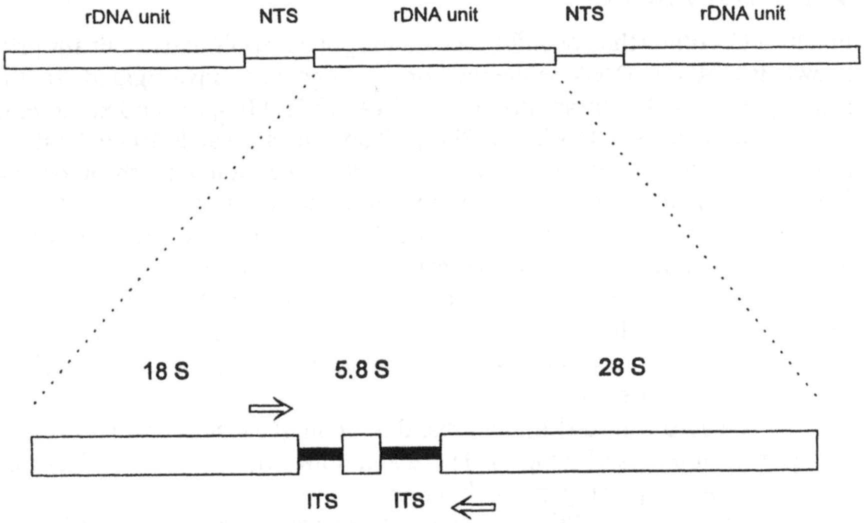

Figure 3 Organization of the eukaryotic ribosomal DNA genes. Several hundred copies of the ribosomal genes are present in a clustered manner. The genes are separated by nontranscribed spacer (NTS) regions, which contain the sequences for rDNA expression. The primary transcript of an rDNA gene is processed to one copy each of 18S, 5.8S, and 28S rRNA. The regions separating these RNAs are called internal transcribed spacers (ITS). Using primers complementary to strongly conserved regions in the 18S and 28S genes (arrows), the highly variable ITS regions of different (sub) species can be amplified by PCR. Analysis of the products reveals the genetic differences.

copies of rDNA are present per genome, only very small amounts of chromosomal DNA are necessary for these amplifications.

The fragments resulting from such PCR reactions can be directly analyzed on agarose gels for differences in length of the NTS or ITS regions [86, 87]. A further discrimination can be obtained by digesting the PCR products with a number of restriction endonucleases and analyzing the products [88].

The highest detail is obtained by direct sequencing of the PCR products [86,89,90], which will detect every single base-pair difference of the amplified fragment between samples. Sequencing of PCR-amplified ribosomal DNA genes is widely used in evolutionary genetics to establish phylogenetic relationships among fungi [91,92]. As more of these sequence data are generated and put into the data bases, the sensitivity and resolution of the method will be further increased by the possibility of devising class-, family-, genus-, and possibly even species-specific primers [93].

6. RAPD ANALYSIS

Finally, PCR offers the possibility of creating polymorphisms without any prior knowledge of the DNA sequences of the organisms investigated. In this technique, termed arbitrary primed PCR (AP/PCR) [94] or random amplification of polymorphic DNA (RAPD) [95] one short (usually 10 nucleotides) primer of arbitrary sequence is used in a PCR reaction with chromosomal DNA. This usually results in the amplification of one or more distinct DNA fragments, although occasionally a primer might not give products at all. Surprisingly, there seems to be no correlation between the number of bands obtained and the genome size of the organism DNA used as template. Each new primer will result in a different band pattern. The patterns produced are highly polymorphic, allowing discrimination between isolates of one species if sufficient primers are screened.

As the acronym RAPD suggests, this technique is very fast in discriminating large numbers of samples. There are, however, several considerations to be made when applying this technique:

1. Sample purity. PCR experiments are very prone to contamination with amplifiable sequences. Since RAPD patterns can be produced from any DNA source, it is of course of the utmost importance that samples or specimen are not contaminated. Especially with samples that have been collected from the field (e.g., from soil, from patients, from infected plants or crops), extreme care has to be taken.

2. Reproducibility. The efficiency and specificity of PCR reactions is very much dependent on the reaction conditions. Slight variations in temperature [96], concentration of $MgCl_2$ [97], primer [98], or contaminating RNA [99] can completely alter the resulting band pattern. Even changing the type of thermostable DNA polymerase may alter the RAPD pattern [100]. Especially with RAPD, in which priming is arbitrary, small changes in the conditions may lead to completely different results. This may set limitations to reproducibility, since the specifications of PCR thermocyclers are highly variable, even if they are the same type and brand [101–103]. It is therefore recommended [104] that the same thermocycler always be employed and that as many control samples as possible be included.

3. Interpretation. First of all one has to bear in mind that RAPD bands are dominant; in diploid organisms both "alleles" must be absent to get a different pattern. Secondly, differences in band patterns can only be interpreted as individual bands being present or absent, since there is absolutely no relation between the individual bands of one pattern. Thirdly, bands of identical size are not necessarily of identical sequence. A fourth consideration to make is that absence of a particular band in a RAPD pattern does not necessarily imply that the target sequence is no longer present. It might just

be that another target sequence is amplified much more efficiently, thus masking the first one. This is especially important if one wants to use RAPD markers to create phylogenetic trees [105].

Taken its limitations into account, RAPDs can be used in a similar way to RFLPs. Most applications of RAPD analysis reported so far are in the differentiation of fungal strains or isolates [34,61,106–111]. However, the use of RAPD markers in parasexual analysis in *P. roqueforti* [112] and in sexual crossings of *Agaricus bisporus* [113] are indicative of the potential this method has in the field of fungal genetics.

REFERENCES

1. Clutterbuck, A. J. (1993). *Aspergillus nidulans*. In: *Genetic Maps* (S. J. O'Brien, ed.). Cold Spring Harbor Laboratory Press, Cold Spring Harbor, NY, pp. 371–384.
2. Mortimer, R. K., C. R. Contopoulou, J. S. King (1993). Genetic and physical maps of *Saccharomyces cerevisiae*. In: *Genetic Maps*, Ed. 11. (S. J. O'Brien, ed.). Cold Spring Harbor Laboratory Press, Cold Spring Harbor, NY, pp. 336–356.
3. Perkins, D. H. (1993). *Neurospora crassa* genetic maps. In: *Genetic Maps* (S. J. O'Brien, ed.). Cold Spring Harbor Laboratory Press, Cold Spring Harbor, NY, pp. 311–320.
4. Sybenga, J. (1992). *Cytogenetics in Plant Breeding*. Monographs on theoretical and applied genetics 17. Springer-Verlag, Berlin, Heidelberg.
5. Raju, N. B. (1980). Meiosis and ascospore genesis in *Neurospora*. *Eur. J. Biol. 23*: 208–223.
6. Fincham, J. R. S., P. R. Day, A. Radford (1979). *Fungal Genetics*, 4th Ed. Blackwell Scientific Publications, Oxford.
7. Schwartz, D. C., C. R. Cantor (1984). Separation of yeast chromosome-sized DNAs by pulsed field gradient gel electrophoresis. *Cell 37*:67–75.
8. Carle, G. F., M. V. Olson (1984). Separation of chromosomal DNA molecules from yeast by orthogonal-field-alternation gel electrophoresis. *Nucl. Acids Res. 12*:5647–5664.
9. Carle, G. F., M. V. Olson (1985). An electrophoretic karyotype for yeast. *Proc. Natl. Acad. Sci. USA 82*:3756–3760.
10. Lai, E., B. W. Birren, S. M. Clark, M. I. Simon, L. Hood (1989). Pulsed field gel electrophoresis. *Biotechniques 7*:34–42.
11. Anand, R. (1986). Pulsed field gel electrophoresis: a technique for fractionating large DNA molecules. *Trends Genet. 2*:278–283.
12. Swart, K., A. J. M. Debets, E. F. Holub, C. J. Bos, R. F. Hoekstra (1994). Physical karyotyping: genetic and taxonomic applications in Aspergilli. In: *The Biology of* Aspergillus. Proceedings of FEMS/BMS symposium no. 69 (K. A. Powell, J. Peberdy, eds.). Plenum, New York.
13. Skinner, D. Z., A. D. Budde, S. A. Leong (1991). Molecular karyotype analysis of fungi. In: *More Gene Manipulations in Fungi.* (J. W. Bennett and L. L. Lasure, eds.). Academic Press, San Diego.

14. Chu, G., D. Vollrath, R. W. Davis (1986). Separation of large DNA molecules by contour-clamped homogeneous electric field. *Science 234*:1582–1585.
15. Orbach, J. M., D. Vollrath, R. W. Davis, C. Yanofsky (1989). An electrophoretic karyotype of *Neurospora crassa. Mol. Cell. Biol. 8*:1469–1473.
16. Debets, A. J. M., E. F. Holub, K. Swart, H. W. J. van den Broek, C. J. Bos (1990). An electrophoretic karyotype of *Aspergillus niger. Mol. Gen. Genet. 224*:264–268.
17. Debets, A. J. M., C. J. Bos (1986). Isolation of small protoplasts of *Aspergillus niger. Fungal Genet. Newsl. 33*:24.
18. Verdoes, J. C., M. R. Calil, P. J. Punt, et al. (1994). The complete karyotype of *Aspergillus niger*: the use of introduced electrophoretic mobility variation of chromosomes for gene assignment studies. *Mol. Gen. Genet. 244*:75–80.
19. Maniatis, T., E. F. Frisch, J. Sambrook (1982). *Molecular Cloning*: *A Laboratory Manual*. Cold Spring Harbor Laboratory, Cold Spring Harbor, NY.
20. Brody, H., J. Carbon (1989). Electrophoretic karyotype of *Aspergillus nidulans. Proc. Natl. Acad. Sci. USA 86*:6260–6263.
21. Brody, H., J. Griffith, A. J. Cuticchia, J. Arnold, W. E. Timberlake (1991). Chromosome-specific recombinant DNA libraries from the fungus *Aspergillus nidulans. Nucl. Acids Res. 19*:3105–3109.
22. Debets, F., K. Swart, R. F. Hoekstra, C. J. Bos (1993). Genetic maps of eight linkage groups of *Aspergillus niger* based on mitotic mapping. *Curr. Genet. 23*: 47–53.
23. Debets, A. J. M., K. Swart, E. F. Holub, T. Goosen, C. J. Bos (1990). Genetic analysis of *amdS* transformants of *Aspergillus niger* and their use in chromosome mapping. *Mol. Gen. Genet. 222*:284–290.
24. Verdoes, J. C., P. J. Punt, J. M. Schrickx, H. W. van Verseveld, A. H. Stouthamer, C. A. M. J. J. van den Hondel (1993). Glucoamylase overexpression in *Aspergillus niger*: molecular genetic analysis of strains containing multiple copies of the *glaA* gene. *Transgenic Res. 2*:84–92.
25. Bos, C. J., S. M. Slakhorst, A. J. M. Debets, K. Swart (1993). Linkage group analysis in *Aspergillus niger. Appl. Microbiol. Biotechnol. 38*:742–745.
26. Walz, M., U. Kück (1991). Polymorphic karyotypes in related *Acremonium* strains. *Curr. Genet. 19*:73–76.
27. Smith, A. W., K. Collis, M. Ramsden, H. M. Fox, J. F. Peberdy (1991). Chromosome rearrangements in improved cephalosporin C-producing strains of *Acremonium chrysogenum. Curr. Genet. 19*:235–237.
28. Lodder, S., K. Gull, D. Wood (1993). An electrophoretic karyotype of the cultivated mushroom—*Agaricus bisporus. Curr. Genet. 24*:496–499.
29. Mäntylä, A. L., K. H. Rossi, S. A. Vanhanen, M. E. Penttilä, P. L. Suominen, K. M. H. Nevalainen (1992). Electrophoretic karyotyping of wild-type and mutant *Trichoderma longibrachiatum* (*reesi*) strains. *Curr. Genet. 21*:471–477.
30. Miao, V., H. D. VanEtten (1989). Non-transmittance through meiosis of a gene for pisatin demethylase in *Nectria haematococca* MP VI is correlated with a change in electrophoretic karyotype. *J. Cell Biochem. Suppl. 13E*:22.

31. Taylor, J. L., I. Borgmann, G. Séguin-Swartz (1991). Electrophoretic karyotyping of *Leptosphaeria maculans* differentiates highly virulent from weakly virulent isolates. *Curr. Genet. 19*:273–277.
32. McCluskey, K., D. Mills (199). Identification and characterization of chromosome length polymorphisms among strains representing fourteen races of *Ustilago hordei. Mol. Plant–Microbe Interact. 3*:366–373.
33. Fekete, C., Nagy, A. J. M. Debets, L. Hornok (1993). Electrophoretic karyotypes and gene mapping in eight species of the *Fusarium* sections *Arthrosporiella* and *Sporotrichiella. Curr. Genet. 24*:500–504.
34. Megnegneau, B., F. Debets, R. F. Hoekstra (1993). Genetic variability and relatedness in the complex group of black Aspergilli based on random amplification of polymorphic DNA. *Curr. Genet. 23*:323–329.
35. Keller, N. P., T. E. Cleveland, D. Bhatnagar (1992). Variable electrophoretic karyotypes of members of *Aspergillus* section *Flavi. Curr. Genet. 2*:371–375.
36. McDonald, B. A., J. P. Martinez (1991). Chromosome length polymorphisms in a *Septoria tritici* population. *Curr. Genet. 19*:265–271.
37. Cooley, R. N., C. E. Caten (1991). Variation in electrophoretic karyotype between strains of *Septoria nodorum*.
38. Kayser, T., J. Wöstemeyer (1991). Electrophoretic karyotype of the zygomycete *Absidia glauca*: evidence for differences between mating types. *Curr. Genet. 19*: 279–284.
39. Jones, R. N., H. Rees (1982). *B-chromosomes.* Academic Press, London.
40. Mills, D., K. McCluskey (1990). Electrophoretic karyotypes of fungi: the new cytology. *Mol. Plant–Microbe Interact. 6*:351–357.
41. Miao, V. P., S. F. Covert, H. D. VanEtten (1991). A fungal gene for antibiotic resistance on a dispensable ("B") chromosome. *Science 254*:1773–1776.
42. Masel, A. W., N. Struyk, C. L. McIntyre, J. A. G. Irwin, J. M. Manners (1993). A strain-specific cyclin homolog in the fungal phytopathogen *Colletotrichum gloesporioides. Gene 133*:141–145.
43. Kusters–van Someren, M. A., R. A. Samson, J. Visser (1991). The use of RFLP analysis in classification of the black Aspergilli: reinterpretation of the *Aspergillus niger* aggregate. *Curr. Genet. 19*:21–26.
44. Meyer, W., R. Morawetz, T. Börner, C. P. Kubicek (1992). The use of DNA-fingerprint analysis in the classification of some species of the *Trichoderma* aggregate. *Curr. Genet. 21*:27–30.
45. Raeder, U., P. Broda (1985). Rapid preparation of DNA from filamentous fungi. *Lett. Appl. Microbiol. 1*:17–20.
46. Goosen, T., G. Bloemheuvel, C. Gysler, D. A. de Bie, H. W. J. van den Broek, K. Swart (1987). Transformation of *Aspergillus niger* using the homologous orotidine-5'-phosphate-decarboxylase gene. *Curr. Genet. 11*:499–504.
47. Ashktorab, H., R. J. Cohen (1992). Facile isolation of genomic DNA from filamentous fungi. *Biotechniques 13*:198.
48. Cenis, J. L. (1992). Rapid extraction of fungal DNA for PCR amplification. *Nucl. Acids Res. 20*:2380.

49. Moody, S. F., B. M. Tyler (1990). Use of nuclear DNA restriction fragment length polymorphisms to analyze the diversity of the *Aspergillus flavus* group—*A. flavus, A. parasiticus* and *A. nomius. Appl. Environ. Microbiol. 56*:2453–2461.
50. Förster, H., T. G. Kinscherf, S. A. Leong, D. P. Maxwell (1989). Restriction fragment length polymorphisms of the mitochondrial DNA of *Phytophthora megasperma* isolated from soybean, alfalfa and fruit trees. *Can. J. Bot. 67*:529–537.
51. Manicom, B. Q., M. Bar-Joseph, A. Rosner, H. Vigodsky-Haas, J. M. Kotze (1987). Potential applications of random DNA probes and restriction fragment length polymorphisms in the taxonomy of *Fusaria. Phytopathology 77*:669–672.
52. Manicom, B. Q., M. Bar-Joseph, J. M. Kotze (1990). A restriction fragment length polymorphism probe relating vegetative compatibility groups and pathogenicity in *Fusarium oxysporum* f. sp. *dianthi. Phytopathology 80*:336–339.
53. Koch, E., K. Song, T. C. Osborn, P. H. Williams (1991). Relationship between pathogenicity and phylogeny based on restriction fragment length polymorphism in *Leptosphaeria maculans. Mol. Plant-Microbe Interact. 4*:341–349.
54. McDonald, B. A., J. P. Martinez (1990). DNA restriction fragment length polymorphisms among *Mycosphaerella graminicola* (anamorph *Septoria tritici*) isolates collected from a single wheat field. *Phytopathology 80*:1368–1373.
55. McDonald, B. A., J. P. Martinez (1990). Restriction fragment length polymorphisms in *Septoria tritici* occur at a high frequency. *Curr. Genet. 17*:133–138.
56. Brown, J. K. M., M. O'Dell, C. G. Simpson, M. S. Wolfe (1990). The use of DNA polymorphisms to test hypotheses about a population of *Erysiphe graminis* f. sp. *hordei. Plant Pathol. 39*:391–401.
57. Tanksley, S. D. (1993). Linkage map of tomato (*Lycopersicon esculentum*). In: *Genetic Maps* (S. J. O'Brien, ed.). Cold Spring Harbor Laboratory Press, Cold Spring Harbor, NY, pp. 639–660.
58. Hulbert, S. C., T. W. Ilott, E. J. Legg, S. E. Lincoln, E. S. Lander, R. W. Michelmore (1988). Genetic analysis of the fungus *Bremia lactucae*, using restriction fragment length polymorphisms. *Genetics 120*:947–958.
59. Christiansen, S. K., H. Giese (1990). Genetic analysis of the obligate parasitic barley powdery mildew fungus based on RFLP and virulence loci. *Theor. Appl. Genet. 79*:705–712.
60. Metzenberg, R. L., J. Grotelueschen (1993). *Neurospora crassa* restriction polymorphism map. In: *Genetic Maps* (S. J. O'Brien, ed.). Cold Spring Harbor Laboratory Press, Cold Spring Harbor, NY, pp. 321–331.
61. Francis, D. M., D. A. St. Clair (1993). Outcrossing in the homothallic oomycete *Pythium ultimum*, detected with molecular markers. *Curr. Genet. 24*:100–106.
62. Tyler, J. W. (1986). Fungal evolutionary biology and mitochondrial DNA. *Exp. Mycol. 10*:259–269.
63. Moody, S. F., B. M. Tyler (1990). Restriction enzyme analysis of mitochondrial DNA of the *Apsergillus flavus* group: *A. flavus, A. parasiticus* and *A. nomius. Appl. Environ. Microbiol. 56*:2441–2452.
64. Förster, H., P. Oudemans, M. D. Coffey (1990). Mitochondrial and nuclear DNA diversity within 6 species of *Phytophthora. Exp. Mycol. 14*:18–31.

65. Förster, H., M. D. Coffey (1993). Molecular taxonomy of *Phytophthora megasperma* based on mitochondrial and nuclear DNA polymorphisms. *Mycol. Res. 97*: 1101–1112.
66. Bates, M. R., K. W. Buck, C. M. Brasier (1993). Molecular relationships of the mitochondrial DNA of *Ophiostoma ulmi* and the NAN and EAN races of *O. novo-ulmi* determined by restriction fragment length polymorphisms. *Mycol. Res. 97*:1093–1100.
67. Kistler, H. C., P. W. Bosland, U. Benny, S. A. Leong, P. H. Williams (1987). Relatedness of strains of *Fusarium oxysporum* from crucifers measured by examination of mitochondrial and ribosomal DNA. *Phytopathology 77*:1289–1293.
68. Kistler, H. C., U. Benny (1989). The mitochondrial genome of *Fusarium oxysporum. Plasmid 22*:86–89.
69. Jacobson, D. J., T. R. Gordon (1990). The variability of mitochondrial DNA as an indicator of the relationship between populations of *Fusarium oxysporum* f. sp. *melonis. Mycol. Res. 94*:734–744.
70. Collins, R. A., B. J. Saville (1990). Independent transfer of mitochondrial chromosomes and plasmids during unstable vegetative fusion in *Neurospora. Nature 345*:177–179.
71. Earl, A. J., G. Turner, J. H. Croft, et al. (1981). High frequency transfer of species specific mitochondrial DNA sequences between members of the Aspergillaceae. *Curr. Genet. 3*:221–228.
72. Girardin, H., J. P. Latge, T. Srikantha, B. Morrow, D. R. Soll (1993). Development of DNA probes for fingerprinting *Aspergillus fumigatus. J. Clin. Microbiol. 31*: 1547–1554.
73. Meyer, W., A. Koch, C. Niemann, B. Beyermann, J. T. Epplen, T. Börner (1991). Differentiation of species and strains among filamentous fungi by DNA fingerprinting. *Curr. Genet. 19*:239–242.
74. Hamer, J. E., N. J. Talbot, M. Levy (1993). Genome dynamics and pathotype in the rice blast fungus. In: *Advances in Molecular Genetics of Plant–Microbe Interactions* (E. W. Nester, D. P. S. Verma, eds.). Kluwer Academic Publishers, Amsterdam, pp. 299–311.
75. Levy, M., J. Romao, M. A. Marchetti, J. E. Hamer (1991). DNA fingerprinting with a dispersed repeated sequence resolves pathotype diversity in the rice blast fungus. *Plant Cell 3*:95–102.
76. Hamer, J. E. (1991). Molecular probes for rice blast disease. *Science 252*:632–633.
77. Leong, S. A., D. W. Holden (1989). Molecular genetic approaches to the study of fungal pathogenesis. *Annu. Rev. Phytopathol. 27*:463–481.
78. Hamer, J. E., S. Givan (1990). Genetic mapping with dispersed repeated sequences in the rice blast fungus: mapping the *SMO* locus. Mol. *Gen. Genet. 223*: 487–495.
79. Valent, B., F. G. Chumley (1991). Molecular genetic analysis of the rice blast fungus, *Magnaporthe grisae. Annu. Rev. Phytopathol. 29*:443–467.
80. Romao, J., J. E. Hamer (1992). Genetic organization of a repeated DNA sequence family in the rice blast fungus. *Proc. Natl. Acad. Sci. USA 89*:5316–5320.

81. Foster, L. M., K. R. Kozak, M. G. Loftus, J. J. Stevens, I. K. Ross (1993). The polymerase chain reaction and its application to filamentous fungi. *Mycol. Res. 97*: 769–781.
82. Goodwin, D. C., S. B. Lee (1993). Microwave miniprep of total genomic DNA from fungi, plants, protists and animals for PCR. *Biotechniques 15*:438.
83. Aufauvre-Brown, A., C. M. Tang, D. W. Holden (1993). Detection of gene disruption events in *Aspergillus* transformants by polymerase chain reaction direct from conidiospores. *Curr. Genet. 24*:177–178.
84. Rollo, F., R. Salvi, P. Torchia (1990). Highly sensitive and fast detection of *Phoma tracheiphila* by polymerase chain reaction. *Appl. Microbiol. Biotechnol. 32*:572–576.
85. Crampin, A. C., R. C. Matthews (1993). Application of the polymerase chain reaction to the diagnosis of candidosis by amplification of an HSP-90 gene fragment. *J. Med. Microbiol. 39*:233–238.
86. White, T. J., T. Bruns, S. Lee, J. Taylor (1990). Amplification and direct sequencing of fungal ribosomal RNA genes for phylogenetics. In: *PCR Protocols, A Guide to Methods and Applications* (M. A. Innis, D. H. Gelfand, J. J. Sninsky, T. J. White, eds.). Academic Press, San Diego, pp. 315–323.
87. Buchko, J., G. R. Klassen (1990). Detection of length heterogeneity in the ribosomal DNA of *Pythium ultimum* by PCR amplification of the intergenic region. *Curr. Genet. 18*:203–205.
88. Saperstein, D. A., J. M. Nickerso (1991). Restriction fragment length polymorphism analysis using PCR coupled to restriction digests. *Biotechniques 10*:488–489.
89. Bevan, I. S., R. Rapley, M. R. Walker (1992). Sequencing of PCR-amplified DNA. *PCR Meth. Appl. 1*:222–228.
90. Meltzer, S. J. (1993). Direct sequencing of polymerase chain reaction products. *PCR Protocols 15*:137–141.
91. Bruns, T. D., R. Vilgalys, S. M. Barns, et al. (1992). Evolutionary relationships within the Fungi: analyses of small subunit rRNA sequences. *Mol. Phylogen. Evol. 1*:231–243.
92. Simon, L., J. Bousquet, R. C. Levesque, M. Lalonde (1993). Origin and diversification of endomycorrhizal fungi and coincidence with vascular land plants. *Nature 363*:67–69.
93. Simon, L., M. Lalonde, T. D. Bruns (1992). Specific amplification of 18S fungal ribosomal genes from vesicular-arbuscular endomycorrhizal fungi colonizing roots. *Appl. Environ. Microbiol. 58*:291–295.
94. Welsh, J., M. McClelland (1990). Fingerprinting genomes using PCR with arbitrary primers. *Nucl. Acids Res. 18*:7213–7218.
95. Williams, J. G. K., A. R. Kubelik, K. J. Livak, J. A. Rafalski, S. V. Tingey (1990). DNA polymorphisms amplified by arbitrary primers are useful as genetic markers. *Nucl. Acids Res. 18*:6531–6535.
96. Yu, F. K., K. P. Pauls (1992). Optimization of the PCR program for RAPD analysis. *Nucl. Acids Res. 20*:2606.
97. Innis, M. A., D. H. Gelfand (1990). Optimization of PCRs. In: *PCR Protocols, A Guide to Methods and Applications* (M. A. Innis, D. H. Gelfand, J. J. Sninsky, T. J. White, eds.). Academic Press, San Diego, pp. 3–12.

98. Muralidharan, K., E. K. Wakeland (1993). Concentration of primer and template qualitatively affects products in random amplified polymorphic DNA PCR. *Biotechniques 14*:362.
99. Yoon, C. S., D. A. Glawe (1993). Pretreatment with RNase to improve PCR amplification of DNA using 10-Mer primers. *Biotechniques 14*:908.
100. Schierwater, B., A. Ender (1993). Different thermostable DNA polymerases may amplify different RAPD products. *Nucl. Acids Res. 21*:4647–4648.
101. Hoelzel, R. (1990). The trouble with PCR machines. *Trends Genet. 6*:237–238.
102. Linz, U. (1990). Thermocycler temperature variation invalidates PCR results. *Biotechniques 9*:286.
103. Stamm, S., B. Gillo, J. Brosius (1991). Temperature recording from thermocyclers used for PCR. *Biotechniques 10*:430.
104. Orrego, C. (1990). Organizing a laboratory for PCR work. In: *PRC Protocols, A Guide to Methods and Applications* (M. A. Innis, D. H. Gelfand, J. J. Sninsky, T. J. White, eds.). Academic Press, San Diego, pp. 447–454.
105. Clark, A. G., C. M. S. Lanigan (1993). Prospects for estimating nucleotide divergence with RAPDs. *Mol. Biol. Evol. 10*:1096–1111.
106. Crowhurst, R. N., B. T. Hawthorne, E. A. H. Rikkerink, M. D. Templeton (1991). Differentiation of *Fusarium solani* f. sp. *cucurbitae* races 1 and 2 by random amplification of polymorphic DNA. *Curr. Genet. 20*:391–396.
107. Goodwin, P. H., S. L. Annis (1991). Rapid identification of genetic variation and pathotype of *Leptosphaeria maculans* by random amplified polymorphic DNA assay. *Appl. Environ. Microbiol. 57*:2482–2486.
108. Guthrie, P. A. I., C. W. Magill, R. A. Frederiksen, G. N. Odvody (1992). Random amplified polymorphic DNA markers: a system for identifying and differentiating isolates of *Colletotrichum graminicola. Phytopathology 82*:832–835.
109. Jones, M. J., L. D. Dunkle (1993). Analysis of *Cochliobolus carbonum* races by PCR amplification with arbitrary and gene-specific primers. *Phytopathology 83*: 366–370.
110. Ouellet, T., K. A. Seifert (1993). Genetic characterization of *Fusarium graminearum* strains using RAPD and PCR amplification. *Phytopathology 83*:1003–1007.
111. Loudon, K. W., J. P. Burnie, A. P. Coke, R. C. Matthews (1993). Application of polymerase chain reaction to fingerprinting *Aspergillus fumigatus* by random amplification of polymorphic DNA. *J. Clin. Microbiol. 31*:1117–1121.
112. Durand, N., P. Reymond, M. Fèvre (1993). Randomly amplified polymorphic DNAs assess recombination following an induced parasexual cycle in *Penicillium roqueforti. Curr. Genet. 24*:417–420.
113. Khush, R. S., E. Becker, M. Wach (1992). DNA amplification polymorphisms of the cultivated mushroom *Agaricus bisporus. Appl. Env. Microbiol. 58*:2971–2977.

6

Chromosomes, Mitosis, and Meiosis

Benjamin C. K. Lu
University of Guelph, Guelph, Ontario, Canada

1. INTRODUCTION

Since the introduction of *Neurospora crassa* to genetic research by Shear and Dodge [195], and the subsequent production of auxotrophic mutants deficient in vitamin biosynthesis by Beadle and Tatum [13], fungi have moved to the center stage of genetic research. Many new insights have been obtained through studies of fungal genetics. Examples include one-gene–one-enzyme hypothesis, tetrad analysis, gene conversion, postmeiotic segregation, and allelic complementations. In recent years, eukaryotic gene expression, cell cycle controls, centromere organizations, the genetics of homologous pairing and its relation to the assembly of synaptonemal complexes, etc., are prominent areas of advancement.

If fungi are gaining respect as suitable genetic organisms, it is imperative that we understand the organization of their genetic materials, such as the chromosomes, their karyotypes, and their behavior in meiosis and mitosis. After all, the chromosome behavior in meiosis is the foundation of Mendelian segregation, independent assortment, linkage, and crossing over. Thus Section 2 deals briefly with the mitochondrial genome, Section 3 deals with the chromosomes, their internal organization, and their functions (these include the nucleolar organizer, the centromere, the telomeres). Section 4 deals with mitosis in the vegetative mycelium, and section 5 deals with meiotic processes. Key stages of mitosis in vegetative mycelia have been revealed by light and electron microscopy, albeit only in a few species. For meiosis, chromosome behavior will be described and illustrated at key stages for Ascomycetes and for Basidiomycetes. Some reinterpretation of previous observations will also be made. Discussions will be given to the question of homologous pairing before the assembly of the synaptonemal complex for which idea abound [108, 127]. As to the coverage of the literature, omission is unavoidable; it is certainly not intentional.

Some key concepts of fungal chromosomes are as follows:

1. The mitochondrial genome varies in size even among closely related species.
2. The size of the haploid genome of fungi can be determined with different methods and may have a size of 15–60 Mb (megabase pairs) and can consist of 10% or more repetitive DNA.
3. The nucleolus, the site of rRNA biosynthesis, increases 10- to 15-fold in size in meiotic cells and participates in the formation of the synaptonemal complex.
4. The chromosomes are similarly organized into nucleosomes as those of higher eukaryotes and contain a localized centromere-kinetochore complex from which the spindle microtubules originate. The centromere regions are being studied by molecular techniques.
5. The telomeres, the ends of chromosome arms, maintain chromosome integrity; they have unique repeat sequences that are important for replication and perhaps also for chromosome pairing.
6. Karyotype analysis can be done with light microscopy, electron microscopy, and pulse field electrophoresis.
7. Variations of chromosome structure and numbers have been recognized and studied.
8. Mitosis follows a normal process; paired kinetochores are connected to opposite poles by spindle microtubules. Chromosome movement at metaphase–anaphase is staggered giving them a scattered or two-track appearance.
9. The meiotic S phase (DNA replication) differs from the mitotic S phase and precedes karyogamy in most cases.
10. Meiotic stages are described for Ascomycetes and Basidiomycetes.
11. Diverse programs of ascus development are adopted by different pseudohomothallic species.
12. Assembly of the synaptonemal complex depends on homologous pairing and the initiation of recombination process; the question of homology search remains speculative.

13. Initiation of recombination during prophase of meiosis I depends on single-strand nicks and gaps or double-strand breaks in homologous non-sister chromatids, and the synapsis is stabilized by the synaptonemal complex.

2. ORGANIZATION OF GENETIC MATERIALS IN THE MITOCHONDRIA

The mitochondrial genomes of fungi vary widely in size, the smallest being 17.6 kilobase pairs (kbp), and the largest 121 kbp. The majority of known species studied have their genome size fall between 30 and 80 kb; only a few species have their genome size smaller than 30 or larger than 80 kb. With few exceptions, all fungal mitochondrial genomes are physically circular. From sequencing data, some 50% to over 80% of mitochondrial DNA (mtDNA) sequences lack a coding function, and are probably nonfunctional. This argument is consistent with the observations that some closely related species of the same genus have their mtDNA contents differ by two- to threefold [80,84 for review].

A few mtDNAs have been completely sequenced. There are a small number of open reading frames (ORFs); some have been identified as genes coding for components of the electron transport chain. These are cytochrome c oxidase (cox) subunits I, II, and III; ATPase subunit 6 (apt6); and apocytochrome b component of ubiquinol cytochrome c reductase (cob) [51,79,80, 210]. There are also open but unidentified reading frames (URFs). For example, a total of 15 URFs have been found in *Aspergillus nidulans* [29]. There are intergenic A+T–rich spacers, and optional introns, even in the smallest genome of *S. pombe*, genes are separated by spacers, quite in contrast to the animal mtDNA, which lacks spacers [see reviews 80,217]. In addition, there are structural RNA genes. These include small (S-rRNA) and large (L-rRNA) ribosomal RNA genes and at least 24 mitochondrial tRNA genes. There is considerable homology between the mitochondrial and *E. coli* rRNAs supporting the eubacterial origin of mitochondria [80,81,102]. It has been proposed that mitochondria originated in evolution as bacterial endosymbionts that were ultimately integrated into a host cell and lost their independent existence [80].

3. ORGANIZATION OF GENETIC MATERIALS IN THE NUCLEUS

Fungi are classified as lower eukaryotes, and their genome organization is typical of an eukaryotic system. The chromosomes are composed of DNA double helices, histones, and nonhistone nuclear proteins. In a haploid set, as in most vegetative nuclei, one chromosome is the nucleolar chromosome which carries the nucleolar organizer and its associated nucleolus. The chromosomes are enclosed in a nuclear envelope which is made up of an annulated double-membrane system.

A. Quantity of DNA per Haploid Genome

The amount of DNA per haploid genome is important and useful information as it may reflect the genomic complexity of the organisms, at least among the organisms in the low end of the scale. The fungi, in general, fall between bacteria, such as *Escherichia coli* and insects such as *Drosophila melanogaster*. If *E. coli* is assigned a scale of 1, the yeast *Saccharomyces cerevisiae* will be 4, *Neurospora crassa* and *Coprinus cinereus* will be 9–10, and *Drosophila melanogaster* will be 40. Data from estimates of some key fungi are presented in Table 1.

There are highly repeated sequences and they represent about 15% of *C. cinereus* genome with an average size of 40–60 kilobases, and 10–20% of *N. crassa*; this has been demonstrated by a biphasic C_0t curve [28,60,61]. When the repeated sequences of *C. cinereus* were isolated and analyzed, the $c_0t_{0.5}$ is 75 times faster than the DNA of *E. coli* (genome size 2.7×10^9 daltons), indicating the size of each copy of repeated DNA as 3.2×10^7 ($2.7 \times 10^9/75$) daltons or 50 kilobases. There are 110 or so copies of these repeated DNA

Table 1 DNA content per haploid genome of selected fungi

Species	Daltons	Base pairs	References
Coprinus cinereus	2.4×10^{10}	3.7×10^7	Dutta [60]
Neurospora crassa	1.8×10^{10}	2.8×10^7	Krumlauf and Marzluf [104]
	2.2×10^{10}	3.4×10^7	Dutta and Ojha [61]
	2.7×10^{10}	4.2×10^7	Duran and Gray [58]
	2.8×10^{10}	4.3×10^7	Horowitz and MacLeod [94]
	2.8×10^{10}	4.3×10^7	Orbach [155]
Aspergillus nidulans	1.7×10^{10}	2.6×10^7	Timberlake [207]
Saccharomyces cerevisiae	0.9×10^{10}	1.4×10^7	Bicknell and Douglas [17]
	1.0×10^{10}	1.5×10^7	Lohr et al. [109]
Schizosaccharomyces pombe	0.9×10^{10}	1.38×10^7	Orbach [155]
Ustilago maydis	1.4×10^{10}	2.2×10^7	Esposito and Holliday [69]
Puccinia graminis f. sp. *tritici*	3.8×10^{10}	5.8×10^7	Backlund and Szabo, 1991

sequences per haploid cell, only a small fraction of which should belong to the mitochondrial DNA. By saturation hybridization of ribosomal and transfer RNAs with the redundant DNA fraction, some or most of the redundant DNA represents the ribosomal cistrons and the transfer RNA genes. The remainder have not been characterized. There is little or no redundant DNA in *S. cerevisiae* and in *Aspergillus nidulans* [207]. There are also highly conserved long interspersed repetitive elements (LINEs) and short interspersed repetitive elements (SINEs) in eukaryotic genomes [111,158,209].

B. Nucleolus

The nucleolus is the largest object in a fungal nucleus; it is easily observed by light microscopy in mitotic and meiotic cells. It is attached to the nucleolar organizer region of the nuclear chromosome. Examples are chromosome-2 in *N. crassa* [131,166,167,199], chromosome-10 in *C. cinereus* [163]. The nucleolus is small in mycelia and in fruit body tissue cells, but increased 10- to 15-fold in volume at the time of karyogamy in meiotic cells. In the ascus of *N. crassa*, the nucleolus is associated with a pair of satellites that are always located at the opposite side of, and connected through, strands of attenuated chromatin (probably equivalent to the nucleolonema) with the nucleolar organizer. This is particularly evident with the DNA-specific acriflavin fluorescent stain [170].

As in higher eukaryotes, the nucleolus is seen under the electron microscopy to have prominent compartments: the fibrillar component, the granular components, and sometimes nucleolar vacuoles. The fibrillar and granular components are often intermixed, and there is a nucleolus dense body which is formed only in meiotic cells (Fig. 1a). Partition of granular components may also appear. For example, under the influence of cycloheximide, an inhibitor of protein synthesis, the granular components are seen scattered in the peripheral area around the fibrillar dense body. Under long exposures to cycloheximide, a nucleolar vacuole is prominent and no granular components are visible, suggesting that the production of the granuflar components has ceased and the existing ones have been exported to the cytoplasm [120].

The nucleolus is the site of rRNA synthesis and processing; the end products are ribonucleoprotein (RNP) particles, which are seen by electron microscopy to be granular particles. The nucleolar organizer is composed of tandem repeats (between 100 and 200 copies) [59,186] of rRNA transcription units which are separated by nontranscribed spacer sequences. Each transcription unit contains genes for 18S, 5.8S, and 25–28S rRNAs. Cloning of such a sequence, such as from *C. cinereus*, allows detailed mappings of 18S, 5.8S, and 26S rRNA genes [44] and their locations. In most higher eukaryotes, the 5S rRNA genes are located elsewhere in the genome. This is also true for some fungi, such as *N. crassa* [135,192], and the yeast *Schizosaccharomyces pombe*

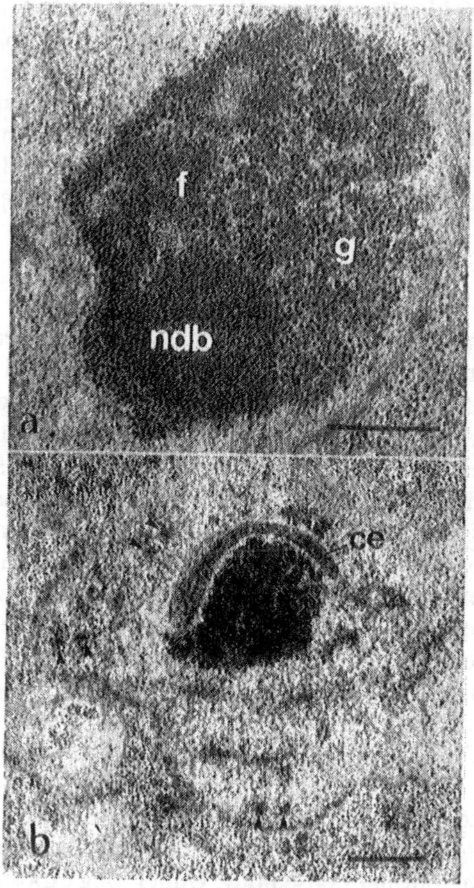

Figure 1 (a) The nucleolus of *Coprinus cinereus* at meiotic prophase showing the nucleolus dense body (ndb) and compartmentalization of fibrillar (f) and granular (g) components. The granular particles are similar in size and appearance to those in the cytoplasm. (b) Effect of cycloheximide treatment on the nucleolus and assembly of the synaptonemal complex of *C. cinereus*: water agar containing cycloheximide was applied to the mushroom cap at 4 h after initiation of karyogamy, and the results are: accumulation of assembled central elements (ce) in the nucleolus dense body; pairing of homologous lateral components without the central element (arrowed); and the nucleolus devoid of granular components. Bar = 0.5 μm. Reproduced from Lu [120] with permission of The Company of Biologists.

[130]. For others, the 5S rRNA genes are located within the rDNA repeat units; examples include *Coprinus* spp., and *S. cerevisiae* [15,44]. The transcription of the 5S gene is in the same orientation as the 26S and 18S rRNA genes in *C. cinereus*, *C. micaceus*, *C. atramentarius*, *Flammulina velutipes*, and *Agaricus bisporus*, but is in the inverted orientation in a bipolar species of *C. comatus* [44]. The phylogenetic relationship between these two species is not clear.

The ribosomal DNA internal transcribed spacers (ITS) are highly divergent. Examples may be found in the phytopathogenic fungi *Fusarium sambucinum* [150], and also in *Verticillium albo-atrum* Reinke and Berth and *V. dahliae* Kleb that cause wilt disease in plants [148]. Species-specific sequences have been identified in the internal transcribed spacers (ITS 1 and ITS 2) for the two *Verticillium* species. These differences permitted the synthesis of oligonucleotides that hybridized differentially with the rDNA of the two species and allowed for an efficient, fungus-specific amplification of either DNA sequence by a polymerase chain reaction (PCR). The PCR assay is an effective diagnostic tool for detecting the presence and species of fungus, and for quantification of the fungal infection in the plant tissues [96,148].

Apart from the biosynthesis of rRNAs, the nucleolus may have other functions. During early meiosis, the nucleolus is the home for assembly of the central element. This was first suggested by Westergaard and von Wettstein [215] in their studies with *Neottiella rutilans*, and later confirmed by Lu [120] using the inhibition of cycloheximide to dissect assembly of the synaptonemal complex (SC). Lu [120] found that at certain time in zygotene cycloheximide can prevent transport of the central elements from the nucleolus to the chromosomal sites where the two lateral components are perfectly aligned. As a consequence, all central elements are accumulated in the nucleolus-dense body (Fig. 1b). The accumulation of the central elements in the nucleolus-dense body is also found in yeast homozygous for the temperature sensitive *cdc4* mutation under restrictive temperature [93], and for *rad50S* non-null mutation that fails to assemble the SCs [3].

C. Basic Nucleosomes

The fungal chromosomes are organized in exactly the same way as in the higher eukaryotes, from DNA into nucleosome repeats which contain core particles and linkers. The nucleosome repeats collapse into 10–30 μM nucleohistone fibers which can be observed by thin-sectioned electron microscopy. The core particle is made up of 140 bp of DNA duplex and a histone core consisting of two each of the histones H2A, H2B, H3, and H4. When chromatin extracts from yeast *S. cerevisiae*, *N. crassa*, or *Aspergillus nidulans* were given a limited digest with mycococcal nuclease, a typical nucleosomal ladder was obtained; the repeat sizes of these fungi are 160, 170, and 155 bp, respectively [103,109,

110, 140, 149, 208]. The extensive digest of these chromatins gives a single 140-bp unit, identical to that of higher eukaryotes. Thus, the linker region contains about 15–30 bp DNA duplex which may be bound by an H1 or equivalent.

D. Centromere

Each chromosome contains a localized centromere around which a primitive kinetochore is organized, though not structurally differentiated like those in

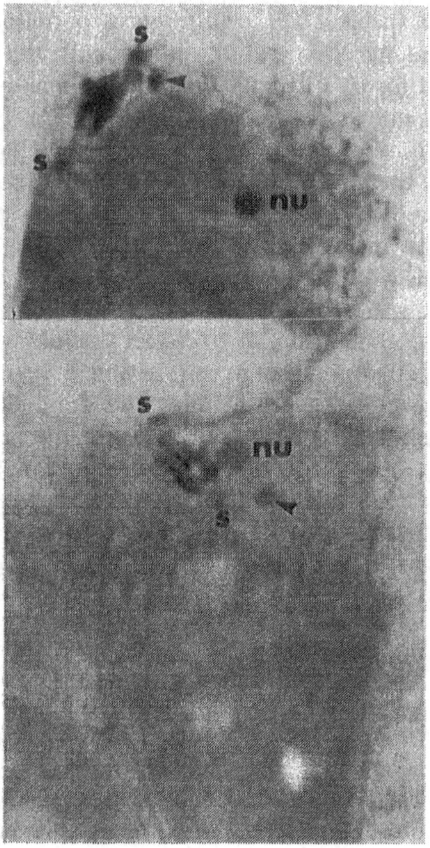

Figure 2 The spindle mechanism of meiotic metaphase of *C. cinereus*. As a result of dikaryotic-monokaryotic mating, aneuploid basidia are produced. The extra bivalents (arrowed) do not congress at the equatorial region. s, Spindle pole body; nu, nucleolus. Bar = 10 μm.

the higher eukaryotes. The kinetochore is the nucleating site for the spindle microtubules. There appears to be only one microtubule per chromosome of *S. cerevisiae* [162]; the same may be true in the higher fungi [2,86,114,115,133, 145,211,220; see Kubai [105] for review]. The number of centromeres per haploid cell for any given species is highly controlled; addition of extra cloned copies of centromeres is toxic to haploid yeast [70,185]. It is possible that the centromere-binding protein CFB3 is limited to one molecule per chromosome [107]. This is consistent with the cytological observation that extra chromosomes (e.g., in dikaryotic–monokaryotic matings) in *C. cinereus* fail to congress at meiotic metaphase I (Fig. 2; B. C. Lu, unpublished observations).

The centromere sequences have been cloned and extensively analyzed in *S. cerevisiae* (Fig. 3) [21,23,49]. The centromere sequences can be cloned either by chromosome walking from centromere-linked genes (e.g., for isolation of CEN III and CEN XI) [21,22], or by direct selection for mitotic stability [88]. The selection scheme is ingenius and deserves a mention. The cloning vector YRp14 carries *ARS*1 for autonomous replication, *URA*3 as a reporter

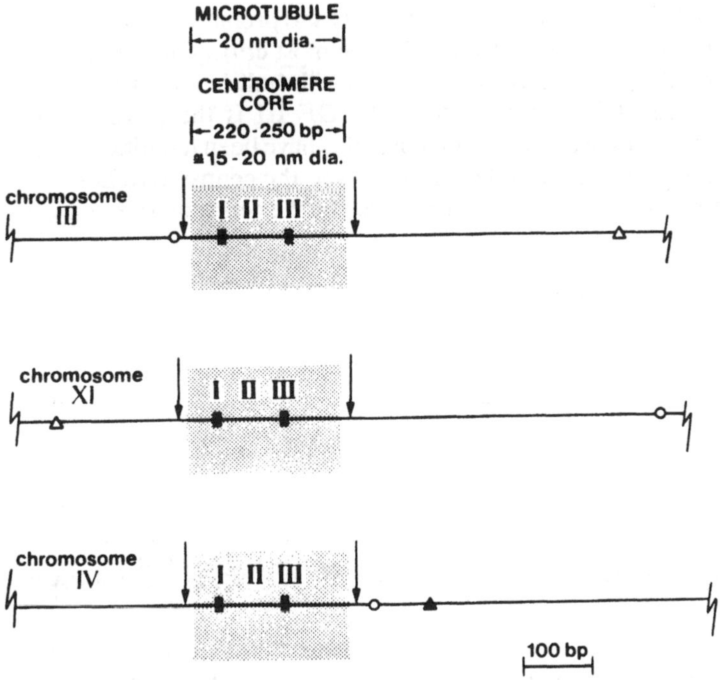

Figure 3 Structure of centromeres of *S. cerevisiae*, reproduced from Bloom [23] by copyright permission of the authors and The Rockefeller University Press.

gene for transformation, and *SUP*11 which is a mutant tyr-tRNA that can read UAA termination codon (i.e., an ocher suppressor). This vector, when not carrying a centromere sequence, will, in a haploid yeast, replicate multiple copies that segregate in a strong mother bias. The daughter cells receive no plasmid and will be selected against while the mother cells die of *SUP*11 toxicity due to excessive production of the mutant tyr-tRNAs. Only when the vector carries a yeast centromere sequence will it confer mitotic stability. Thus, with a few minor exceptions, Hieter et al. [88] were able to isolate a large number of yeast centromere sequences.

The centromere sequence is highly conserved among all yeast chromosomes, but is also species-specific; the yeast CEN sequences do not function as centromeres when introduced on plasmids into cells of other fungi [33]. The 220- to 250-bp sequence contains three functionally distinct centromere DNA elements (CDEs) as shown in Figure 4. CDE I and III are short and highly conserved sequences of 8 and 25 bp, respectively. These are sequence-specific protein-binding sites [6,31,107]. CDE I is the least critical, because its deletion causes little effect on mitotic and meiotic functions. CDE II is an A+T–rich central region of 78–86 bp; deletion of all or part of it leads to increased nondisjunction of chromosomes in mitosis and premature separation of sister chromatids in meiosis [72]. CDE III is the most critical one; a point mutation of the central C to T in the inverted repeat will abolish the centromere function and structure [99]. It is possible that CDE III is the recognition site for centromere-binding proteins. Such proteins have been identified, and they are centromere sequence–specific [107]. In fact, the centromere DNA sequence of the native chromatin is highly protected by the centromere-binding proteins

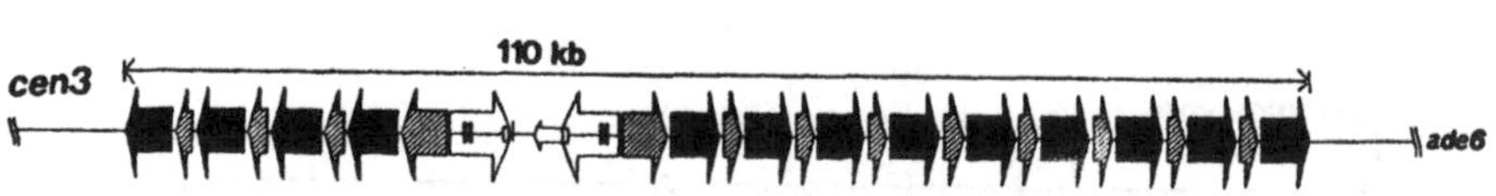

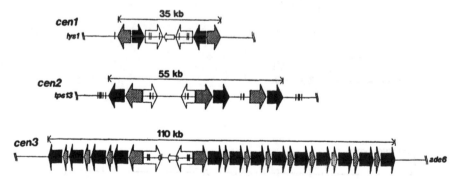

Figure 4 Structure of centromeres of S. pombe. Filled arrows, dg repeats; shaded arrows, dh repeats; large open arrows, imr sequences; small arrows, tm sequences; vertical lines, tRNA genes. Reproduced from Murakami et al. [147] with permission of Springer-Verlag.

against nuclease digest [23]. Microtubules do not react with DNA directly, but microtubule-associated proteins do, suggesting that the "kinetochore" works between centromere DNA sequences and the spindle microtubules [21]. As shown in Figure 3, the 220- to 250-bp sequence measures about 20 nm, which is probably large enough for insertion of only one microtubule as observed by Peterson and Ris [162].

The centromere DNA sequence of *S. cerevisiae* is very small and primitive compared with that of the fission yeast *S. pombe.* In the fission yeast, the functional centromeres are roughly 100–1000 times larger than that of *S. cerevisiae*, and contain repeated sequences identified as dg and dh [47,50,147]. Although these repeated sequences vary in number and size among different chromosomes, the sequence homology among them is 97–99%. As shown in Figure 4, the *cen1* is only 35 kb and contains only two dg-dh repeats, the *cen2* is 55 kb and contains three repeats, while the *cen3* of the smallest chromosome is 100 kb long and contains 13 repeats. The second class of repeated sequences is the innermost repeat (*imr*) sequence, which flanks the central core. The *imr* are inverted repeats, and chromosome-specific, and contain tRNA genes; i.e., the *imr* derived from *cen1*, *cen2*, and *cen3* do not cross-hybridize except the tRNA genes they contain (K. Takahashi et al., quoted by Murakami et al. [147]). These tRNA genes are not transcribed, and the significance of their presence is unknown. The third class of sequence is the central core sequence, which alone does not have the centromere function. Plasmids that contain portions or all of the core and only limited inverted repeat sequences lack centromere function entirely. The repeated elements dg and dh are particularly important for correct segregation of chromosomes in meiosis I, specifically with respect to holding sister chromatids together during metaphase I. In this case, they may function like the pericentric heterochromatin in the higher eukaryotes. The repeated elements alone have no centromere function, as plamids that contain only a repeat unit (either from the right or from the left arm) or a repeat unit and a portion of the central core lack centromere function. On the other hand, plasmid that contains the entire right arm of the inverted repeat (15.5 kb), a small portion of the left arm (3.5 kb), and the entire central core (5–7 kb) is stable in mitosis, segregates 2+:2– through meiosis, but still does not exhibit full function, as it shows a high degree of sister chromatid separation in meiosis I [50]. Thus, the inverted repeat structure including the *imr* sequences is essential for chromosome segregation in mitosis and meiosis as it occurs in all three chromosomes of *S. pombe* [50,147]. While *imr* sequences occur in all chromosomes, they are not homologous, and they contain different sets of tRNA genes. These differences in sequence specificity and number of repeated elements in each chromosome may serve as recognition sites for individual chromosomes during mitosis and meiosis.

Like *S. pombe*, the centromere DNA sequence of *N. crassa* is much more complex than that of yeast. The centromere locus from linkage group VII of *N. crassa* has also been cloned, characterized, and physically mapped. It is contained within a 450-kb region between *qa-2+* and *met-7+*, and it contains A+T–rich repetitive sequences [45]. The A+T content of the DNA in this region is at least 67%, and there are 73 distinct *PatI* sites and clusters of sites. The same reptitive sequences are found in all seven linkage groups. The centromere region is recombination-deficient; the recombination frequency is only 0.2% of the estimated average over the entire genome.

Although heterochromatic knobs have been described for *Neurospora* [199] and the presence of highlfy repeated sequences has been recognized, the presence of pericentric heterochromatin having highly reiterated satellite sequences like those found in *Drosophila* has not been established in fungi. It is possible that the repeated sequences found in *S. pombe* and *Neurospora* may be functionally akin to the heterochromatin of the higher eukaryotes.

E. Chromomeres

The chromomeres are visible in meiotic chromosomes at pachytene of *N. crassa* using orcein stain [8,11]. McClintock made sketches showing chromomere patterns and their variations. Some of these unpublished sketches are reproduced in an article by Perkins [159], who noted that "drawings of this type are more informative than photographs, the usefulness of which is severely limited because focal depth is minimal at the high magnification required for working with such small chromosomes." Observations of chromomere patterns are not as successful in other fungi.

F. Telomeres

The telomeres are the ends of a chromosome. They provide protection to the termini of a chromosome so that different chromosome ends do not fuse. When chromosomes are broken, such as by radiation, the broken ends are sticky and they tend to rejoin with another sticky end. As a consequence, chromosome rearrangements, such as translocations and inversions, occur. Clearly, the broken ends are different from the telomeres. Like higher eukaryotes, the telomeres of fungal chromosomes may be associated with the inner membrane of the nuclear envelope. This is implied by observations showing the association of the ends of synaptonemal complexes with the inner membrane of the nuclear envelope during meiotic prophase [92,113,178,221]. It is conceivable that the nuclear envelope provides a two-dimensional surface for chromosomes to glide around in search for homologous pairing partners during meiosis. This movement is necessary in fungi when meiotic prophase follows karyogamy; two separate sets of chromosomes occupying two different nuclear domains need to come together for homologous pairing.

The telomere has yet another important function: to safeguard the replication of linear chromosomes. The understanding of the telomere sequence first came to light from cloning and sequence analysis of extra chromosomal rDNA from *Tetrahymena*, which contains repeated sequence of $(C_4A_2)n$ or (TTGGGG)n [19,20]. When this Tetrahymena rDNA sequence was added to a linear plasmid carrying a yeast CEN sequence, it increased mitotic stability to the plasmid in yeast cells. When the telomere sequence was analyzed, it carried the yeast-specific telomere repeats [194]. It appears that there is a specific telomerase (or telomere terminal transferase), and the first to be discovered is that of Tetrahymena [82,83] that can extend a couple of repeats of 5'(TTGGGG)3' per cell cycle; the extended TG strand can fold back on itself to serve as a primer for replication of the AC complementary strand. Because of this foldback mode of end replication, the telomere sequence produced is an inverted repeat. When a circular plasmid carrying an artificially created telomere inverted repeat was transformed into haploid yeast, the inverted repeat structure was recognized and resolved into linear minichromosomes by yeast cells. This observation provides strong circumstantial evidence that an inverted repeat is a natural intermediate in telomere replication [205].

The telomere sequence of *Neurospora crassa* has also been cloned and sequenced by Schechtman [187,188]. This was achieved by chromosome walking from the closest gene *his*-6 to the right end of linkage V (VR) in the standard wild-type Oak Ridge background. The sequence is a tandem repeat of the hexanucleotide TTAGGG. Interestingly, this sequence is shared by human and many other mammals [146].

Apart from the telomere repeat sequences, which are shared by all chromosomes, there are telomere-associated sequences, such as X and Y in yeast; the most characteristic feature of telomere-associated sequences is their variability (see Zakian [218] for review). Details of their organization remain to be investigated. If their organization were shown to be chromosome-specific, one might conjecture that these subtelomeric sequences might provide some ancillary form of homology recognition during chromosome pairing.

G. Karyotype Analysis

The chromosome cytology has been an important link to the genetic systems of eukaryotes since the turn of the century. Karyotype analysis may be related to cytotaxonomy, speciation, and evolution. Within a single genetic system, it may be related to linkage groups, the number and distributions of chiasmata, aneuploidy, polyploidy, and chromosome rearrangements, such as translocations, inversions, duplications, and deletions.

Traditionally, chromosomes of higher eukaryotes were stained with a basic dye such as carmine, orcein, hematoxylin, Feulgen, or Giemsa stain, and observed by light microscopy. The first clear demonstration of chromosomes of *N. crassa* in meiosis was made by McClintock [131] using aceto-orcein and acetocarmine in a squash technique. McClintock established n=7 for *N. crassa*; she also numbered them 1 to 7, with chromosome 1 being the longest, chromosome 2 the nucleolar chromosomes, and so on. She certainly laid the groundwork for fungal cytogenetics.

Studying chromosomes of fungi in mitosis and meiosis had its share of difficulties and controversies. For meiotic chromosome counts before 1950 (see review by Olive [152]), most species of Basidiomycetes investigated were reported to have either two or four chromosomes. Olive observed that "reports of odd number of chromosomes in the Basidiomycetes are rare, as are reports of haploid numbers greater than 8." The erroneous reports arose in part because of small sizes of the basidia and because of high density of basophilic cytoplasmic ribosomes that picked up more stain than the chromosomes. In actual fact, more than 12 bivalents have been found in several basidiomycetes [92,112,121,126,163]. Chromosome counts in the Pyrenomycetes have not been handicapped because of the size of the ascus. With a good protocol, detailed analysis of pachytene chromosomes can be achieved (see the methodology below).

H. Methodology of Karyotype Analysis

Three basic techniques have been used to determine chromosome numbers: 1.) conventional light microscopy using a variety of stains; 2.) electron microscopy of the synaptonemal complex; and 3.) pulsed-field gel electrophoresis of intact chromosome-size DNA molecules. Each technique has its advantages and disadvantages.

Conventional light microscopy has been a standard procedure for karyotype analysis of higher eukaryotes. The protocols and the choice of stains vary according to the cell types. The favorite stains are acetocarmine, aceto-orcein, Feulgen and Giemsa, and iron hematoxylin. For fungal chromosomes, the meiotic cells are the material of choice. The initial success was obtained with the ascomycetes where the ascus is very large. The aceto-orcein [7–9,11,131], the acetocarmine [42,199,211], and the Feulgen and Giemsa [183,184] have been used with success. The introduction of the time-controlled hot HCl hydrolysis to remove cytoplasmic stain improves the visibility and clarity of the chromosomes. Additional techniques have been developed and introduced to fungal nuclear cytology through the years. The propiono–iron hematoxylin squash technique was first developed for *Gelasinospora calospora* and *Coprinus* spp. [114,115,126] and later adopted to other species [123,168,175,219]. This procedure proved to be extremely effective for staining chromosomes, nucleoli, spindles, and spindle pole bodies, and it has been used extensively to examine wild-type, developmental mutants, and aneuploids of *Neurospora* spp. (for reviews see Raju [167,171,172] and Perkins et al. [161]). However, the hematoxylin stain is inferior to aceto-orcein in resolving the chromomere details of pachytene bivalents [159], and it is quite useless to make permanent slides as the staining quality is destroyed in the process. Several DNA-specific fluorochromes have been tested for fungal cytology, and in acriflavine with epifluorescence microscopy was found to be most promising [168,170]; it gives unusual clarity of chromosome details not only for pachytene but also for diplotene bivalents (Fig. 5), the latter not usually being resolvable by other stains. Acriflavine-stained pachytene chromosomes show a characteristic chromomere pattern like those observed in orcein-stained preparations. Acriflavine is also superior to hematoxylin and acetocarmine in another respect: it does not stain the nucleolus and the spindle pole bodies, so it eliminates the potential ambiguity in chromosome counts. The nucleolus organizer region shows as an attenuated strand. Although fading of fluorescence is a common problem with any fluorescent stain, it can be overcome to some degree with reducing agents such as mercaptoethanol or dithiothreitol.

Karyotype analysis of the Basidiomycetes is more difficult, because the nuclei and meiocytes are much smaller than those of the Ascomycetes. Although the chromosomes can be stained with either acetocarmine or propiono–iron hematoxylin after time-controlled hydrolysis of the cytoplasmic ribosomes, resolution of individual chromosomes in intact basidia is quite impossible. Only when the protoplasts are "hammered" out of the cell wall will the chromosomes be spread enough for karyotype analysis (see below). Even at diakinesis, the highly condensed bivalents tend to become associated and unambiguous chromosome counts are difficult, because the size of each chromosome is almost below the resolution power of the optics (0.1–0.2 μm).

Karyotype analysis can also be achieved by electron microscopy. During pachytene of meiotic prophase I, each bivalent forms a synaptonemal complex (SC). This three-dimensional reconstruction of a complete nucleus from serial sections allows tracing of all bivalents using the SC as a marker. This technique was first developed by Moens and colleagues [137,138] and has since been used for a number of fungi [30,35,73,92,178]. The use of this technique is limited to meiosis; it is applicable, however, to any species in which tripartite synaptonemal complexes are found. Beyond accurate karyotype analysis, this technique also yields important information such as attachment of the SC ends to the nuclear envelope, chromosome rearrangements (e.g., translocation), and re-

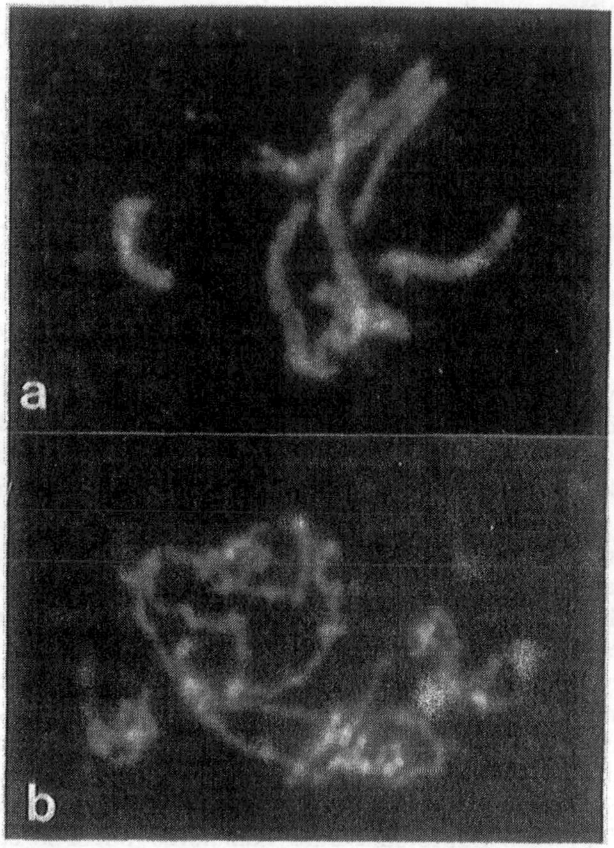

Figure 5 Meiotic chromosomes of *N. crassa* as shown by acriflavine staining and epifluorescence microscopy: (a) pachytene, (b) diplotene; arrows point to the rDNA region and satellites. Courtesy of N. B. Raju.

combination nodules and their distributions (to be discussed later). This technique, however, is time-consuming and requires expert hands. To circumvent this difficulty, a surface spread technique was developed by Moses and his colleagues initially for animal systems [57,143]. A surface spread for SC was first achieved in *S. cerevisiae* after the cell wall was removed by enzymatic means [56]; it has been used extensively to study various meiotic mutants (see below). A surface spread protocol was later developed using mechanical breakage of meiotic cells first for *C. cinereus* [163]; the improved protocol was developed for *C. cinereus* [164] and for *N. crassa* [120a]. Examples are shown in Figures 6 and 7. For *Coprinus*, where meiosis is synchronous, karyotype analysis is definitely simplified; several spreads can be obtained in an afternoon's work. A protocol is attached at the end of the chapter.

In recent years, the molecular karyotyping by pulsed-field gel electrophoresis (PFGE) has been developed whereby intact chromosomes migrate through a pulsed field in an agarose gel matrix with a velocity dependent on their size and three-dimensional structure [191]. PFGE can separate large DNA molecules up to 2 megabasepairs. This will allow researchers to determine both chromosome number and their sizes. A number of "improved" techniques have subsequently been developed, namely, the orthogonal-field-alternation gel electrophoresis (OFAGE) system [34], the contour-clamped homogeneous electric field (CHEF) gel electrophoresis, and the transverse alternating-field electrophoresis (TAFE) [48,71]. The result is a resolution of genome into chromosomal bands, each of which may represent one and sometimes more than one chromosome. Southern hybridization with a known cloned gene will allow identification of a given band to a linkage group. For

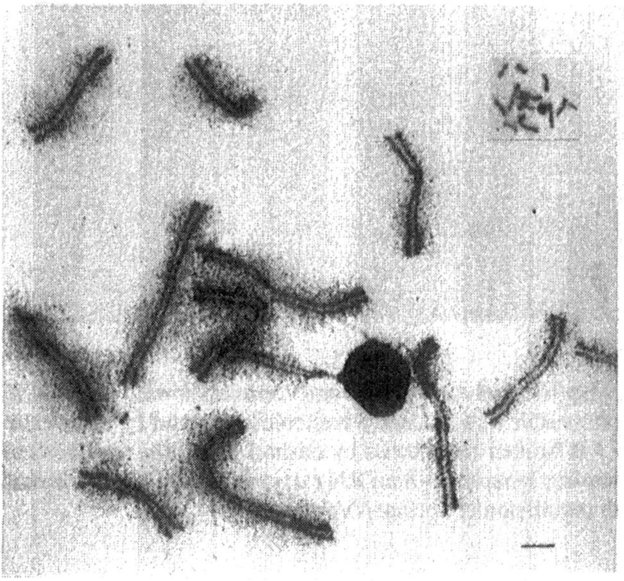

Figure 6 Synaptonemal complexes from *C. cinereus* showing 13 SCs with light microscope image insert of the same nucleus. Bar = 1 μm. Reproduced from Pukkila et al. [164] with permission of John Wiley & Sons, Inc.

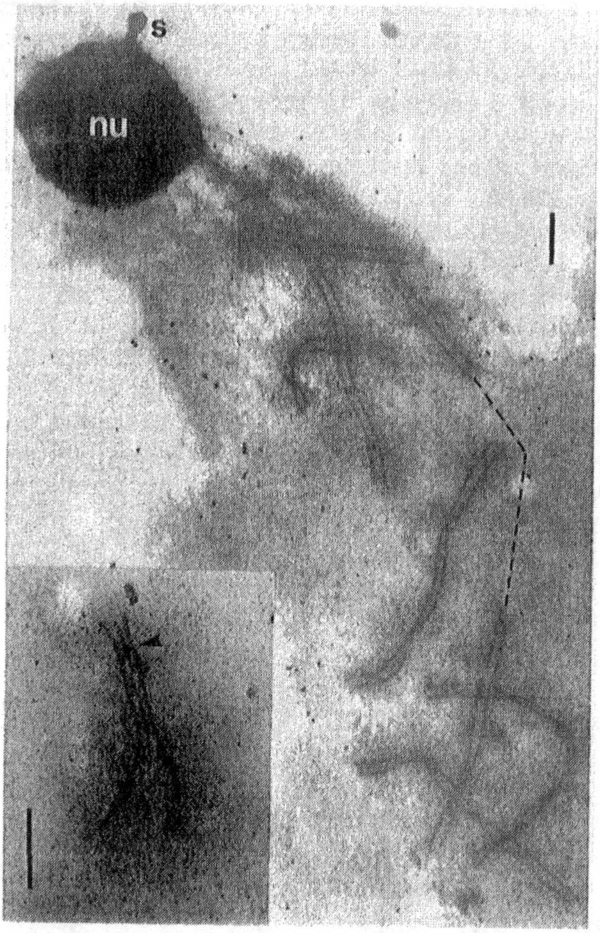

Figure 7 Synaptonemal complexes from *N. crassa* showing seven SCs. Recombination nodules are visible in all SCs, the nucleolus (nu) and its associated satellite (s); chromosome 1 is broken (connected by dashed line). The insert is chromosome 7 showing incomplete synapsis with an RN (arrowed). Bar = 1 μm. Reproduced from Lu [120a] with permission of Springer-Verlag.

detailed information, readers should consult the paper Molecular karyotype analysis of fungi by Skinner et al. [200].

I. Variations of Chromosome Structure and Numbers

As in higher eukaryotes, changes in chromosome structure and numbers have been recognized and studied. It is beyond the scope of this chapter to discuss them in detail. Readers should consult an excellent review by Perkins and Barry on cytogenetics of *N. crassa* [160]. Cytologically, chromosome rearrangements can be studied by light microscopy [10,131] and with great clarity by three-dimensional reconstruction of synaptonemal complexes of *N. crassa* and *C. cinereus* [26,75,92] or by surface spread electron microscopy [163] (B. C. Lu, unpublished). An example is shown in Figure 8. The approximate location of translocations and inversions can be mapped at early pachytene when pairing is precise. Bojko [26] found that synaptic adjustments also occur in *N. crassa* when the inversion loop is eliminated, much as is found in mice [144].

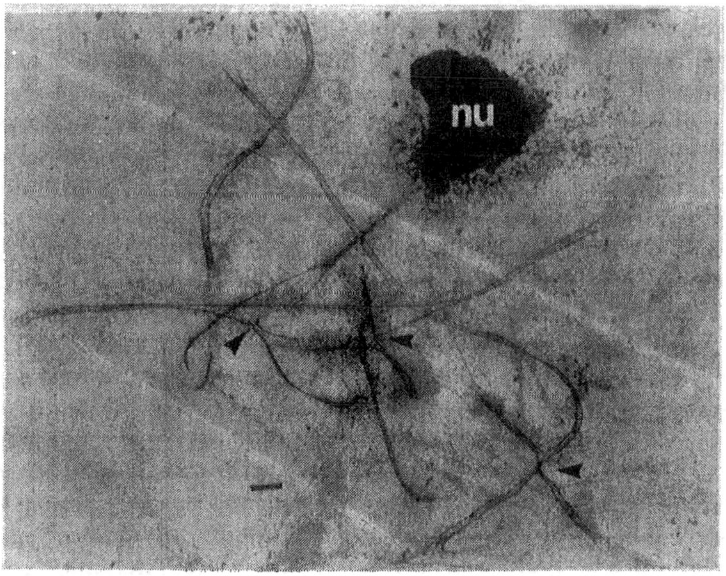

Figure 8 Synaptonemal complexes of *N. crassa*, heterozygous for translocations in acoy-T(I;II)4637 al-1; T(IV,V)R2355, cot-1; T(III;VI)1, ylo-1 involving chromosomes 1 and 6; 2 (associated with the nucleolus (nu) and 5; and 3 and 4; three translocation points are marked by arrows. Bar = 1 μm.

Polyploidy also occurs in fungi. *Cyathus stercoreus* strain 1305 collected by Brodie is an autotetraploid [112a]. In *C. cinereus*, an autotriploid was found by accident as a result of dikaryon–monokaryon mating [178]. In fact, autotriploid basidia occasionally occur in normal dikaryotic fruit bodies probably because of mistakes at the last mitosis when one nucleus fails to enter the clamp connection resulting in the formation of a three-nucleated basidium. In this case, the three lateral components are paired together. A complete set of triple SCs has been observed among normal diploid meioses (Pukkila and Lu, unpublished).

4. MITOSIS IN VEGETATIVE MYCELIUM

The study of mitosis in the vegetative mycelium has its problems and controversies from the beginning. Most of these may be attributed to the nature of the cell types, and how investigators interpreted their observations. The mycelia are narrow tubes with septal pores through which protoplasmic streaming occurs; often the nuclei are forced to become linear threads when they pass through the septal pores, and there is very little space to allow nuclei to spread as in the case of animal or plant cells when they are stained and squashed for cytological observations. Compounding these problems is the occurrence of nuclear division taking place when the protoplasm is streaming as in *Neurospora* (unpublished observations). In most earlier observations, what was observed, under the light microscope, were nuclear filaments, beaded tracks, two-track configurations, and "bilobed" sister chromatin masses at anaphase, with no hint of any chromosomes or spindle mechanism. These observations may contribute to erroneous concepts of "amitosis" or mitosis employing mechanisms very different from those operating in conventional mitosis [53,76,132,180, 181]. A good deal of emphasis of these studies was focused on negative observations such as no evidence of a "classical" metaphase plate, absence of a "classical" spindle apparatus, etc. An excellent review is provided by Kubai [106].

The two-track configuration is the most commonly observed at metaphase–anaphase of mitosis of vegetative mycelium of both Ascomycetes and Basidiomycetes, including species such as *Fusarium* [2], *Poria* [193], and yeast [162]. In fact, the two-track configuration is also common at meiotic metaphase–anaphase in the basidium—e.g., *Coprinus*. For the above species where two-track configurations were observed, the spindle mechanism has been clearly demonstrated by electron microscopy, and it is definitely unexceptional (see review by Kubai [108]). The best explanations for the two-track image observed at metaphase–anaphase is offered by Kubai [108], who suggested that anaphase chromosomes become arranged around the periphery of the central spindle, lying at various positions staggered along the spindle length. In light microscopy, the optical sectioning of the rather cylindrical chromatin distribution produces the two-track images. This description is supported by electron microscopy of meiotic metaphase–anaphase in the basidium of *Coprinus* where two-track images are also commonly observed; chromosomes are arranged around a bundle of the central spindle microtubules [114,115,206].

Electron microscopy has established that mitosis is intranuclear in some groups of fungi, while in others regional dissolution of the nuclear envelope occurs [2,86,132,145] (see also Kubai [105,106] for reviews).

Aist and Williams's [2] observations of mitosis in the vegetative mycelium of *Fusarium oxysporum* by a combined phase contrast microscopy and electron microscopy have provided the clearest picture of mitotic events. The time required for the stages of mitosis are prophase, 70 sec; metaphase, 120 sec; anaphase–telophase, 125 sec; for a total of 5.5 min. At metaphase under phase live, a thin spindle is formed between two spindle pole bodies and the chromosomes (unresolvable) are grouped at the equatorial region. This image is comparable to those observed in meiosis and mitosis in the ascus of *Gelasinospora* and *Neurospora* [114,167], and in the basidia of *Coprinus* [115,126]. A typical spindle mechanism has also been demonstrated by electron microscopy using serial sections [2,86]. Aist and Williams [2] showed that kinetochores occur in pairs, each kinetochore of which is connected to the opposite pole by a single spindle microtubule (Fig. 9a). A similar image has been observed in *Armillaria mellea* and *Poria* [145,193] and in *S. cerevisiae* [162]. At the EM level, there is no difference between the image of mitotic spindle apparatus in the vegetative cells (Fig. 9a) and those in meiotic cells (Fig. 9b). Thus, metaphase congregation of chromosomes at the equatorial region and their association with a spindle are identical to the conventional mitosis. It is true that the chromosome movement at metphase-anaphase is staggered or nonsynchronous. The staggered movement of chromosomes is also seen in asci and in basidia. The absence of a classical prometaphase is most likely due to limited space; given space, a typical prometaphase and metaphase congregation of chromosomes is seen clearly in the postmeiotic mitosis (division III) in the ascus (Fig. 9c) and in the ascospores (division IV) of *Gelasinospora* and *Neurospora* [114,167]. When *Neurospora* conidia are grown in liquid medium containing 3.22 M ethylene glycol, they grow without cell division, forming giant spheres with multiple enlarged nuclei. In these giant cells, all stages of mitosis including prometaphase and metaphase configurations have been observed [169].

5. MEIOTIC PROCESSES

Since McClintock [131] introduced the aceto-orcein squash technique to study meiosis of *Neurospora*, many papers have been devoted to this subject in a large number of fungi. Among these are *Neurospora* [9,167,172,199], *Sordaria* [42], *Gelasinospora* [114], *Cochliobolus* [95], *Venturia* [54], *Hypomyces* [63], *Hypoxy-*

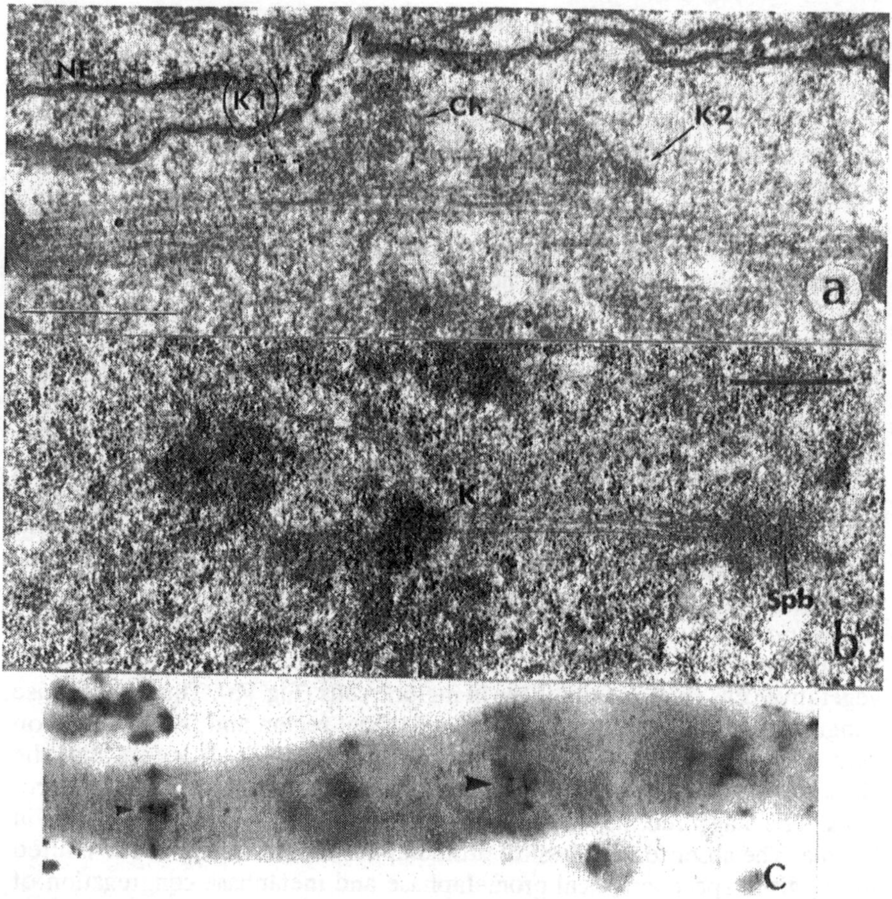

Figure 9 Spindle mechanism in mitosis and meiosis: (a) Mitotic metaphase in *F. oxysporium*: two sister chromatids (ch) are each connected to the pole by a single microtubule inserted at the kinetochore (K1, K2). Bar = 0.5 μm. Reproduced from Aist and Williams [2] by copyright permission of Rockefeller University Press. (b) Meiotic metaphase I of *C. cinereus*. A chromosome is clearly shown to be connected to the Spb by a microtubule inserted at each of its two kinetochores (k). Bar = 0.5 μm. Reproduced from Lu [114,115] with permission of The Company of Biologists. (c) Spindle mechanism in postmeiotic mitosis in *G. calospora*, showing classical (small arrow) and not so congressed metaphase (large arrow) configurations. Bar = 10 μm. Reproduced from Lu [114,115] with permission of Springer-Verlag.

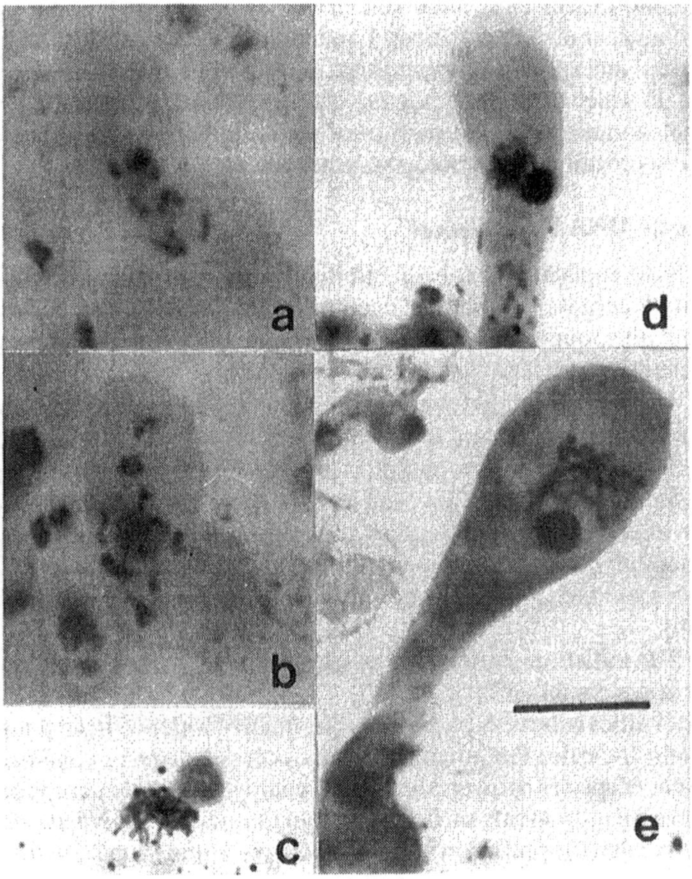

Figure 10 (a, b) Young ascus of *N. crassa*. Chromosomes are highly contracted shortly after karyogamy but before homologous pairing. Reproduced from Lu and Galeazzi [123] with permission of the publisher. (c–e) Young ascus of *G. calospora*. Synizetic knot is observed hammered out of the ascus (c) or in the intact ascus (d) at the end of zygotene before the synapsed bivalents open into pachytene configuration (e). Bar = 10 μm. Reproduced from Lu [114,115] with permission of Springer-Verlag.

lon [183], *Ascobolus* [211,219,226], *Podospora* [198], *Aspergillus* [64], *Cyathus* [112,113], and *Coprinus* [115,126]. Earlier studies on meiosis of fungi have been reviewed by Olive [153] and by Heywood and Magee [87]. In general, the meiotic processes of fungi are identical to those of higher eukaryotes, with perhaps minor differences.

Meiotic processes include one round of DNA replication and two rounds of nuclear divisions: meiosis I and meiosis II. Meiosis I is a reductional division, and meiosis II is an equational division. The major interest lies with meiosis I, in which occur homologous pairing of chromosomes, DNA nicking and double-strand breaks, formation of the synaptonemal complex, crossing over, and segregation of homologous centromeres.

A. Meiotic DNA Replication

Meiotic DNA replication has been habitually named premeiotic S phase. This is inaccurate because it implies an event prior to, and not a part of, meiosis. In actual fact, this round of DNA replication is meiotic in nature and should be named meiotic S phase for the following reasons:

1. The time required for complete DNA replication in meiotic S phase is 8–10 times longer than that of mitotic S phase.
2. The control of the initiation of DNA replication is different for meiosis and mitosis. In *Coprinus cinereus*, where meiosis is synchronous, the initiation of DNA replication is subject to arrest by high temperature and light in meiotic S but not in mitotic S phase [124]. In addition, a mutation was found to abolish meiotic S phase, but the same mutation has no effect on mitotic S phase [98].
3. The initiation of meiotic S phase signals the commitment to meiotic pathway and recombination [5,68,197].

When does meiotic S phase occur in fungi? Evidence from a number of experiments indicates that meiotic S phase occurs before karyogamy, at least in a number of species. Rossen and Westergaard [184] used microphotometric absorbance measurements of Feulgen-stained nuclei in *Neottiella rutilans* to demonstrate that the prefusion nuclei in the ascus initial have 2C value of DNA each, while the fusion nucleus in the ascus has 4C value of DNA. The same was demonstrated for *Sordaria fimicola* [16], *Neurospora crassa* [97], and *Schizophillum commune* [35]. Similarly, completion of meiotic S before karyogamy was demonstrated by ^{32}P incorporation in *Coprinus cinereus* [124]. The same conclusion was reached by Oishi et al. [151] using fluorescence of propidium iodide–stained nuclei in heterokaryotic fruit bodies of *C. macrorhizus* (= *C. cinereus*). However, exceptions have been reported in monokaryotic fruiting bodies of *C. macrorhizus* [151] and in homokaryotic strain *C. patouillardii* [12, 58]. Further studies of these exceptions are needed.

B. Meiosis in Ascomycetes

Meiosis in Ascomycetes may be represented by *Neurospora*, *Gelasinospora*, and *Sordaria*. By a sequential crozier formation from a single pair of nuclei, an array of asci of advancing stages are developed from the penultimate cells.

The length of the ascus varies from 30 μm at zygotene to 100 μm at full pachytene to 170 μm at spore-forming stage. Because of the small size of the ascus, the prefusion nuclei and the early postfusion nucleus are difficult to observe. In her preliminary study of *Neurospora* chromosomes, McClintock [131] described highly condensed chromosomes before homologous pairing. This is shown in Figure 10a. There is also a synizesis stage at the end of zygotene, like that seen in corn meiosis (often neglected by cytologists) when all bivalents are bundled together near the nucleolus. This configuration is called synizetic knot (Fig. 10b) before it eventually opens into full pachytene. The photographic documentation of the complete meiosis is shown in Figure 11.

Leptonema–Zygonema

Leptonema is not well defined in this group of fungi, and it is difficult to study as the ascus is small. It is a stage after karyogamy and before homologous pairing of chromosomes. Under the light microscope, the chromosomes are highly contracted. This is quite different from most organisms where chromosomes at leptotene are fully extended. Under the electron microscope, a unique electron-dense structure appears on each unpaired chromosome; these structures are called the axial cores, and they are formed only after karyogamy. At zygotene, the homologous cores start to move around to find each other and form loose pairs; often the pairing starts from the ends. This event is clearly demonstrated by three-dimensional reconstruction of serial sections by electron microscopy [221] in *Sordaria macrospora* (Fig. 12) and by surface spread of *N. crassa* (Fig. 13) [120a]. Since these chromosomes are highly contracted before homologous pairing, they must have elongated when the homologous axial cores are completely aligned (Fig. 13) as they become the lateral components before the deposition of the central elements to form a tripartite synaptonemal complex. This may be considered early zygotene. Late zygotene shows almost complete assembly of the SC. During zygotene, the nucleolus increases in volume. This is clearly evident in *Gelasinospora calospora* [114].

Pachynema–Diplonema

At pachytene, homologous pairing is complete, and bivalents can be resolved by light microscopy. These bivalents continue to elongate until they reach the full pachytene (Figs. 5, 11b). The most exciting structure discovered by electron microscopists is the tripartite synaptonemal complex (SC). This is a ribbonlike structure composed of two parallel lateral components (LC), each of which is about 20 nm thick and 40–50 nm wide, flanking the central element (CE), about 20 nm thick and 20 nm wide. The distance between the two LCs is 120 nm. The space between the two LCs is the central region in which fine fibers perpendicular to the ribbon appear to connect the LCs to the CE. In some

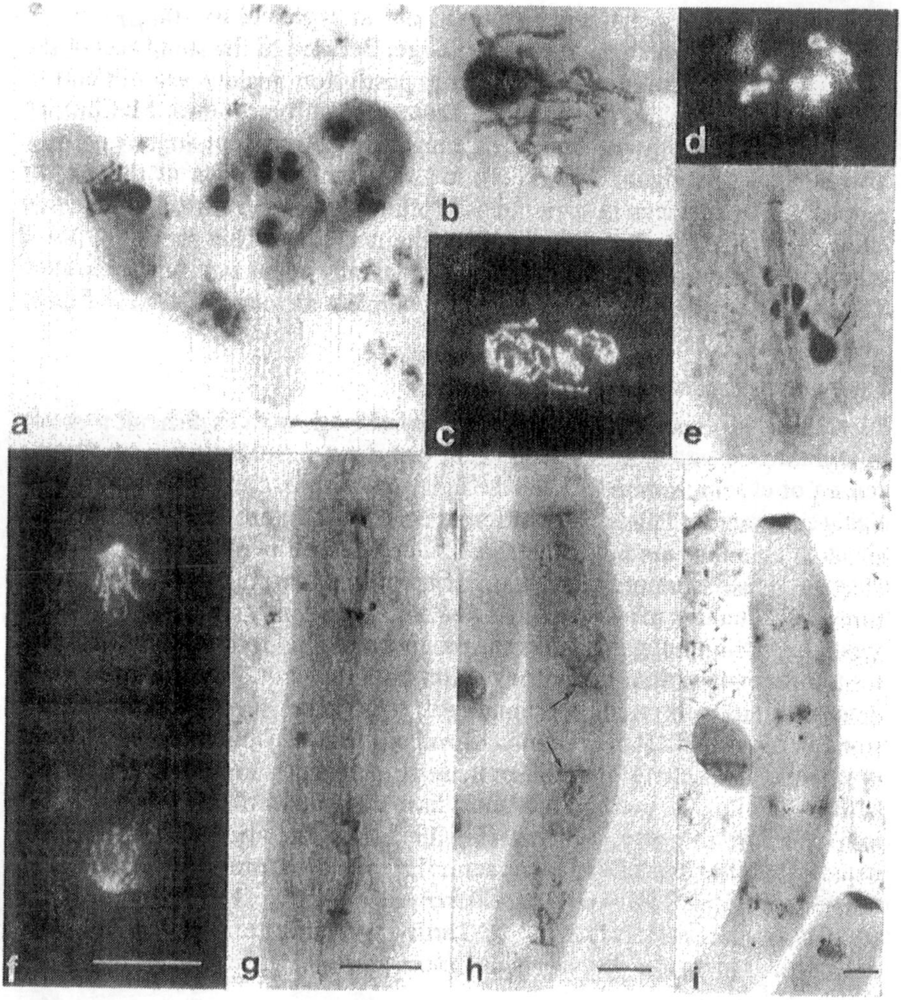

Figure 11 Ascus development in *Neurospora* spp. Staining was iron–hematoxylin [166,175] except for c, d, and f, where the DNA-specific fluorochrome acriflavine was used [170]. (a) Crazier and young ascus, showing apical, penultimate, and stalk cells. (b) Pachytene; the paired chromosomes are much extended and the nucleolus is at its maximum size. (c) Diplotene chromosomes. (d) Diakinesis showing all seven condensed bivalents. (e) Metaphase I. The nucleolus (arrow) is still attached to the nucleolus organizer chromosome. (f) Interphase I. (g) Telophase II. The two second-division spindles are aligned in tandem, parallel to the ascus wall. (h) Late interphase II.

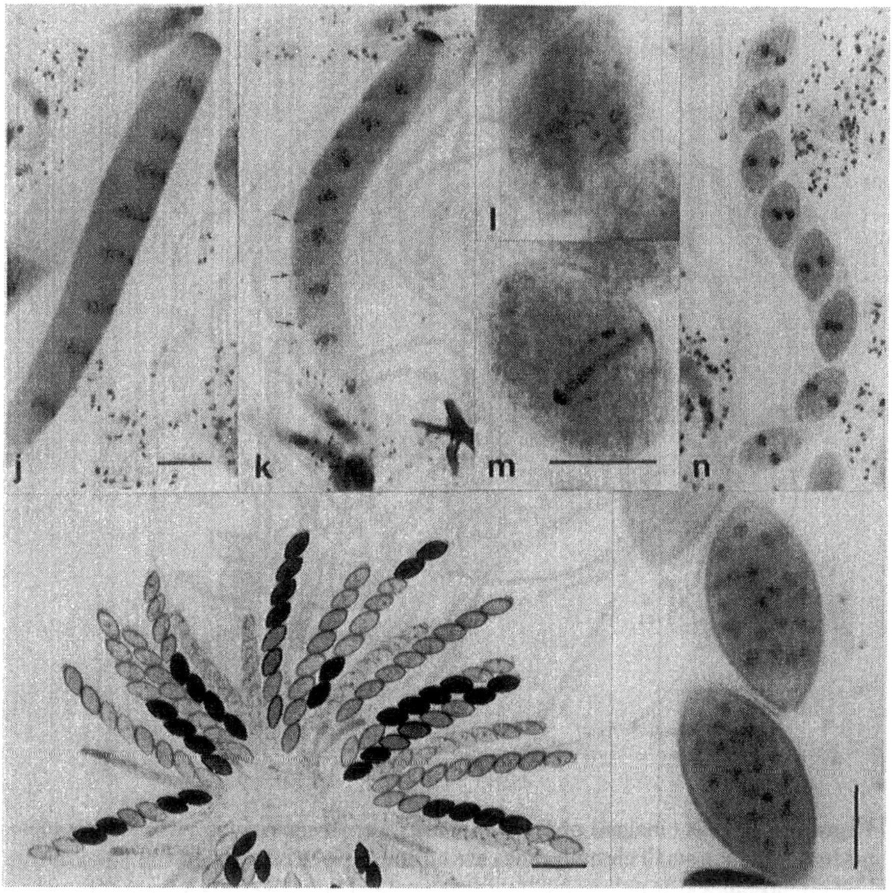

Legend cont. The enlarged double SPB plaques (arrows) are formed at this stage. (i) Anaphase III. All four spindles are oriented across the ascus. (j) Interphase III. The sister nuclei from opposite sides of the ascus line up in single file with all SPB plaques facing the same side of the ascus; spore delimiting outlines are discernible. (k) Uninucleate ascospores have just been delimited. SPB plaques (arrows) usually occupy the lower end of each ascospore, relative to the ascus base. (l) Metaphase IV in an immature ascospore. The polar view shows all seven chromosomes. (m) Telophase IV. The spindle, old nucleolus, and nuclear envelope are all visible. (n) Binucleate immature ascospores at interphase IV. (o) A rosette of maturing asci showing first- and second-division segregations. Bar = 50 μm. (p) Multinucleate mature ascospores from a mutant (*per-1*). Bar = 10 μm. Reproduced from Raju [172] with permission of *Mycological Research*.

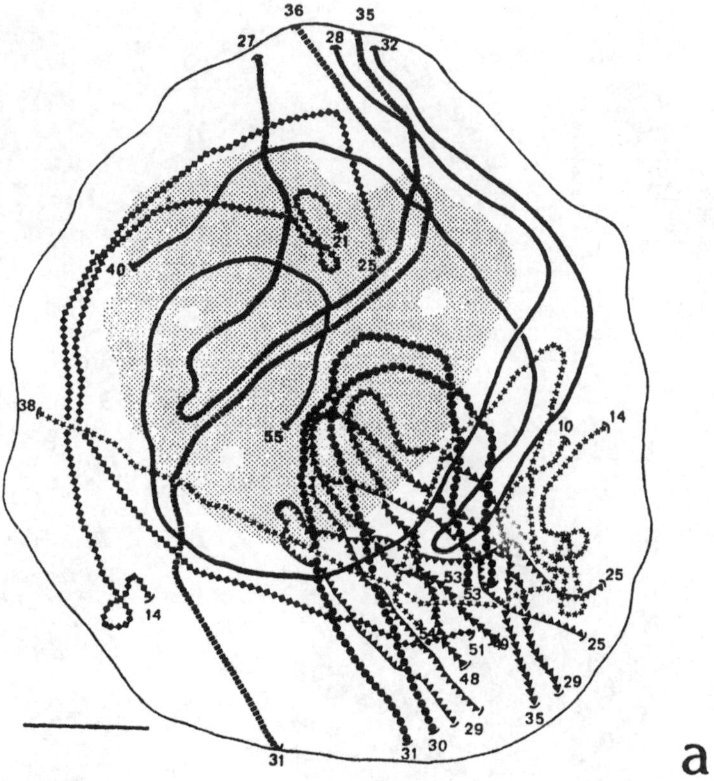

Figure 12 Early meiosis of *S. macrosporus*. (a) Reconstruction of late leptotene nucleus when the small chromosomes are all roughly paired whereas the midsize and long chromosomes are only partially paired over their length (e.g., the two chromosome 4s are paired at one end at sections 35 and 36 but are diametrically opposed at the other end at sections 27 and 31). (b) Reconstruction of early zygotene nucleus when synapsis starts mostly from one or both telomeres, but some starts interstitially. Bar = 1 μm. Reproduced from Zickler [221] with permission of Springer-Verlag.

cases, the LCs appear alternating thick and thin bands in precision regularity as exemplified by those of *Neottiella*, but in most cases they appear as solid electron-dense structure. The formation of the SC is achieved by two steps: 1.) the pairings of the homologous axial cores that give rise to the LCs, and 2.) the synapsis and deposition of the CEs, which are first assembled in the nucleolus and then transported to the chromosome sites, to bring about the tripartite structure of the synaptonemal complex. These SCs can be studied by a three-

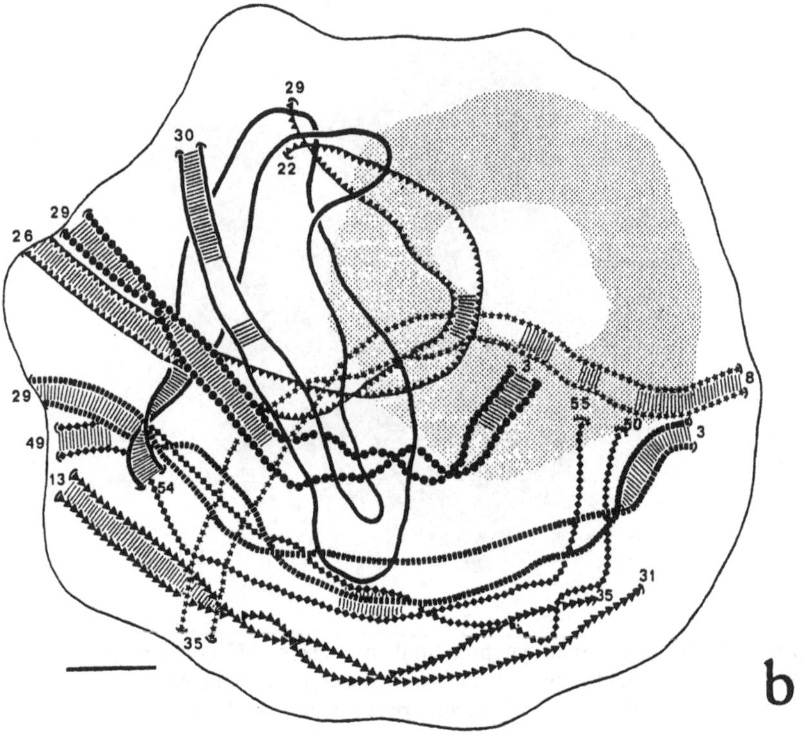

dimensional reconstruction of serial sections of a complete meiotic nucleus and by surface spread electron microscopy. Another structure associated with the SCs is the recombination nodule (RN). This is an oval structure that sits on top of the central element in the central region of the SC (Figs. 7, 8). As the name implies, this organelle may be involved in recombination. More detailed discussion of the SCs and the RNs will follow later in this chapter.

Recombination nodules can be counted and their distribution mapped at zygotene and pachytene stages [25,75,120a,221]. There appear to be two kinds of nodules in N. crassa, early and late. Early nodules appear at zygotene and increase in number until a dramatic reduction occurs at zygotene–pachytene transition. Thereafter, they are steadily eliminated until they disappear by diplotene. Late nodules are also present during zygotene. Their number doubles at the zygotene–pachytene transition and stays at this level until

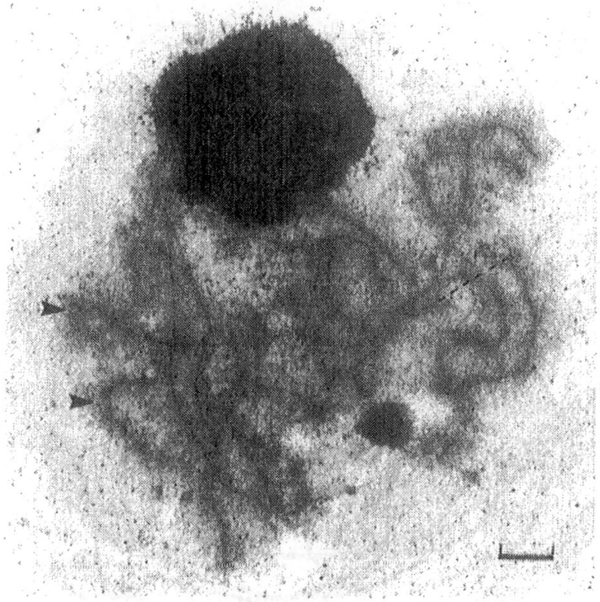

Figure 13 Surface spread synaptonemal complexes of *N. crassa* at early zygotene. Parallel pairing is evident; the broken chromosome 1 is connected by dotted line. Bar = 1 μm. Reproduced from Lu [120a] with permission from Springer-Verlag.

diplotene. The early nodules appear to be randomly distributed, but the distribution of the late nodules is nonrandom, and shows interference [25]. Nonrandom distribution of RNs is also true in *S. macrospora* [221,223].

Diplonema–Diakinesis

Diplotene chromosomes are quite diffused and often difficult to resolve. It is quite possible that this is an active stage for transcription as the meiocytes prepare for the subsequent two division events, analogous to the lampbrush stage of amphibian oocytes. The diplotene chromosomes can be resolved more clearly by acriflavine fluorescence microscopy than by any conventional stains [170]. At early diplotene, the lateral components of the SCs start to pull apart from each other before they become disintegrated; the only areas that remain paired are always associated with an RN [120a]. These are considered to be chiasmata [221]. From diplotene to diakinesis, chromosomes contract dramatically; chromosome count is possible, when there is ample space in the ascus for chromosomes to spread. Chiasmata can be counted in favorable spreads.

Division I to III and Spore Delimitation

At metaphase I, the chromosomes are very condensed and are congressed at the equatorial region; they are so condensed and compacted that individual chromosomes cannot be resolved. The nucleolus, somewhat reduced in size, is still attached to a chromosome, the nucleolar chromosome. The spindle fibers are clearly visible with the hematoxylin stain, and they connect the homologous kinetochores of each bivalent to a pair of SPBs; here the sister kinetochores behave as one. The metaphase I spindle is oriented along the long axis of the ascus. At late anaphase I, the chromosomes have reached the pole and are still connected by the pole-to-pole central spindle fibers. The disintegrating nucleolus may still be seen in the cytoplasm. Segregation of homologous centromeres occurs at this stage (the first division segregation), and proper disjunction of homologs is dependent on chromosome pairing and chiasma formation. It appears that maintenance of bivalents by one or more chiasmata per chromosome pair at metaphase–anaphase I is essential for correct disjunction. Asynaptic mutants generally fail to achieve correct disjunction.

After a brief interphase I, when all chromosomes are decondensed, the two haploid nuclei enter division II in synchrony. The metaphase II spindles are also oriented along the long axis of the ascus. The spindle microtubules are now connecting the sister kinetochores to the opposite poles. The two second-division spindles are usually tandem; in some cases, they are overlapping and often parallel such as *N. tetrasperma*, depending on the species. The first postmieotic mitosis is of particular interest. All phases of a typical mitosis—prophase, prometaphase, metaphase, anaphase, and telophase—can be observed with clarity. At prophase III, the rectantular SPB plaques have divided (the division of SPB probably occurs at late interphase II), but the daughter SPB plaques are still attached together, in a V configuration, to one side of the nucleus. At prometaphase III, the daughter SPB plaques migrate to the opposite poles flanking the highly condensed chromosomes; chromosome count can be achieved at this stage with certainty. At metaphase (division III), all four spindles normally orient across the ascus, like ladder rungs [172]. Typical metaphase plates are observed in favorable preparations [114]. Because of the orientation of the spindles in this division, and because of the space limitation by the ascus, the central spindles at late anaphase are often curved when they elongate. At the end of division III, the daughter nuclei are on the opposite side of the ascus. However, all eight interphase nuclei line up in a single file with their enlarged SPB plaques facing to one side of the ascus wall [42,114,167]. This is followed by the delimitation of ascospores, the mechanics of which has been reviewed by Beckett [14]. It appears that the SPB plaques/microtubule complexes may be involved in the process.

Spindle Pole Body

The spindle pole bodies (SPBs) and the spindle fibers are clearly stained with great clarity with the propiono–iron hematoxylin stain [114,172]. During the course of the four nuclear divisions, the SPBs change their size dramatically. They grow in size from 1 to 1.5 μm in divisions I and II, and reach maximum in size of 2–2.5 μm in metaphase III; the SPB is reduced to the original size in the spores. The size change during division III may be attributed to the formation of a plaquelike structure associated with the SPBs to form SPB plaques [167,172]. During the first meiotic division, they become visible under light microscopy at metaphase I, and they appear to be crescent-shaped and about 1–1.5 μm long, and to have spindle fibers attached to their side. At the interphase before the first postmeiotic mitosis (division III), the SPBs are closely associated with their respective nuclei and appear to be single and rectangular in shape. At prophase III, each SPB plaque has divided into two daughter SPB plaques, and they may be positioned at an angle or in a straight line to each other with their ends closely associated. At prometaphase III, the daughter SPBs move to the opposite pole, and they give the clearest image. They may appear as rod-shaped in side view, or rectangular in face view. It is possible that the spindle fibers are attached to a fixed side.

Diverse Programs of Ascus Development in Different Pseudohomothallic Species

In filamentous Ascomycetes, species may be homothallic, heterothallic, or pseudohomothallic. Homothallic species are self-compatible and fruiting bodies are generated under nutritional stress without requirement of matings between different strains. Examples of homothallic species are *Aspergillus nidulans, Gelasinospora calospora,* and *Sordaria fimicola*. Heterothallic species are self-incompatible and fruiting bodies are generated only upon matings between compatible strains. In these Ascomycetes, a simple monofactorial system is the rule; the mating type may be *A* or *a*, as in *Neurospora crassa* and its relatives. Pseudohomothallic species represent a special class of fungi in which the basic heterothallism is the rule, but due to diverse cytological programs, two compatible nuclei are enclosed in the same ascospore, thus generating four-spored asci. Examples are found in *N. tetrasperma, G. tetrasperma, P. tetraspora,* and *P. anserina.*

Detailed cytological descriptions have been documented [55,177], and explanatory diagrams are reproduced here (Fig. 14). As pointed out by Raju and Perkins [177], reprogramming of ascus development during evolution of the pseudohomothallic species has typically involved reducing the spore

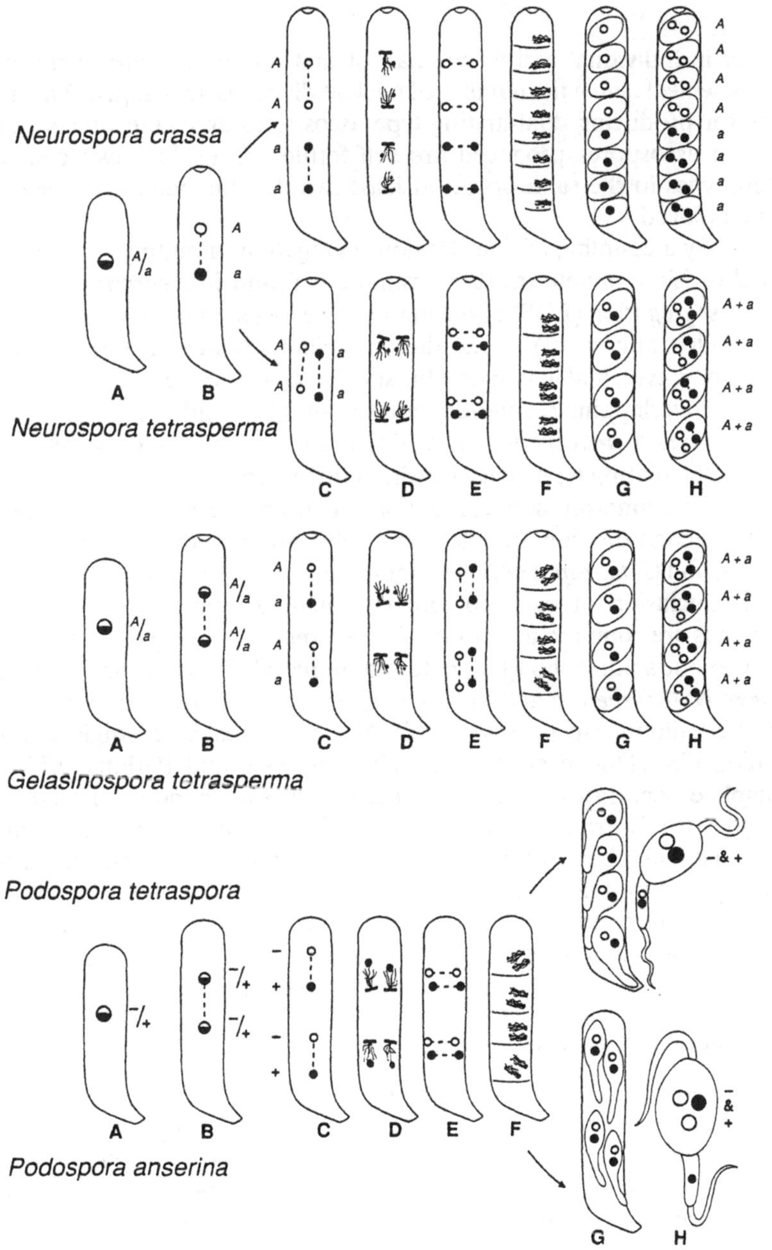

Figure 14 Schematic diagram of ascus development in *N. crassa* and in four pseudohomothallic Pyrenomycetes. Reproduced from Raju and Perkins [177] with permission of John Wiley & Sons, Inc.

number in individual asci to one-half of that found in related heterothallic species, while leaving the number of nuclear divisions unchanged. The mechanisms for producing dual-mating-type ascospores are quite precise; almost 100% of ascospores produced are self-fertile. To enclose two compatible mating types in the same spore, at least two developmental strategies have been employed.

1. By a coupling of first-division segregation of mating type locus with second-division spindle overlap: example is found in *Neurospora*. The first division segregation (FDS) of the mating type locus coupled with nonoverlapping (tandem alignment) second-division spindles is found in the heterothallic species, *N. crassa*, and in homothallic species, *G. calospora*, whereas FDS coupled with overlapping second-division spindles is found in the pseudohomothallic species *N. tetrasperma*. The FDS is ensured by lack of recombination between the mating type locus and the centromere.

2. By second-division segregation of mating type, coupled with nonoverlapping second-division spindles: examples are found in *P. tetraspora*, *P. anserina*, and *G. tetrasperma*. The alignment of third-division spindles in pairs and the inclusion of nuclei of opposite mating type in each of the four ascospores are common to both the *Neurospora* and *Podospora* strategies.

Genetic studies show that heterothallic condition is dominant in crosses between *N. tetrasperma* and the eight-spored *N. sitophila* and that the eight-spored condition can be restored in *N. tetrasperma* by a single dominant mutation [55] (Dodge et al., 1950; cited by Raju and Perkins [177]) with incomplete penetrance because the second-division spindles in a cross heterozygous for E are somewhat variable, overlapping in some and tandem arrangement in others [171]. The converse is not true; a single-gene mutation in the eight-spored heterothallic species could not in one step have given rise to a four-spored condition. Other changes are needed. Raju and Perkins [177] suggested that four-sporedness is a derived condition from its ancestral eight-spored condition and its control is complex and multigenic.

C. Meiosis in Basidiomycetes

For meiosis in basidiomycetes, *Coprinus* is chosen because it has a naturally evolved synchronous meiosis [115,119,126]. Its fruiting body contains approximately 100 gills with an estimated 10 million basidia, of which 85% are in synchrony in meiotic processes [173]. Two stages can be used to monitor the development: one is karyogamy, during which the basidium changes from binucleate to uninucleate, and the other is metaphase I, when the condensed and congressed metaphase chromosomes are highly visible. Since all gills are identical, a few gills may be removed hourly with care (without killing the fruiting body) for cytological studies of meiotic progression. Thus timing of meiotic stages can be obtained: 4 h for karyogamy, 5 h for pachytene, 4.5 h for diplotene, 2 h for metaphase I through telophase II, 2 h for sterigma formation [173], and an additional 3 h for spores to mature. The timing may be different for different geographical strains. The system also allows cytological and genetic studies on the same fruiting body, such as stage-specific effect of temperatures, radiation, or chemicals on genetic recombination [117,122,125, 174,179]. *C. cinereus* is unique in another respect. The initiation of meiotic DNA replication is subject to arrest by combined high temperature and continuous light regime [118,124]; the control of meiosis is possible when the production of a large number of synchronized fruiting bodies is desirable. The meiotic S phase occurs before karyogamy [124,151]. This pathway is not mandatory, however, for a mutant that abolishes meiotic S can proceed through meiotic prophase I but cannot enter metaphase I [98]. A complete photographic documentation of meiosis in *Coprinus* is shown in Figure 15.

As in the Pyrenomycetes, the chromosomes of *Coprinus* are already organized before karyogamy and the nucleoli increase in volume (Fig. 15a). After karyogamy, two nucleoli fuse immediately. *Coprinus* chromosomes appear to be fully extended (Fig. 15b) prior to homologous pairing at zygotene (Fig. 15c). The basidia grow to double their volume. This is unlike many filamentous ascomycetes where ascus grows 10- to 20-fold from karyogamy through the end of meiosis. Nevertheless, the complete alignment of homologues is the same (Fig. 15c). The axial cores are formed very shortly after karyogamy; then they are aligned between the homologous pairs [115]. This is followed by synapsis and the completion of the tripartite synaptonemal complex. The complete set of SCs by surface spread is shown in Figure 6. The pachytene may last from 5 to 8 h, while the diplotene lasts from 4 to 1 h depending on the strains and the light regime used. During diplotene, the chromosomes are diffused, and chiasmata may be observed although a reliable chromosome count is difficult [126]. A pair of SPBs can be observed at pachytene (Fig. 7) and diplotene (Fig. 15g,h) with light and electron microscopy [164, 174]. Metaphase I through telophase II occur rapidly within 2 hours for the population of 10 million basidia. At metaphase I, chromosomes are congressed at the equatorial region. The chromosome configuration is quite variable, from a typical metaphase plate to two-track configuration; the latter is common. The spindle lies perpendicular to the long axis, and near the top of the basidium. Observations with electron microscopy shows that the nuclear volumn is reduced to approximately 1.5 μm in diameter and the condensed chromosomes are closely associated with the nuclear envelope [115]. At this stage, the SPBs are located on each pole outside the nuclear envelope. When the spindle microtubules are organized, the SPBs pull away to the opposite end of the basidium and the nuclear envelope is dissociated, leaving a massive amount

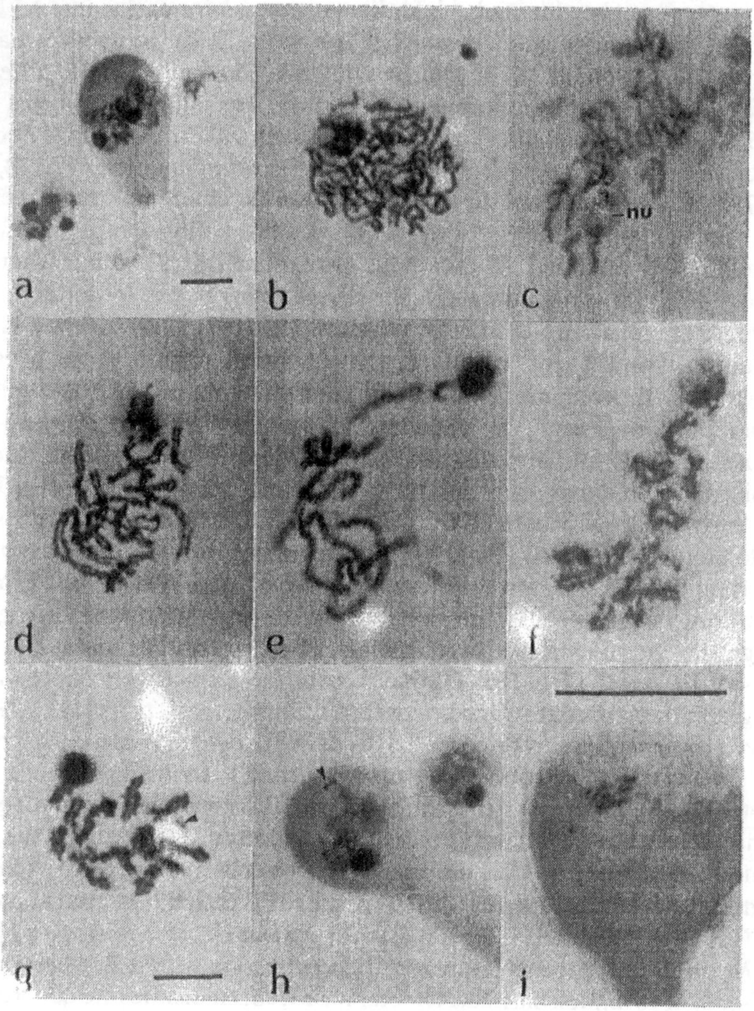

Figure 15 Meiosis of *C. micaceus*. (a) Young basidium before karyogamy. (b) Leptotene after karyogamy. Chromosomes are well organized but not paired; the chromosomes are hammered out of the basidium. Note, the chromosome and the nucleolar volumes appear much increased when released from the restriction of the basidium as compared with the image in a. (c) Early zygotene, where loose pairing occurs. Note the parallel alignment of the homologues. (d) Near pachytene. (c) Pachytene. (f) Diplotene. Chromosomes are lampbrushed. (g) Late diplotene; a pair of duplicated spindle pole bodies (SPB, arrowed) are visible. (h) Diplotene in an intact basidium showing a pair of SPB (arrowed). (i) Metaphase I where chromosomes are congressed at the equatorial region between two SPBs. Bar = 5 μm. Reproduced from Lu and Raju [126] with permission of Springer-Verlag.

of membranous vesicles, which presumably be used for the reorganization of new nuclear envelopes. Because of the smallness of space, the change from metaphase I to anaphase I is difficult to ascertain. When the nucleus is squashed for light microscopy, the contracted chromosomes tend to display a wide variation of forms including a two-track and some typical metaphase configurations. The chromosome movement at anaphase I and II is staggered; this leads to stringlike configurations. There is a brief interphase, during which new nucleoli are organized, before the second meiotic division. The second meiotic division appears to be intranuclear, and even the SPBs are internalized. At the end of second division, the SPBs are evaginated and are positioned just outside the nuclear envelope [206]. Two hours after completion of meiosis, four sterigmata are formed from the apex of the basidium from which basidiospores are formed. Each nucleus migrates into a developing spore, with the SPB leading the move [206].

D. The Synaptonemal Complex (SC)

The discovery of the SC in meiocytes of crayfish as a unique structure of pachytene bivalents by Moses [141,142] has added a new avenue to the analysis of homologous chromosome pairing. This is particularly significant in light of the findings that the SC is found in female meiosis of *Drosophila melanogaster*, where recombination occurs, and not in males, where meiosis is achiesmatic [136]. For fungi, the SCs were first found in *Neottiella rutilans* [213] and *C. cinereus* [113,114], and there have been many subsequent observations in various species, including *Armillaria, Neurospora, Ascobolus, Podospora, Sordaria*, etc. [74,157,221] (see Westergaard and von Wettstein [215] and von Wettstein et al. [216] for reviews). Exceptions are found in *Aspergillus* and the fission yeast, where complete SCs are not found [4,62]. The formation of the SC is meiosis-specific, and its presence is correlated with the time of genetic recombination. There is a fourfold increase in meiotic recombination by cold temperature treatment on *Coprinus* fruiting bodies; this increase is possible only when the SC is present at the time of treatment [117,118,122].

The temporal analysis of the formation of SCs was first described in the synchronous meiotic system of *C. cinereus* [115]. Shortly after karyogamy, single cores are formed on chromosomes. These are brought to align loosely when the homologous chromosomes pair (this is defined as *pairing*). Finally, axial cores become the lateral components of the tripartite SC when central elements are added to complete synapsis (now defined as *synapsis*). The temporal analysis was extended later to show that protein synthesis is necessary for complete assembly of the SC using an inhibitor of protein synthesis, such as cycloheximide [120]. This experimental approach accentuates different steps involved in the formation and dissolution of the SC. The axial cores are formed within an hour of karyogamy, and their subsequent pairing and assembly into SCs have been confirmed. The most dramatic finding in this experiment is the accumulation of central elements in the nucleolus-dense body and their failure to transport to the chromosomal sites where lateral components are paired. This occurs only within a narrow time period, i.e., when cycloheximide was applied to the fruiting cap 4 h after the beginning of karyogamy. That this phenomenon represents an intermediate step in the morphogenesis of the SC is supported by the finding of similar accumulation of central elements in the nucleolus in the yeast *cdc*4 and *rad*50S mutant strains [3,93]. The temporal analysis of the SC formation was also described for the wild-type yeast *S. cerevisiae* taking advantage of synchronized meioses in SK1 strain [3,156]. However, because the time spent in meiosis is short in the yeast, the formation and pairing of axial cores and their assembly into tripartite SCs occur in concert. Separation of pairing and synapsis is possible when time permits. Examples of pairing before synapsis are found in *Coprinus* (Fig. 15c), in *Sordaria* (Fig. 12), and in *Neurospora* (Fig. 13).

The three-dimensional reconstruction of axial cores and synaptonemal complexes by electron microscopy has increased the resolution of meiotic chromosomes. Examples include *N. crassa* [25,73,75,120a], *S. macrospora* [221], *Schizophyllum commune* [35], *Puccinia gramini* [24], *C. cinereus* [92, 164,178], and in yeast *S. cerevisiae* [30], while the study by Zickler [221] of ascomycete *S. macrospora* is one of the best in showing the temporal sequence of events of homologous pairing. The three-dimensional reconstruction not only resolves the formation of the SC, but also reveals the recombination nodules, their numbers, and their distributions.

More recently, the surface spread whole-mount electron microscopy has been developed for fungal synaptonemal complexes that adds a further dimension to the analysis of chromosomes in meiosis. This has been achieved in *Saccharomyces* [56], in *Coprinus* [163,164], and in *Neurospora* [120a]. Examples are shown in Figures 6–8. The silver nitrate stains lateral components, central elements, and recombination nodules in *Neurospora*, but only lateral components in *Coprinus* and *Saccharomyces*; there is no explanation for the differences.

With surface spread electron microscopy, accurate and detailed temporal analysis of the SC formation in relation to other meiotic events can be achieved. Among the most informative studies is the work of Padmore et al. [156] using the synchronized yeast SK1 strain. The first event of interest is the meiotic DNA replication. The analysis of propidium iodide–stained samples in a fluorescence-activated cell sorter show that the bulk of meiotic DNA replication is completed before other meiotic events are initiated. Approximately 1 h after the end of DNA replication (4.3 h after the initiation of meiosis) appear nuclei with SC precursors having short axial cores and partial tripartite structures. These two substages may be equated to leptotene and zygotene, and they occupy about 30 min each. The next substage is the full pachytene when SCs are fully assembled, and it occupies about 1–1.3 h. The end of pachytene is followed within 15 min by disappearance of the SCs and the appearance of metaphase I spindles at 6.8 h. Other meiotic events of interest are transient site-specific double-strand breaks (DSBs) and the formation of mature recombinant molecules. The DSBs are found in meiotic recombination hot spots, such as *HIS*4-*LEU*2, *ARG*4, *THR*4, and *DED*81-*DED*82 regions [32, 78,156,203]. The meiosis-induced DSBs may be "pairing sites" for homologous chromosomes in meiosis [78]. DSBs can occur prior to or concomitant with the first appearance of visible tripartite SC structure, and they disappear during the stage when SCs are forming and elongating and are gone by the beginning of pachytene. Mature reciprocal recombinant molecules, defined by restriction site polymorphisms, appear at or just after the end of pachytene. The above observations provide a clear picture of the sequence of events that occur in meiotic prophase I in the yeast *S. cerevisiae*. It is quite possible that similar sequence of events occurs in higher fungi.

E. Recombination Nodules

Recombination nodules (RNs) are dense, oval (or near spherical) structures associated with the synaptonemal complex during late zygotene to early diplotene. They are located above and adjacent to the central element of the SC. The term recombination nodule was coined by Carpenter [36–38] after her extensive analysis of the SCs of *Drosophila melanogaster* females, both the wild-type and the recombination defective mutants (e.g., *mei*9, *mei*218, and *mei*41). She discovered that the number and distribution of RNs closely match those of genetic exchanges. She also found by EM autoradiography that the locations of the RNs are the sites of repair DNA synthesis [39]. She suggested that the RN may be an enzyme complex for genetic recombination [40]. She described two types of nodules—ellipsoidal and spherical. The ellipsoidal appear first in high numbers, followed by the spherical in reduced numbers. Based on her analyses of *mei*9 and *mei*218, in which gene conversion events and exchange events can be uncoupled, Carpenter suggested that the ellipsoidal nodules may be correlated with the gene conversion events, and the spherical nodules may be correlated with the exchange events. Since then, the RNs have been analyzed in a variety of higher eukaryotes. Where data of genetic exchanges (or chiasmata) are available, the results appear to be the same; the RN frequency and their chromosomal locations always match those of chiasma frequency and their chromosomal locations. Indeed, both the recombination nodules and chiasma show nonrandom chromosomal distribution and positive interference, which is an established genetic phenomenon. The RNs have been analyzed in a number of fungi [25,30,35,75,178,221]. When the chiasmata can be counted, as in the case of *Sordaria*, their numbers seem to agree with the number of RNs; there are 18–19 chiasmata at diplotene–diakinesis, and there are 18–19 RNs at pachytene and diplotene SCs. In general, the data on exchange frequency and their distributions are insufficient. Nevertheless, the RN distribution is definitely nonrandom, and positive interference is demonstrated. In *Neurospora*, Bojko [25] described two types of RNs; the early nodule (spherical) is randomly distributed while the late one (ellipsoidal) is nonrandomly distributed. The morphological distinction is not so obvious in the surface spread preparations [120a]; most RNs appear to be distally located. In *Coprinus*, the number of RNs is high at midzygotene, and they exhibit a random distribution; these numbers are reduced almost by half (from 46 to 26) at mid-late pachytene, and they exhibit a nonrandom distribution. Similar results are found in *Schizophyllum* and *Sordaria*. Again, the predominant locations are the telomere regions [35,178,221].

F. The Question of Homology Search

All observations point to the consensus that the SC and the meiotic level of recombination are intimately related. The SC is required for the observed meiotic levels of genetic recombination. It is possible that the SC functions as a "cast" or a zipper to hold homologues together long enough to allow recombination processes to take effect [119]. Mutants that are defective in the assembly of SCs are also defective in meiotic recombination. The converse is not true; mutants that are defective in meiotic recombination may have normal SCs, as recombination requires complex enzymatic actions. Examples include *mei*9 of *Drosophila melanogaster* [38], *asy*2-17 of *S. macrospora* [223], and *rad*52 of *S. cerevisiae* [77]. Questions have been raised whether 1.) SCs come before the initiation of recombination, 2.) the initiation of recombination comes before the assembly of SCs, or 3.) the two events are parellel.

From various observations, it is clear that the synthesis of protein components and their assembly into axial cores and central elements are the early steps. Without them, meiotic recombination is also defective. In *D. melanogaster*, females homozygous for the mutation c(3)G fail to form SCs, and meiotic recombination is drastically reduced; the gametes produced are mostly inviable [201]. In yeast, diploid homozygous for *hop*1 also fails to form SCs and is defective in meiotic crossovers and gene conversions, but mitotic and intrachromosomal recombinations are not affected. The *hop*1 mutation affects yeast meiosis in a manner analogous to that of the c(3)G mutation of *Drosophila* [89]. Thus the *HOP*1 gene in *S. cerevisiae* is important for the SC assembly, and it codes for a protein component of the lateral element [90].

Closely related to *hop*1 is *red*1, which fails to assemble discrete axial cores, although some short segments of tripartite structure may be present, and the recombination machinery is intact [182]. The *hop*1 is epistatic to *red*1, suggesting that they belong to the same pathway and that *RED*1 operates at a later step. In fact, a diploid yeast homozygous for a *hop*1-628 allele can be suppressed by an overexpression of the *RED*1 gene [91]. In addition to *hop1* and *red1*, another mutant *zip*1 has been found. The diploid yeast homozygous for zip1 assembles full-length axial cores that are paired but not synapsed. The anti-ZIP1 antibodies are localized along the length of the SC except for the unsynapsed area; the ZIP1 is probably a component of the central region of the SC [204]. It is possible that *HOP*1, *RED*1, and *ZIP*1 control the assembly of SCs in association with synapsis.

Between the assembly of axial cores and tripartite SCs lies the problem of homologous pairing and synapsis (see Loidl [108] for review). This is the least understood step, and there may be a number of genetic functions involved. Strains carrying mutations affecting this braod step are known as asynaptic mutants. Examples include, in addition to those mentioned above, *mei*1, *mei*2, and *mei*3 of *N. crassa* (46,123,176,189]; *spo*11, *mei*4, *rad*50 null allele, *rad*50S, *rad*51, *mer*1, *mer*2, and *dcm*1 of *S. cerevisiae* [3,18,66,67,77, 100,128,134,196]; *spo*44, *spo*76, and *spo*77 of *S. macrospora* [139,222]; and *rad*3, *rad*9, and *rad*12 of *Coprinus* [164,224,225]. More mutants are probably needed to map the complete process. Most of these mutants, when homozygous, exhibit reduced meiotic recombination; some have incomplete axial core assembly (e.g., *rad*3-1, *rad*9-1 of *Coprinus*); and all have partial tripartite structures, except *mer1*, which shows all (10% of the nuclei) or none (90% of the nuclei) of SCs. The *rad*9-1 is of particular interest where the tripartite structures are limited to telomeres [225]. It is puzzling that some asynaptic mutants (*mei*1, *mei*2, *rad*50) exhibit pleiotrophic effect on spindle mechanism.

What controls homologous pairing or what is homology search? The protein components of SCs have no property of homology. The only molecule in the chromosomes having the property of homology is the DNA. The idea of direct DNA interaction before homologous pairing has been advanced recently [41,202]. The DNA molecules, when single-stranded, are capable of finding their complementary sequences as shown in DNA renaturation studies [27]. The larger the number of repeat sequences, the faster they find each other. In higher eukaryotes, there are many highly reiterated sequences located in strategic locations of the chromosomes, such as pericentric heterochromatin, the nucleolar organizer regions, long and short interspersed mono- and dinucleotide repetitive elements (LINE/SINE), and telomeric and subtelomeric sequences. These regions may provide initial contacts for homology search. That these sequences play some role in homologous pairing is demonstrated recently in *Drosophila* by Hawley et al. [85]. These authors demonstrated that pericentric heterochromatin plays a primary role in ensuring the pairing and proper segregation of achiamate homologs. The pericentric heterochromatin is not known in fungal genomes, but up to 20% repeated sequences have been demonstrated in *Neurospora* and in *Coprinus* [28,60,61], and heterochromatin knobs have been observed in *Neurospora* [199]. The closest thing to the pericentric heterochromatin is hinted by the organization of centromere repeated sequences of the fission yeast where three chromosomes have three distinct centromere patterns [147]. The LINE/SINE may be involved in the organization of the SC and the initiation of meiotic recombination [158,190,209]. Cytologically, the first event observed after karyogamy is the fusion of the nucleoli, suggesting that the nucleolar organizer regions may be involved in the early contact. There are telomeric and subtelomeric repeated sequences in yeast [219]. Since the telomeres are inserted on the nuclear envelope, they are like casters rolling on a two-dimensional surface in search for homologous contacts; in most cases, pairing appears to start from the telomeres [92,221] and "zippering up" toward the centromere regions; although interstitial initiation, or "buttoning up," has been observed [74]. The initial contacts may not be sufficient to hold the two homologues together like those observed in presynaptic alignments where the two axial cores are held at a distance [115,120a,221], further actions need to happen. The distance in the presynaptic alignment may represent the outer zones of the chromatin where DNA loops of the homologues may interact. For this interaction, single-stranded tails or gaps or double-strand breaks may be needed as demonstrated by the in vitro D-loop studies [52,165,212]. Furthermore, if the nicking enzymes make random nicks and gaps on nonsister chromatids, many of these single-stranded regions will not interact because they are not homologous or do not have extended homology; only a small number will fall within the same locus (i.e., coincidental nicks), in which extended homologous, single-stranded sequences are exposed to form a joint heteroduplex. Insufficient length of homology will cause this interaction to fail. Thus, when mutants are unable to make these DNA substrates for the initiation of recombination, such as *spo*11, *rad*50 null mutants, or are mutants defective in RecI- or RecA-like protein [52,101], such as *dcm*1, *rad*51 of yeast [18,196] and *mei*3 of *Neurospora* [46], pairing is not possible and the assembly of the SC fails. On the other hand, when *rad*50 non-null mutants, such as *rad*50S, double-stranded breaks are made, some defective pairing is possible. The above argument brings the initiation of recombination and the homologous pairing into an intimate relationship as two separate pathways that converge to a successful synapsis and the assembly of the SCs. Failing in one or the other pathway will result in asynapsis. This converging pathway was suggested by Malone et al. [129].

If the homology search is dependent on single-stranded DNA substrates and RecA-like protein functions, which alone can achieve gene conversion events, then why is the SC necessary? The homologous pairing of DNA substrates mediated by the RecA-like proteins represents prokaryotic and mitotic recombination pathway. Such reaction is chance-dependent, transient, and unstable. Only when these DNA substrates are held or zippered together long enough will the level of meiotic recombination be achieved. Thus, the SCs are zippers to stabilize homologous interactions, as recombination is a time-dependent event.

The completion of synapsis requires the transport of central elements and recombination nodules to the paired lateral elements. It is possible that only when homologous pairing is achieved will the transport of central lements and recombination nodules be permitted.

APPENDIX: SPREADING THE SYNAPTONEMAL COMPLEX OF *COPRINUS*

Preparation of Fruiting Cultures

Coprinus cinereus, as well as several related species, has a naturally evolved synchronous meiosis [114,115]. The meiotic chromosomes are highly observable, and all stages are well defined [126,173], the synaptonemal complex can be easily spread [164], and the development of fruiting body is controllable by laboratory routines [114,115,122,124]. Fruiting primordial development is light-dependent, and it takes 3 days to reach meiotic phase.

To allow a reasonable prediction of the time of initiation of karyogamy, the following routine can be used. Dikaryotic mycelium is inoculated onto an agar medium in crystallizing dishes (50 × 90 cm) containing glucose, malt extract, and yeast extract, and incubated in a 35°C incubator in total darkness for 5 days, during which the mycelium will completely cover the surface of the medium. On day 6, the cultures are transferred to a low-temperature incubator fitted with a 15 watt, cool-white fluorescent light with the temperature set at 25°C, and the light regime set for a 16-h light and 8-h dark cycle. The time when light cycle begins (and light intensity) has a direct influence on the time of meiosis, which should be determined beforehand under a set of laboratory conditions. In my laboratory, the light cycle begins at 1100 h, and karyogamy starts usually around 0300 h on day 10 or shortly after that; synaptonemal complexes can be spread between 1100 and 1600 h.

Preparation of Slides

Slides should be of high quality, free of flaws and chips which can be detected under a low-power microscope; some brands are better (e.g., VWR) than others. Wash (never soak) in hot water with a dish-washing detergent (e.g., Dove), rinse well in distilled or deionized water, soak in 95% ethanol, wipe dry with tissue paper (e.g., Kimwipe), and polish with nonlinting lens tissue (e.g., Ross Optical Lens Tissue). Store in a slide box until use.

For light and electron microscopy, clean and polished slides are coated with a plastic solution (0.9% polystyrene from a Falcon optilux brand Petric dish w/v in chloroform). Dip a clean slide in the plastic solution (in a fume hood, preferably with the airflow turned down for even coating) for 1 sec and quickly withdraw vertically with one side running against the container (this will remove the streak on the edge). Drain briefly on a paper towel vertically, then lay it down flat when it is nearly dry.

Fixatives and Solutions

Make a stock solution of 0.05 M sodium borate (1.91 g in 100 mL double-distilled water) and store in a refrigerator. For use, dilute 2 parts of stock to 3 parts of double-distilled water to make 0.02 M, then adjust pH to 9.22 with 0.5 N NaOH (about 4–6 drops); filter through a Millipore filter (0.2 μm).

Fixative I. Make a 4% paraformaldehyde fixative (4 g in 100 mL dd water) in a 200-mL beaker, heat and stir to 40°C, and add 10 N NaOH, drop by drop (about 18 drops) until solution is clear. Let cool, adjust pH to 5–7.0 with full-strength HCl (drop by drop very carefully), then bring up the pH to 8.2 with 0.02 M borate buffer (pH 9.22). Filter through Millipore (0.2 μm) and store in a refrigerator. Check pH each day before use as pH drops by oxidation; the workable pH is 7.8–8.2. This fixative is good for 14 days.

Fixative II. For spreading, take 50 mL of the 4% paraformaldehyde fixative made above, and add 150 μL of a 10% SDS stock solution to make a 0.03% SDS. This solution is kept at room temperature, and the pH appears to be stable.

Solutions

Photoflo rinse solution. 0.4% Kodak photoflo 200 adjusted to pH 8.2 with borate buffer (0.02 M, pH 9.22). Use separate solutions for spreading and staining; change frequently, as it becomes contaminated, contributing to dirty slides.

Staining solutions. Solution A: 40% silver nitrate in double-distilled water, filtered using 0.2-μm filter. Keep in a brown dropping bottle. Solution

B, a gel developer: dissolve 2 g gelatin in 100 mL double-distilled ater at 40°C. When cool, add 1 mL of formic acid and filter through a Millipore filter (0.2 μm). Keep in brown dropping bottle.

Spreading the Synaptonemal Complex for Light and Electron Microscopy

1. Cut out a small piece of gill tissue from a mushroom at pachytene with a razor blade and put it on a slide. Dissect four or five gills without the outer skin in a drop of cold fixative I, and place them in a 1.5-mL microtube (Applied Scientific). Add a drop (about 50 μL) of cold fixative I for a total of 4 min. The gills are disrupted to release nuclei using a plastic pellet pestle (Kontes 749515-000) in 10 up-down actions using a cordless drill (6.0–7.2 v) at full speed. Take great care that the 50-μL suspension remains in the tube (just tapping it down will help). The tube is placed on ice while preparing for spreading dish.
2. Prepare a spreading dish—glass dish (2 cm in diameter) coated with polystyrene plastic to make it hydrophobic. Fill the spreading dish with 0.5% NaCl to a slight convex full (about 2 mL); sweep the surface clean with tissue paper twice; be sure the surface remains convexed. If the surface is too convex the spread could be lost during the pick-up.

   Use an Eppendorf pipette to pick up 5 μL disrupted basidial suspension. Push it out to form a hanging drop, and touch it on the surface of the spreading solution as gently as possible.

   Immediately pick up the spread by touching a plastic coated slide (face down horizontally) against the convex spreading solution. The circular area should become wet indicating successful pick up.
3. Cover the circular area with a few (12) drops of fixative II and allow to spread for 20–40 min. Mark the side of the slide with a felt-tip marking pen with a water-resistant ink to show the area to be stained later. Gently pour off the fixative and dip the slide in distilled water and then into a 0.4% Kodak photoflo solution. Drain quickly, and allow to air-dry.

Silver Staining

4. Prepare the staining solution on a coverslip (22 × 22 mm) by adding 2 drops of 40% silver nitrate and 1 drop of a gel developer; mix well. Pick up the stain by the slide to be stained, and invert slide carefully in a quick action so that the stain does not run off. Incubate it on top of a hot plate at 55°C for 2–2.5 min.
5. The stain and the coverslip are washed off by distilled water using a squeeze bottle. After the coverslip falls off, continue washing for a few more seconds (take care not to damage the plastic support membrane). Dip it in

0.4% Kodak photoflo solution (pH 8.2), and let air-dry. The slide is now ready for light microscope examination.

Preparation for Electron Microscopy

6. When good spreads of SC are found, the locations of SC are marked with a felt-tipped marking pen using a water-resistant ink so as to guide grid laying. Take great care not to break the supporting membrane.

7. Carefully cut off the plastic supporting membrane from the edges of the slide with a razor blade, and float it off the slide very carefully on clean dd water. Avoid water leaking through to the top surface. Lay a slot grid on the desired spot (to protect the grid from running off during pickup, coat the grid with a thin film of a superglue on three sides, except the area held by the forceps, immediately before laying. Never touch the forceps with the glue.)

8. Use parafilm to pick up the membrane and the grids on it and let it air-dry in a dust-free, partially covered container. When dry, cut out carefully around the grid with a thin needle so it can be picked up without tearing the membrane.

REFERENCES

1. Aist, J. R. The mitotic apparatus in fungi, *Ceratocystis fagacearum* and *Fusarium oxysporum*. *J. Cell Biol.* *40*:120–135 (1969).
2. Aist, J. R., and P. H. Williams. Ultrastructure and time course of mitosis in the fungus *Fusarium oxysporum*. *J. Cell Biol.* *55*:368 (1972).
3a. Backlund, J. E., L. J. Szabo. Genome size and repetitive DNA content of *Puccinia graminis* f.sp. *tritici*. *Phytopathology 81* (Abstr.) (1991).
3. Alani, E., R. Padmore, N. Kleckner. Analysis of wild-type and rad50 mutants of yeast suggests an intimate relationship between meiotic chromosome synapsis and recombination. *Cell 61*:419–436 (1990).
4. Bahler, J., T. Wyler, J. Loidl, J. Kohli. Unusual nuclear structures in meiotic prophase of fission yeast: a cytological analysis. *J. Cell Biol.* *121*:241–256 (1993).
5. Baker, B. S., A. T. C. Carpenter, M. S. Esposito, R. E. Esposito, L. Sandler. The genetic control of meiosis. *Annu. Rev. Genet.* *10*:53–134 (1976).
6. Baker, R. E., M. Fitzgerald-Hayes, T. C. O'Brien. Purification of the yeast centromere binding protein CP1 and a mutational analysis of its binding site. J. *Biol. Chem.* *264*:10843–10850 (1989).
7. Barry, E. G. Cytological techniques for meiotic chromosomes in *Neurospora. Neurospora Newsl.* *10*:12 (1966).
8. Barry, E. G. Chromosome aberrations in Neurospora and the correlation of chromo- somes and linkage groups. *Genetics 55*:21 (1967).
9. Barry, E. G. The diffuse diplotene stage of meiotic prophase in N*eurospora. Chromosoma 26*:119–129 (1969).
10. Barry, E. G., J. F. Leslie. An interstitial pericentric inversion in *Neurospora. Can. J. Genet. Cytol.* *24*:693–703 (1982).

11. Barry, E. G., D. D. Perkins. Position of linkage group V markers in chromosome 2 of *Neurospora crassa*. *J. Hered.* *60*:120 (1969).
12. Bayman, P., O. R. Collins. Meiosis and premeiotic DNA synthesis in a homothallic *Coprinus*. *Mycologia* *82*:170–174 (1990).
13. Beadle, G. W., E. L. Tatum. Genetic control of biochemical reactions in *Neurospora*. *Proc. Natl. Acad. Sci. USA* *27*:499–506 (1941).
14. Beckett, A. Ascospore formation. In: *The Fungal Spores: Morphogenetic Controls* (ed. G. Turian, H. R. Hohl), pp. 107–129 (1981). Academic Press, Toronto.
15. Bell, G. I., L. J. DeGennaro, D. H. Gelfand, R. J. Bishop, P. Velenzuela, W. J. Rutter. Ribosomal RNA genes of *Saccharomyces cerevisiae*. I. physical map of the repeating unit and location of the regions coding for 5S, 5.8S, 18S, and 25S ribosomal RNAs. *J. Biol. Chem.* *252*:8118–8125 (1977).
16. Bell, W. R., C. D. Therrien. A cytophotometric investigation of the relationship of DNA and RNA synthesis to ascus development in *Sordaria fimicola*. *Can. J. Genet. Cytol.* *19*:359–370 (1977).
17. Bicknell, J. N., H. C. Douglas. Nucleic acid homologies among species of *Saccharomyces*. *J. Bacteriol.* *101*:505–512 (1970).
18. Bishop, D. K., D. Park, L. Xu, N. Kleckner. DMC1: a meiosis-specific yeast homolog of *E. coli* recA required for recombination, synaptonemal complex formation, and cell cycle progression. *Cell* *69*:439–456 (1992).
19. Blackburn, E. H., M. L. Budarf, P. B. Challoner, et al. DNA termini in ciliate macronuclei. *Cold Spring Harbor Symp. Quant. Biol.* *47*:1195–1207 (1983).
20. Blackburn, E. H., J. W. Szostak. The molecular structure of centromeres and telomeres. *Annu. Rev. Biochem.* *53*:163–194 (1984).
21. Bloom, K. S., J. Carbon. Yeast centromere DNA is in a unique and highly ordered structure in chromosomes and small circular minichromosomes. *Cell* *29*:305–317 (1982).
22. Bloom, K. S., M. Fitzgerald-Hayes, J. Carbon. Structural analysis and sequence organization of yeast centromeres. *Cold Spring Harb. Symp. Quant. Biol.* *47*:1175–1185 (1983).
23. Bloom, K. S., E. Amaya, J. Carbon, L. Clarke, A. Hill, E. Yeh. Chromatin conformation of yeast centromeres. *J. Cell Biol.* *99*:1559–1568 (1984).
24. Boehm, E. W. A., J. C. Wenstrom, D. J. McLaughlin. An ultrastructural pachytene karyotype for *Puccinia graminis* f. sp. *tritici*. *Can. J. Bot.* *70*:401–413 (1992).
25. Bojko, M. Two kinds of "recombination nodules" in *Neurospora crassa*. *Genome* *32*:309–317 (1989).
26. Bojko, M. Synaptic adjustment of inversion loops in *Neurospora crassa*. *Genetics* *124*:593–598 (1990).
27. Britten, R. J., D. E. Kohne. Repeated sequences in DNA. Science 161:529–540 (1968).
28. Brooks, R. R., P. C. Huang. Redundant DNA of N*eurospora crassa*. *Biochem. Genet.* *6*:41–49 (1972).
29. Brown, T. A., R. W. Davies, J. A. Ray, R. B. Waring, C. Scazzochio. The mitochondrial genome of *Aspergillus nidulans* contains reading frames homologous to human URFs 1 and 4. *EMBO J.* *2*:427–435 (1983).

30. Byers, B., L. Goetsch. Electron microscopic observations of the meiotic karyotype of diploid and tetraploid *Saccharomyces cerevisiae*. *Proc. Natl. Acad. Sci. USA 72*: 5056–5060 (1975).
31. Cai, M., R. W. Davis. Purification of a yeast centromere binding protein that is able to distinguish single base-pair mutations in its recognition site. *Mol. Cell. Biol. 9*: 2544–2500 (1989).
32. Cao, L., E. Alani, N. Kleckner. A pathway for generation and processing of double-strand breaks during meiotic recombination in *S. cerevisiae*. *Cell 61*:1089–1101 (1990).
33. Carbon, J. Yeast centromeres: structure and function. *Cell 37*:351–353 (1984).
34. Carle, G. F., M. V. Olson. Separation of chromosomal DNA molecules from yeast by orthogonal-field-alternation gel electrophoresis. *Nucl. Acids Res. 12*:5647–5664 (1984).
35. Carmi, P., P. B. Holm, S. W. Rasmussen, J. Sage, D. Zickler. The pachytene karyotype of *Schizophyllum commune* analyzed by three dimensional reconstructions of synaptonemal complexes. *Carlsberg Res. Commun. 43*:117–132 (1978).
36. Carpenter, A. T. C. Electron microscopy of meiosis in *Drosophila melanogaster* females. II. The recombination nodule—a recombination-associated structure at pachytene? *Proc. Natl. Acad. Sci. USA 72*:3168–3189 (1975).
37. Carpenter, A. T. C. Synaptonemal complex and recombination nodules in wild-type *Drosophila melanogaster* females. *Genetics 92*:511–541 (1979).
38. Carpenter, A. T. C. Recombination nodules and synaptonemal complex in recombination-defective females of *Drosophila melanogaster*. *Chromosoma 75*:259–292 (1979).
39. Carpenter, A. T. C. EM autoradiographic evidence that DNA synthesis occurs at recombination nodules during meiosis in *Drosophila melanogaster* females. *Chromosoma 83*:59–80 (1981).
40. Carpenter, A. T. C. Meiotic roles of crossing-over and of gene conversion. *Cold Spring Harbor Symp. Quant. Biol. 49*:23–29 (1984).
41. Carpenter, A. T. C. Gene conversion, recombination nodules, and the initiation of meiotic synapsis. *BioEssays 6*:232–236 (1987).
42. Carr, A. J. H., L. S. Olive. Genetics of *Sordaria fimicola*. II. Cytology. *Am. J. Bot. 45*:142–150 (1958).
43. Cassidy, J. R., D. Moore, B. C. Lu, P. J. Pukkila. Unusual organization and lack of recombination in the ribosomal RNA genes of *Coprinus cinereus*. *Curr. Genet. 8*: 607–613 (1984).
44. Cassidy, J. R., P. J. Pukkila. Inversion of 5S ribosomal RNA genes within the genus *Coprinus*. *Curr. Genet. 12*:33–36 (1987).
45. Centola, M., J. Carbon. Cloning and characterization of centromeric DNA from *Neurospora crassa*. *Mol. Cell. Biol. 14*:1510–1519 (1994).
46. Cheng, R., T. I. Baker, C. E. Cords, R. J. Radloff. mei3, A recombination and repair gene of *Neurospora crassa*, encodes a RecA-like protein. *Mutat. Res. DNA Repair 294*:223–234 (1993).
47. Chikashige, Y., N. Kinoshita, Y. Nakaseko, et al. Composite motifs and repeat symmetry in *S. pombe* centromeres: direct analysis by integration of Not1 restriction sites. *Cell 57*:740–751 (1989).

48. Chu, G., Vollrath, D., R. W. Davis. Separation of large DNA molecules by countour-clamped homogeneous electric fields. *Science 234*:1582–1585 (1986).
49. Clarke, L., J. Carbon. The structure and function of yeast centromeres. *Annu. Rev. Genet. 19*:29–56 (1985).
50. Clarke, L., M. P. Baum. Functional analysis of a centromere from fission yeast: a role for centromere-specific repeated DNA sequences. *Mol. Cell. Biol. 10*:1863–1872 (1990).
51. Clayton, D. A. Transcription of the mammalian mitochondrial genome. *Annu. Rev. Biochem. 55*:573–594 (1984).
52. Cunningham, R. P., C. DasGupta, T. Shibata, C. M. Radding. Homologous pairing genetic recombination: recA protein makes joint molecules of gapped circular DNA and closed circular DNA. *Cell 20*:223–235 (1980).
53. Day, A. W. Genetic implications of current models of somatic nuclear division in fungi. *Can. J. Bot. 50*:1337–1347 (1972).
54. Day, P. R., D. M. Boone, G. W. Keitt. *Venturia inaequalis* (CKE.) Wint. XI. The chromosome number. *Am. J. Bot. 43*:835–838 (1956).
55. Dodge, B. O. Nuclear phenomena associated with heterothallism and homothallism in the ascomycete *Neurospora. J. Arg. Res. 35*:289–305 (1927).
56. Dresser, M. E., C. N. Giroux. Meiotic chromosome behavior in spread preparations of yeast. *J. Cell Biol. 106*:567–573 (1988).
57. Dresser, M. M., M. J. Moses. Synaptonemal caryotyping in spermatocytes of the Chinese hamster (*Cricetulus griseus*). IV. Light and electron microscopy of synapsis and nucleolar development by silver staining. *Chromsoma 76*:1–22 (1980).
58. Duran, R., P. M. Gray. Nuclear DNA, an adjunct to morphology in fungal taxonomy. *Mycotaxon 36*:205–219 (1989).
59. Dutta, S. K. Transcription of non-repeated DNA in *Neurospora crassa. Biochim. Biophys. Acta 324*:482–487 (1973).
60. Dutta, S. K. Repeated DNA sequences in fungi. *Nucl. Acids Res. 1*:1411–1419 (1974).
61. Dutta, S. K., M. Ojha. Relatedness between major taxonomic groups of fungi based on the measurement of DNA nucleotide sequence homology. *Mol. Gen. Genet. 114*:232–240 (1972).
62. Egel-Mitani, M., L. S. Olson, R. Egel. Meiosis in *Aspergillus nidulans*: another example for lacking synaptonemal complexes in the absence of crossover interference. *Hereditas 97*:179–187 (1982).
63. El-Ani, A. S. Ascus development and nuclear behaviour in *Hypomyces solani* f. *cucurbitae. Am. J. Bot. 43*:769–778 (1956).
64. Elliott, C. G. The cytology of *Aspergillus nidulans. Genet. Res. Camb. 1*:462–476 (1960).
65. Engebrecht, J. A., G. S. Roeder. Yeast *mer*1 mutants display reduced levels of meiotic recombination. *Genetics 121*:237–247 (1989).
66. Engebrecht, J. A., G. S. Roeder. *MER*1, a yeast gene required for chromosome pairing and genetic recombination, is induced in meiosis. *Mol. Cell. Biol. 10*:2379–2389 (1990).

67. Engebrecht, J. A., J. Hirsch, G. S. Roeder. Meiotic gene conversion and crossing over: their relationship to each other and to chromosome synapsis and segregation. *Cell 62*:927–937 (1990).
68. Esposito, R. E., M. S. Esposito. Genetic recombination and commitment to meiosis in *Saccharomyces. Proc. Natl. Acad. Sci. USA 71*:3172–3176 (1974).
69. Esposito, R. E., R. Holliday. The effect of 5-fluorodeoxyuridine on genetic replication and somatic recombination in synchronously dividing cultures of *Ustilago maydis. Genetics 50*:1009–1017 (1964).
70. Futcher, B., J. Carbon. Toxic effects of excess cloned centromeres. *Mol. Cell. Biol. 6*:2213–2222 (1986).
71. Gardiner, K., D. Patterson. Transverse alternating electrophoresis. *Nature (Lond.) 331*:371–372 (1988).
72. Gaudet, A., M. Fitzgerald-Hayes. Mutations in *CEN*3 cause aberrant chromosome segregation during meiosis in Saccharomyces cerevisiae. Genetics 121:477–489 (1989).
73. Gillies, C. B. Reconstruction of the *Neurospora crassa* pachytene karyotype from serial sections of synaptonemal complexes. *Chromosoma 36*:119–130 (1972).
74. Gillies, C. B. Synaptonemal complex and chromosome structure. *Annu. Rev. Genet. 9*:91–109 (1975).
75. Gillies, C. B. The relationship between synaptonemal complexes, recombination nodules and crossing over in *Neurospora crassa* bivalents and translocation quadrivalents. *Genetics 91*:1 (1979).
76. Girbardt, M. Ultrastructure of the fungal nucleus. II. The kinetochore equivalent (KCE). *J. Cell Sci. 9*:453–473 (1971).
77. Giroux, C. N., M. E. Dresser, H. F. Tiano. Genetic control of chromosome synapsis in yeast meiosis. *Genome 31*:88–94 (1989).
78. Goldway, M., A. Sherman, D. Zenvirth, T. Arbel, G. Simchen. A short chromosomal region with major roles in yeast chromosome III meiotic disjunction, recombination and double strand breaks. *Genetics 133*:159–169 (1993).
79. Gray, M. W. Mitochondrial genome diversity and the evolution of mitochondrial DNA. *Can. J. Biochem. 60*:157–171 (1982).
80. Gray, M. W. Origin and evolution of mitochondrial DNA. *Annu. Rev. Cell Biol. 5*:25–50 (1989).
81. Gray, M. W., W. F. Doolittle. Has the endosymbiont hypothesis been proven? *Microbiol. Rev. 46*:1–42 (1982).
82. Greider, C. W., E. H. Blackburn. Identification of a specific telomere terminal transferase activity in Tetrahymena extracts. *Cell. 43*:405–413 (1985).
83. Greider, C. W., E. H. Blackburn. The telomere terminal transferase of Tetrahymena is a ribonucleoprotein enzyme with two kinds of primer specificity. *Cell 51*:887–898 (1987).
84. Grossman, L. I., M. E. S. Hudspeth. Fungal mitochondrial genomes. In: *Gene Manipulations in Fungi* (J. W. Bennett, L. L. Lasure, eds.). Academic Press, Toronto, 1985, pp. 65–103.
85. Hawley, R. S., H. Irick, A. E. Zitron, et al. There are two mechanisms of achiesmate segregation in *Drosophila* females, one of which requires heterochromatic homology. *Dev. Genet. 13*:440–467 (1993).

86. Heath, I. B. Mitosis in the fungus *Thraustotheca clavata. J. Cell Biol. 60*:204 (1974).
87. Heywood, P., P. T. Mcgee. Meiosis in protists, some structural and physiological aspects of meiosis in algae, fungi, and protozoa. *Bacteriol. Rev. 40*:190–240 (1976).
88. Hieter, P., D. Pridmore, J. H. Hegemann, M. Thomas, R. W. Davis, P. Philippsen. Functional selection and analysis of yeast centromeric DNA. *Cell 42*:913–921 (1985).
89. Hollingsworth, N. M., B. Byers. *HOP*1: a yeast meiotic pairing gene. *Genetics 121*: 445–462 (1989).
90. Hollingsworth, N. M., L. Goetsch, B. Byers. The *HOP*1 gene encodes a meiosis-specific component of yeast chromosomes. *Cell 61*:73–84 (1990).
91. Hollingsworth, N. M., A. D. Johnson. A conditional allele of the *Saccharomyces cerevisiae HOP*1 gene is suppressed by overexpression of two other meiosis-specific genes: *RED*1 and *REC*104. *Genetics 133*:785–797 (1993).
92. Holm, P. B., S. W. Rasmussen, D. Zickler, B. C. Lu, J. Sage. Chromosome pairing, recombination nodules and chiasma formation in the basidiomycete *Coprinus cinereus. Carlsberg Res. Commun. 46*:305–346 (1981).
93. Horesh, O., G. Simchen, A. Friedmann. Morphogenesis of the synaptonemal complex during yeast meiosis. *Chromosoma 75*:101–115 (1979).
94. Horowitz, N. H., H. Macleod. The DNA content of *Neurospora* nuclei. *Microb. Genet. Bull. 17*:6–7 (1960).
95. Hrushovetz, S. B. Cytological studies of ascus development in *Cochliobolus sativus. Can. J. Bot. 34*:641–651 (1956).
96. Hu, X., R. N. Nazar, J. Robb. Quantification of *Verticillum* biomass in wilt disease development. *Physiol. Mol. Plant Pathol. 42*:23–36 (1992).
97. Iyengar, G. A. S., P. C. Deka, S. C. Kundu, S. K. Sen. DNA synthesis in course of meiotic development in *Neurospora crassa. Genet. Res. 29*:1–8 (1977).
98. Kanda, T., H. Arakawa, Y. Yasuda, T. Takemaru. Basidiospore formation in a mutant of incompatibility factors and in mutants that arrest at meta-anaphase I in *Coprinus cinereus. Exp. Mycol. 14*:218–226 (1990).
99. Kenna, M., E. Amaya, K. Bloom. Selective excision of the centromere chromatin complex from *Saccharomyces cerevisiae. J. Cell Biol 107*:9–15 (1988).
100. Klapholz, S., C. S. Waddell, R. E. Esposito. The role of the *SPO*11 gene in meiotic recombination in yeast. *Genetics 110*:187–216 (1985).
101. Kmiec, E. B., W. M. Holloman. Synapsis promoted by *Ustilago* Rec1 protein. *Cell 36*:593–598 (1984).
102. Kochel, H. G., H. Kuntzel. Nucleotide sequence of the *Aspergillus nidulans* mitochondrial gene coding for the small ribosomal subunit RNA: homology to *E. coli* 16S rRNA. *Nucl. Acids Res. 9*:5689–5696 (1982).
103. Kornberg, R. K. Structure of chromatin. *Annu. Rev. Biochem. 46*:931–954 (1977).
104. Krumlauf, R., G. A. Marzluf. Characterization of the sequence complexity and organization of the *Neurospora crassa* genome. *Biochemistry 18*:3705–3713 (1979).
105. Kubai, D. F. The evolution of the mitotic spindle. *Int. Rev. Cytol. 43*:167–227 (1975).
106. Kubai, D. F. Mitosis and fungal phylogeny. In: *Nuclear Division in the Fungi* (ed. I. B. Heath). Academic Press, New York (1978), pp. 177–229.

107. Lechner, J., J. Carbon. A 240-kd multisubunit complex, CBF-3, is a major component of the budding yeast centromere. *Cell 64*:717–725 (1991).
108. Loidl, J. The initiation of meiotic chromosome pairing: the cytological view. *Genome 33*:759–778 (1990).
109. Lohr, D., J. Corden, K. Tatchell, R. T. Kovacic, K. E. Van Holde. Comparative subunit structure of HeLa, yeast, and chicken erythrocyte chromatin. *Proc. Natl. Acad. Sci. USA 74*:79–83 (1977).
110. Lohr, D., K. E. van Holde. Yeast chromatin subunit structure. *Science 188*:165–166 (1975).
111. Lowenhaupt, K., A. Rich, M. L. Pardue. Nonrandom distribution of long mono- and dinucleotide repeats in *Drosophila* chromosomes: Correlations with dosage compensation, heterochromatin and recombination. *Mol. Cell. Biol. 9*:1173–1182 (1989).
112. Lu, B. C. Chromosome cycles of the basidiomycete *Cyathus stercoreus* (Schw.) de Toni. *Chromosoma 15*:170–178 (1964).
112a. Lu, B. C. Polyploidy in the Basidiomycete *Cyathus stercoreus. Amer. J. Bot. 51*:343–347 (1964).
113. Lu, B. C. Fine structure of the meiotic chromosomes of the fungus *Coprinus lagopus. Exp. Cell Res. 43*:224–227 (1966).
114. Lu, B. C. The course of meiosis and centriole behaviour during the ascus development of the ascomycete *Gelasinospora calospora. Chromosoma 22*:210–226 (1967).
115. Lu, B. C. Meiosis in *Coprinus lagopus*: a comparative study with light and electron microscopy. *J. Cell Sci. 2*:529–536 (1967).
116. Lu, B. C. Genetic recombination in *Coprinus*: II. Its relations to the synaptinemal complexes. *J. Cell Sci. 6*:669–678 (1970).
117. Lu, B. C. Genetic recombination in *Coprinus*. IV. A kinetic study of the temperature effect on recombination frequency. *Genetics 78*:661–677 (1974).
118. Lu, B. C. Meiosis in *Coprinus*. VI. The control of the initiation of meiosis. *Can. J. Genet. Cytol. 16*:155–164 (1974).
119. Lu, B. C. Replication of deoxyribonucleic acid and crossing over in *Coprinus*. In: *Basidium and Basidiocarp, Evolution, Cytology, Function, and Development* (K. Wells, E. K. Wells, eds.). Springer-Verlag, New York (1982), pp. 93–112.
120. Lu, B. C. The cellular program for the formation and dissolution of the synaptonemal complex in *Coprinus. J. Cell Sci. 67*:25–43 (1984).
120a. Lu, B. C. Spreading the synaptonemal complex of *Neurospora crassa. Chromosoma 102*:464–472 (1993).
121. Lu, B. C., H. J. Brodie. Chromosomes of the fungus *Cyathus. Nature (Lond.) 194*:606 (1962).
122. Lu, B. C., S. M. Chiu. Genetic recombination in *Coprinus*. V. Repair synthesis of deoxyribonucleic acid and its relation to meiotic recombination. *Mol. Gen. Genet. 147*:121–127 (1976).
123. Lu, B. C., D. Galeazzi. Light and electron microscope observations of a meiotic mutant of *Neurospora crassa. Can. J. Bot. 56*:2694–2706 (1978).

124. Lu, B. C., D. Y. Jeng. Meiosis in *Coprinus*. VII. The prekarogamy S-phase and the postkaryogamy DNA replication in *C. lagopus*. *J. Cell Sci. 17*:461–470 (1975).
125. Lu, B. C., S. Y. Li. Correlation of endonuclease activity at meiotic prophase and cofactor-induced genetic recombination in the basidiomycete *Coprinus cinereus*. *Genome 30*:380–386 (1988).
126. Lu, B. C., N. B. Raju. Meiosis in *Coprinus*. II. Chromosome pairing and the lamp-brush diplotene stage of meiotic prophase. *Chromosoma 29*:305–316 (1970).
127. Maguire, M. Interactive meiotic systems, chromosome structure and function. In: *Impact of New Concepts* (J. P. Gustufson, R. Appels, eds.). Plenum, New York, 1988, pp. 117–144.
128. Malone, R. E., R. E. Esposito. Recombinationless meiosis in *Saccharomyces cerevisiae*. *Mol. Cell. Biol. 1*:891–901 (1981).
129. Malone, R. E., S. Bullard, M. Hermiston, R. Rieger, M. Cool, A. Galbraith. Isolation of mutants defective in early steps of meiotic recombination in the yeast *Saccharomyces cerevisiae*. *Genetics 128*:79–88 (1991).
130. Mao, J., B. Appel, J. Schaack, H. Yamada, D. Soll. The 5S RNA genes of *Schizosaccharomyces pombe*. *Nucl. Acids Res. 10*:487–500 (1982).
131. McClintock, B. Neurospora. I. Preliminary observations of the chromosomes of *Neurospora crassa*. *Am. J. Bot. 32*:671–678 (1945).
132. McCully, E. K., C. F. Robinow. Mitosis in heterobasidiomycetous yeast. I. *Leucosporidium scottii* (*Candida scottii*). *J. Cell Sci. 10*:857–881 (1972).
133. McLaughlin, D. J. Centrosomes and microtubules during meiosis in mushroom *Boletus rubinellus*. *J. Cell Biol. 50*:737–745 (1971).
134. Menees, T., G. S. Roeder. *MEI4*, a yeast gene required for meiotic recombination. *Genetics 123*:675–682 (1989).
135. Metzenberg, R. L., J. N. Stevens, E. U. Selker, E. Morzycka-Wroblewska. Identification and chromosomal distribution of 5S ribosomal RNA genes in *Neurospora crassa*. *Proc. Natl. Acad. Sci. USA 82*:2067–2071 (1985).
136. Meyer, G. F. A possible correlation between the submicroscopic structure of meiotic chromosomes and crossing over. *Proc. Eur. Reg. Conf. Electron Microsc.* Prague, Czechoslovakia (1964), pp. 461–462.
137. Moens, P. B. The fine structure of meiotic chromosome pairing in natural and artificial *Lilium* polyploids. *J. Cell Sci. 7*:55–64 (1970).
138. Moens, P. M., F. O. Perkins. Chromosome number of a small protist: accurate determination. *Science 166*:1289–1291 (1969).
139. Moreau, P. J. F., D. Zickler, G. Leblon. One class of mutants with disturbed centromere cleavage and chromosome pairing in *Sordaria macrospora*. *Mol. Gen. Genet. 198*:189–197 (1985).
140. Morris, N. R. Nucleosome structure in *Aspergillus nidulans*. *Cell 8*:357–363 (1974).
141. Moses, M. J. Chromosomal structures in crayfish spermatocytes. *J. Cell Biol. 2*: 215–218 (1956).
142. Moses, M. J. The relation between the axial complex of meiotic prophase chromosomes and chromosome pairing in a salamander (*Plethodon cinereus*). *J. Biophys. Biochem. Cytol. 4*:633–638 (1958).

143. Moses, M. J. Synaptonemal complex karyotyping in spermatocytes in the Chinese hamster (*Cricetulus griseus*). I. Morphology of the autosomal complement in spread preparations. *Chromosoma 60*:99–125 (1977).
144. Moses, M. J., P. A. Poorman, T. H. Roderick, M. T. Davisson. Synaptonemal complex analysis of mouse chromosomal rearrangements. IV. Synapsis and synaptic adjustment in two paracentric inversions. *Chromosoma 84*:457–474 (1982).
145. Motta, J. J. Somatic nuclear division in *Armillaria mellea*. *Mycologia 61*:873–886 (1969).
146. Moyzis, R. K., J. M. Buckingham, L. S. Cram, et al. A highly conserved repetitive DNA sequence, $(TTAGGG)_n$, present at the telomeres of human chromosomes. *Proc. Natl. Acad. Sci. USA 85*:6622–6626 (1988).
147. Murakami, S., T. Matsumoto, O. Niwa, M. Yanagida. Structure of the fission yeast centromere *cen*3: direct analysis of the reiterated inverted region. *Chromosoma 101*:214–221 (1991).
148. Nazar, R. N., X. Hu, J. Schmidt, D. Culham, J. Robb. Potential use of PCR-amplified ribosomal intergenic sequences in the detection and differentiation of *Verticillium* wilt pathogens. *Physiol. Mol. Plant Pathol. 39*:1–11 (1991).
149. Noll, M. Differences and similarities in chromatin structure of *Neurospora crassa* and higher eucaryotes. *Cell 8*:349–355 (1976).
150. O'Donnell, K. Ribosomal DNA internal transcribed spacers are highly divergent in the phytopathogenic ascomycete *Fusarium sambucinum* (*Gibberella pulicaris*). *Curr. Genet. 22*:213–220 (1992).
151. Oishi, K., I. Uno, T. Ishikawa. Timing of DNA replication during the meiotic process in monokaryotic basidiocarps of *Coprinus macrorhizus*. *Arch. Microbiol. 132*:372–374 (1982).
152. Olive, L. S. The structure and behavior of fungus nuclei. Bot. Rev. 19:439–586 (1953).
153. Olive, L. S. Nuclear behavior during meiosis. In: *The Fungi: An Advanced Treatise* (ed. G. C. Ainsworth, A. S. Sussman), Vol. 1. Academic Press, New York (1965), pp. 143–161.
154. Orback, M. J., D. Vollrath, R.. W. Davis, C. Yanofsky. An electrophoretic karyotype of *Neurospora crassa*. *Mol. Cell. Biol. 8*:1469–1473 (1988).
155. Orback, M. J. The revised *Neurospora* chromosome sizes. *Fungal Genet. Newsl. 93*:92 (1992).
156. Padmore, R., L. Cao, N. Kleckner. Temporal comparison of recombination and synaptonemal complex formation during meiosis in *S. cerevisiae*. *Cell 66*:1239–1256 (1991).
157. Peabody, D. C., J. J. Motta. The ultrastructure of nuclear division in *Armillaria mellea*: meiosis I. *Can. J. Bot. 57*:1860–1872 (1979).
158. Pearlman, R. E., N. Tsao, P. B. Moens. Synaptonemal complexes from DNase-treated rat pachytene chromosomes contain $(GT)_n$ and LINE/SINE sequences. *Genetics 130*:865–872 (1992).
159. Perkins, D. D. *Neurospora* chromosomes. In: *The Dynamic Genome: Barbara McClintock's Ideas in the Century of Genetics* (N. Federoff, D. Botstein, eds.). Cold Spring Harbor Laboratory Press, Cold Spring Harbor, NY (1992), pp. 33–43.

160. Perkins, D. D., E. G. Barry. The cytogenetics of *Neurospora*. *Adv. Genet.* *19*:133–285 (1977).
161. Perkins, D. D., N. B. Raju, E. G. Barry. A chromosome rearrangement in *Neurospora* that produces segmental aneuploid progeny containing only part of the nucleolus organizer. *Chromosoma* *89*:8–17 (1984).
162. Peterson, J. B., H. Ris. Electron microscopic study of the spindle and chromosome movement in the yeast *Saccharomyces*. *J. Cell Sci.* *22*:219–242 (1976).
163. Pukkila, P. J., B. C. Lu. Silver staining of meiotic chromosome in the fungus, *Coprinus cinereus*. *Chromosoma* *91*:108–112 (1985).
164. Pukkila, P. J., C. Skrzynia, B. C. Lu. The *rad*3-1 mutant is defective in axial core assembly and homologous chromosome pairing during meiosis in the basidiomycete *Coprinus cinereus*. *Dev. Genet.* *13*:403–410 (1992).
165. Radding, C. M. Homologous pairing and strand exchange in genetic recombination. *Annu. Rev. Genet.* *16*:405–437 (1982).
166. Raju, N. B. Meiotic nuclear behavior and ascospore formation in five homothallic species of *Neurospora*. *Can. J. Bot.* *56*:754–763 (1978).
167. Raju, N. B. Meiosis and ascospore genesis in *Neurospora*. *Eur. J. Cell Biol.* *23*: 208–223 (1980).
168. Raju, N. B. Easy method for fluorescent staining of *Neurospora* nuclei. Neurospora *Newsl.* *29*:24–26 (1982).
169. N. B. Raju. Use of enlarged cells and nuclei for studying mitosis in *Neurospora*. *Protoplasm* *121*:87–98 (1984).
170. Raju, N. B. A simple fluorescent staining method for meiotic chromosomes of *Neurospora*. *Mycologia* *78*:901–906 (1986).
171. Raju, N. B. Functional heterothallism resulting from homokaryotic conidia and ascospores in *Neurospora tetrasperma*. *Mycol. Res.* *96*:103–116 (1992).
172. Raju, N. B. Genetic control of the sexual cycle in *Neurospora*. *Mycol. Res.* *96*:241–262 (1992).
173. Raju, N. B., B. C. Lu. Meiosis in *Coprinus*. III. Timing of meiotic events in *C. lagopus* (sensu Buller). *Can. J. Bot.* *48*:2183–2186 (1970).
174. Raju, N. B., B. C. Lu. Meiosis in *Coprinus*. IV. Morphology and behaviour of spindle pole bodies. *J. Cell Sci.* *12*:131–141 (1973).
175. Raju, N. B., D. Newmeyer. Giant ascospores and abnormal croziers in a mutant of *Neurospora crassa*. *Exp. Mycol.* *1*:152–165 (1977).
176. Raju, N. B., D. D. Perkins. Barren perithecia in *Neurospora crassa*. *Can. J. Genet. Cytol.* *20*:41–59 (1978).
177. Raju, N. B., D. D. Perkins. Diverse programs of ascus development in pseudohomothallic species of *Neurospora*, *Gelasinospora*, and *Podospora*. *Dev. Genet.* *15*: 104–118 (1994).
178. Rasmussen, S. W., P. B. Holm, B. C. Lu, D. Zickler, J. Sage. Synaptonemal complex formation and distribution of recombination nodules in pachytene trivalents of triploid *Coprinus cinereus*. *Carlsberg Res. Commun.* *46*:347–360 (1981).
179. Raudaskoski, M., B. C. Lu. The effect of hydroxyurea on meiosis and genetic recombination in the fungus *Coprinus lagopus*. *Can. J. Genet. Cytol.* *22*:41–50 (1978).

180. Robinow, C. F., A. Bakerspigel. Somatic nuclei and forms of mitosis in fungi. In: *The Fungi: An Advanced Treatise* (ed. G. C. Ainsworth, A. S. Sussman), Vol. 1. Academic Press, New York (1965), pp. 119–142.
181. Robinow, C. F., C. E. Caten. Mitosis in *Aspergillus nidulans*. *J. Cell Sci.* *5*:403 (1969).
182. Rockmill, B., G. S. Roeder. Meiosis in asynaptic yeast. *Genetics* *126*:563–574 (1990).
183. Rogers, J. D. *Hypoxylon fuscum*. I. Cytology of the ascus. *Mycologia* *57*:789–803 (1965).
184. Rossen, J. M., M. Westergaard. Studies on the mechanism of crossing over. II. Meiosis and the time of meiotic chromosome replication in the ascomycete *Neottiella rutilans* (Fr.) Dennis. *C. R. Trav. Lab. Carlsberg* *35*:233–260 (1966).
185. Runge, K. W., R. J. Wellinger, V. A. Zakian. Effects of excess centromeres and excess telomeres on chromosome loss rates. *Mol. Cell. Biol.* *11*:2919–2928 (1991).
186. Russell, P. J., S. Wagner, K. D. Rodland, et al. Organization of the ribosomal ribonucleic acid genes in various wild-type strains and wild-collected strains of *Neurospora*. *Mol. Gen. Genet.* *196*:275–282 (1984).
187. Schechtman, M. G. Isolation of telomere DNA from *Neurospora crassa*. *Mol. Cell. Biol.* *7*:3168–3177 (1987).
188. Schechtman, M. G. Characterization of telomere DNA from *Neurospora crassa*. *Gene* *88*:159–165 (1990).
189. Schroeder, A. L., N. B. Raju. *mei*-2, A mutagen-sensitive mutant of *Neurospora* defective in chromosome pairing and meiotic recombination. *Mol. Gen. Genet.* *231*:41–48 (1991).
190. Schultes, N. P., J. W. Szostak. A poly(dA.dT) tract is a component of the recombination initiation site at the ARG4 locus in *Saccharomyces cerevisiae*. *Mol. Cell. Biol.* *11*:322–328 (1991).
191. Schwartz, D. C., C. R. Cantor. Separation of yeast chromosome-sized DNAs by pulsed field gradient gel electrophoresis. *Cell* *37*:67–75 (1984).
192. Selker, E. U., C. Yanofsky, K. Driftmier, R. L. Metsenberg, B. Alzner-DeWeerd, U. L. RajBhandary. Dispersed 5S RNA genes in *N. crassa*: structure, expression and evolution. *Cell* *24*:819–828 (1981).
193. Setliff, E. C., H. C. Hoch, R. F. Patton. Studies on nuclear division in basidia of *Poria latemarginata*. *Can. J. Bot.* *52*:2323–2333 (1974).
194. Shampay, J., J. W. Szostak, E. H. Blackburn. DNA sequences of telomeres maintained in yeast. *Nature* *310*:154–157 (1984).
195. Shear, C. L., B. O. Dodge. Life histories and heterothallism of the red bread-mold fungi of the *Monilia sitophila* group. *J. Agric. Res.* *34*:1019–1042 (1927).
196. Shinohara, A., H. Ogawa, T. Ogawa. Rad51 protein involved in repair and recombination in *S. cerevisiae* is a RecA-like protein. *Cell* *69*:457–470 (1992).
197. Simchen, G., R. Pinon, Y. Salts. Sporulation in *Saccharomyces cerevisiae*: premeiotic DNA synthesis, readiness, and committment. *Exp. Cell Res.* *75*:207–218 (1972).
198. Simonet, J. M., D. Zickler. Mutations affecting meiosis in *Podospora anserina*. I. Cytological studies. *Chromosoma* *37*:327–351 (1972).
199. Singleton, J. R. Chromosome morphology and the chromosome cycle in the ascus of *Neurospora crassa*. *Am. J. Bot.* *40*:124–144 (1953).

200. Skinner, D. Z., A. D. Budde, S. A. Leong. Molecular karyotype analysis of fungi. In: *More Gene Manipulations in Fungi* (eds. J. W. Bennett, L. L. Lasure). Academic Press, New York (1991), pp. 86–103.
201. Smith, P. A., R. C. King. Genetic control of synaptonemal complexes in *Drosophila melanogaster. Genetics 60*:335–351 (1968).
202. Smithies, O., P. A. Powers. Gene conversions and their relation to homologous chromosome pairing. *Philos. Trans. R. Soc. Lond. Ser. B. 312*:291–302 (1986).
203. Sun, H., D. Treco, N. P. Schultes, J. W. Szostak. Double strand breaks at an initiation site for meiotic gene conversion. *Nature 338*:87–90 (1989).
204. Sym, M., J. A. Engebrecht, G. S. Roeder. ZIP1 is a synaptonemal complex protein required for meiotic chromosome synapsis. *Cell 72*:365–378 (1993).
205. Szostak, J. W. Replication and resolution of telomeres in yeast. *Cold Spring Harb. Symp. Quant. Biol. 47*:1187–1194 (1984).
206. Thielke, C. Meiotic divisions in the basidium. In: *Basidium and Basidiocarp, Evolution, Cytology, Function, and Development* (K.. Wells, E. K. Wells, eds.), Springer-Verlag, New York (1982), pp. 75–91.
207. Timberlake, W. E. Low repetitive DNA content in *Aspergillus nidulans. Science (Washington, D.C.) 202*:973–975 (1978).
208. Thomas, J. O., V. Furber. Yeast chromatin structure. *FEBS Lett. 66*:274–280 (1976).
209. Treco, D., N. Arnheim. The evolutionarily conserved repetitive sequence d(TG.AC)n promotes reciprocal exchange and generates unusual recombination tetrads during yeast meiosis. *Mol. Cell. Biol. 6*:3934–3947 (1986).
210. Wallace, D. C. Structure and evolution of organelle genomes. *Microbiol. Rev. 46*: 208–240 (1982).
211. Wells, K. Light and electron microscopic studies of *Ascobolus stercorarius*. I. Nuclear divisions in the ascus. *Mycologia 42*:761–790 (1970).
212. West, S. C., E. Cassuto, P. Howard-Flanders. Postreplication repair in *E. coli*: strand exchange reactions of gapped DNA by recA protein. *Mol. Gen. Genet. 187*:209–217 (1982).
213. Westergaard, M., D. von Wettstein. Studies on the mechanism of crossing over. III. On the ultrastructure of the chromosomes in *Neottiella rutilans* (Fr.) Dennis. *C. R. Trav. Lab. Carlsberg 35*:261–286 (1966).
214. Westergaard, M., D. von Wettstein. Studies on the mechanism of crossing over. IV. The molecular organization of the synaptinemal complex in *Neottiella rutilans* (Cooke) Saccardo (Ascomycetes). *C. R. Trav. Lab. Carlsberg 37*:239–268 (1970).
215. Westergaard, M., D. von Wettstein. The synaptinemal complex. *Annu. Rev. Genet. 6*:71 (1972).
216. von Wettstein, D., S. W. Rassmussen, P. B. Holm. The synaptonemal complex in genetic segregation. *Annu. Rev. Genet. 18*:331–413 (1984).
217. Wolf, K., L. Del Giudice. The variable mitochondrial genome of ascomycetes: organization, mutational alterations, and expression. *Adv. Genet. 25*:185–308 (1988).
218. Zakian, V. A. Structure and function of telomeres. *Annu. Rev. Genet. 23*:579–604 (1989).

219. Zickler, D. Analyse de la meiose du champignon discomycete *Ascobolus immerus* (Pers.). *C. R. Hebd. Seances Acad. Sci. Ser. D 265*:189–201 (1967).
220. Zickler, D. Division spindle and centrosomal plaques during mitosis and meiosis in some Ascomycetes. *Chromosoma 30*:287–304 (1970).
221. Zickler, D. Development of the synaptonemal complex and the "recombination nodules" during meiotic prophase in the seven bivalents of the fungus *Sordaria macrospora* Auersw. *Chromosoma 61*:289–316 (1977).
222. Zickler, D., L. de Lares, P. J. F. Moreau, G. Leblon. Defective pairing synaptonemal complex formation in a *Sordaria* mutant (spo44) with a translocated segment of the nucleolar organizer. *Chromosoma 92*:37–47 (1985).
223. Zickler, D., P. J. F. Moreau, A. D. Huynh, A.-M. Slezec. Correlation between pairing initiation sites, recombination nodules and meiotic recombination in *Sordaria macrospora. Genetics 132*:135–148 (1992).
224. Zolan, M. E., C. J. Tremel, P. J. Pukkila. Production and characterization of radiation-sensitive meiotic mutant of *Coprinus cinereus. Genetics 120*:379–387 (1988).
225. Zolan, M. E., N. Y. Stassen, M. A. Ramesh, B. C. Lu, G. Valentine. Meiotic mutants and DNA repair genes of *Coprinus cinereus. Can. J. Bot.,* in press (1995).
226. Zuk, J., Z. Swietlinska. Cytological studies in *Ascobolus immersus. Acta Soc. Bot. Poloniae 34*:171–179 (1965).

7

Extrachromosomal and Transposable Genetic Elements

Heinz D. Osiewacz
Johann Wolfgang Goethe-Universität, Frankfurt am Main, Germany

1. INTRODUCTION

In eukaryotes, the vast majority of the genome is part of the chromosomes in the nucleus (chromosomal DNA). During nuclear divisions this genetic material is distributed in a well-defined way according to the Mendelian rules. In addition, genetic traits are found either in the nucleus or in the cytoplasm which are not subject to this controlled distribution. However, since the nuclear envelope becomes disintegrated during nuclear division, these elements and the chromosomes are part of the same compartment, at least at particular stages during the cell cycle. Finally, genetic elements are found in the mitochondria of both heterotrophic and autotrophic eukaryotes as well as in the plastids of photoautotrophic plants (e.g., higher plants) (Table 1).

Besides a classification according to their localization, the different genetic traits can be distinguished according to their physical characteristics. Eukaryotic chromosomes represent linear DNA species that are associated with proteins. In mitochondria and plastids, usually several copies of a high-molecular-weight DNA species are found as circular molecules. In only a few

Table 1 Genetic Traits as Part of the Eukaryotic Genome

Genome		
Chromosomal (Mendelian) Traits	Extrachromosomal (non-Mendelian) Traits	
Nuclear/cytoplasmic elements	Mitochondrial genome (chondriome)	Genome of plastids (plastone)
• Set of chromosomes	• High-molecular-weight mtDNA (mtDNA)	• High-molecular-weight ptDNA
• Circular plasmids	• Circular plasmids	• Circular plasmids
• Linear plasmids	• Linear plasmids	• Linear plasmids
• Viruses and virus-related elements	• Viruses and virus-related elements	

cases were linear DNA molecules found to replace these genetic traits. In addition, autonomous, low-molecular-weight DNA and RNA species are part of the genetic complement in many, but not all, fungi. In some cases, these elements are found to be encapsidated by proteins and therefore, although most of them appear to be noninfectious, may be viewed as viruses. In other cases, the corresponding elements are not encapsidated and can be considered to be plasmids in the broader view as it was originally defined by Lederberg [1] as *all extrachromosomal hereditary determinants.* According to this definition all circular and linear low-molecular-weight DNAs as well as double-stranded RNA molecules (dsRNA) are part of this group of genetic elements.

In this chapter different extrachromosomal genetic elements will be introduced and discussed. Special emphasis is put on genetic traits found in fungal mitochondria. In addition, related elements identified outside these organelles are briefly discussed. Experimental approaches are an integral part of this chapter. They are aimed to outline how particular questions can be addressed and what the results of a particular experiment look like. In some cases, a rather detailed description of a typical experiment is presented. In others, a more general outline is given. It will be stressed that the experimental procedures can only be an example and can nowadays be replaced by a large number of alternatives.

As a prerequisite for the understanding of this chapter, and in particular to follow the experimental procedures, some basic knowledge about the different life cycles of fungi and some training in basic microbiology and in molecular biology are required.

2. DISCRIMINATION BETWEEN CHROMOSOMAL AND EXTRACHROMOSOMAL GENETIC TRAITS

In cases in which genetic traits are associated with a particular function, the mutation of the corresponding element may lead to an altered phenotype. The analysis of the corresponding mutant can result in the identification and characterization of the responsible element. In fungi that are accessible to a genetic analysis, one of the first steps in the characterization of a novel mutant is the analysis of the type of inheritance of the genetic trait that gives rise to the altered phenotype. This analysis may answer the question whether the corresponding phenotype is encoded by a chromosomal, Mendelian factor or an extrachromosomal genetic trait.

The outcome of a genetic cross is dependent on the type of cells that are contributing to it. In fungi, there exists a large variety of different modes of sexual propagation. It is not possible to cover these as a whole in this chapter. For a more detailed discussion of sexual reproduction in fungi, the reader is referred to textbooks on fungal biology [e.g., 2]. However, in order to introduce the possibility to distinguish genetically between extranuclear and nuclear genetic traits, a rather general introduction is provided.

Basically, two different modes of fertilization can be distinguished. First, reproduction may be initiated by the fusion of two structurally identical sexual units (e.g., gametes or gametangia) which contribute approximately the same amount of cytoplasm to the fusion product (isogamy). An example for this type of sexual reproduction is the fusion of isogamous gametes of certain fungi or the fusion of haploid cells of *Saccharomyces cerevisiae.* Second, the fusing cells or organs may be structurally different, and the vast majority of the cytoplasm is derived from the larger, "female" gamete or gametangium. Examples are the fusion of gametes or gametangia of different size (anisogamy and oogamy) or the fusion of small *male gametes* and *female gametangia.* The latter mode of fertilization is found in a number of genetically important ascomycetes (e.g., *Neurospora crassa, Podospora anserina*).

A. Isogamy

In isogamous species that are accessible to a genetic analysis the cells that fuse are characterized by a polarity, a particular mating type (mat). In the simplest situation this mating type corresponds to one of two alternative nuclear genetic factors, which may be termed *mat+* or *mat–*. In this situation, two different genetic crosses can be performed between a wild-type and a particular mutant strain. As illustrated in Figure 1, the parent with mating type + may harbor the mutation whereas the mating type – parent may have wild-type characteristics. In the so-called reciprocal cross, the wild-type parent is of mating type +

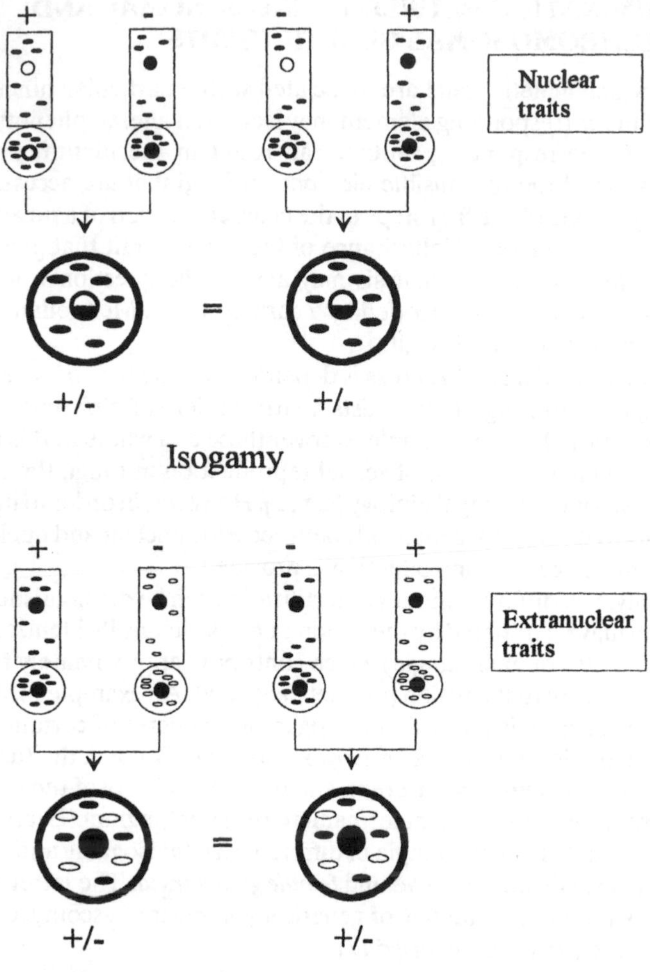

Figure 1 Reciprocal crosses between chromosomal and extrachromosomal mutants and wild-type strain of a species (e.g., *Saccharomyces cerevisiae*) in which the two parents contribute approximately the same amount of cytoplasm to a genetic cross (isogamy). The parents of each cross are either of mating type + or of mating type –. In this scheme the vegetative parts of an organism are indicated by rectangles; generative cells (e.g., gametes), by the larger circles attached to the rectangles. The genetic trait of interest that gives rise to a particular phenotype is indicated by small, solid or open circles in the case of a nuclear factor, or as open or solid ellipsoids in the case of an extranuclear factor.

and the mutant of mating type –. Assuming that the mutation is located in the nucleus (Fig. 1), the outcome of reciprocal crosses is identical. In both cases, the diploid zygotes will contain a mixture of cytoplasm (*hetero*plasmon) derived from both parental strains. After meiosis, the four products (e.g., ascospores) will segregate according to Mendelian rules. Two spores will give rise to a mutant progeny, and two spores will lead to strains with wild-type characteristics. In contrast, no 2:2 segregation is observed in the haploid products of a meiosis when the mutation is located in the cytoplasm. It may, however be that, as in the yeast *Saccharomyces cerevisiae*, the two types of extrachromosomal genetic traits completely separate after a number of subsequent mitotic divisions, giving rise to *homo*plasmons with the genetic traits derived from either one of the two parents.

B. Anisogamy, Oogamy

A difference in the outcome of reciprocal crosses will be observed when the fusing partners contribute different amounts of cytoplasm to the cross. In these cases, almost no cytoplasm in the zygote is derived from the smaller, "male" cells (Fig. 2). As a consequence, the outcome of a cross between strains with different genetic traits in the cytoplasm depends on the genotype of the "female" parent, a situation which is termed "maternal inheritance." In this case, reciprocal crosses lead to differences (reciprocal differences), whereas those between chromosomal mutants do not.

C. Experiments

The genetic analysis of an isogamous organism is basically dealt with in the case study on *Saccharomyces cerevisiae* in this volume. It should be emphasized that it was a genetic analysis of cytoplasmic mutants of this yeast that resulted in the first postulation of genetic factors in mitochondria [3]. This was long before DNA was first discovered to be present in these eukaryotic organelles.

Reciprocal Crosses Between Extrachromosomal Mutants and the Wild-Type Strain of Podospora anserina

A general introduction into the genetic analyses of *Podospora anserina* is presented in the case study on this ascomycete. Since the analysis of the extrachromosomal genetic traits in a particular mutant of this species will be used as an example for the analyses of the different extrachromosomal factors, the results of an initial genetic analyses of this mutant are presented at this point. For details on the life cycle of *P. anserina*, the reader is referred to the case study.

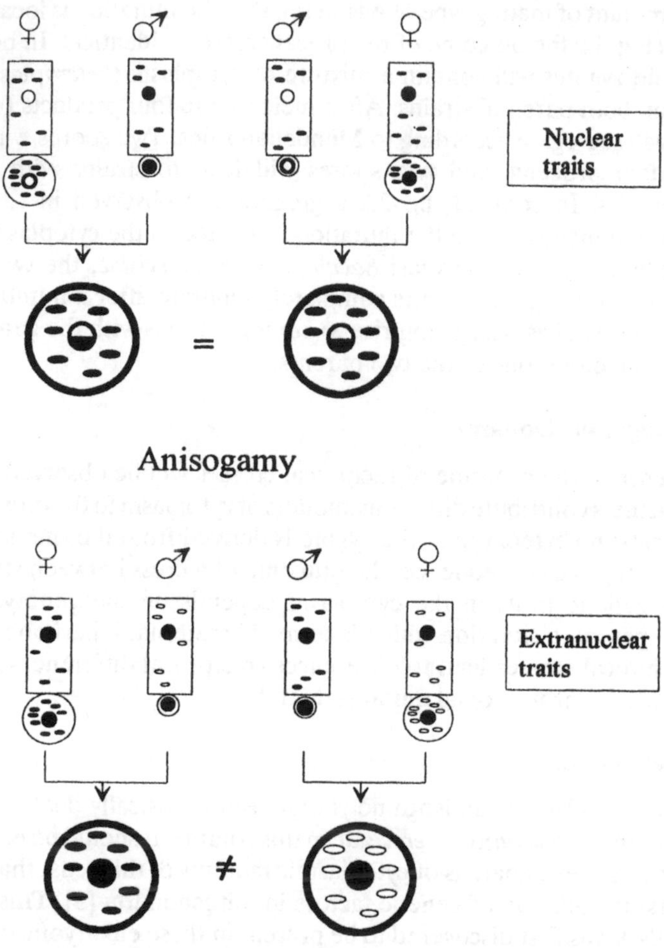

Figure 2 Scheme, reciprocal crosses between chromosomal and extrachromosomal mutants and a wild-type strain of a species with an anisogamous mode of fertilization. Generative cells with different amounts of cytoplasm are indicated by circles of different size. Symbols for genetic factors are the same as in Figure 1.

The two strains used in this analysis are wild-type strain *A* with a life span of 18 days, and mutant *AL2* which was isolated from wild-type strain *A*. Mutant *AL2* is characterized by the life span of 220 days [4].

Method

1. Fertilization of a culture of the wild-type strain *A* with isolated spermatia ("male" gametes) of long-lived mutant *AL2*.
2. Fertilization of a culture of long-lived mutant *AL2* with spermatia of wild-type strain *A* (compare Fig. 3 in the case study on *P. anserina*).

After the formation of perithecia, mononucleate ascospores are isolated from about 100 individual asci of each reciprocal cross. After germination, the life span of cultures derived from the individual ascospores of these crosses is determined.

Results

In this example, the life span of the progeny turned out to be identical to that of the "female" parent. Therefore it can be concluded that the genetic traits responsible for this case of maternal inheritance are most likely located in the cytoplasm. Since in *P. anserina* it is known that the DNA in mitochondria is involved in the control of the life span, these organelles were good candidates to contain the genetic trait leading to an increased life span.

3. EXTRACHROMOSOMAL GENETIC ELEMENTS

A. The Mitochondrial Genome—The Chondriome

In fungi, different types of genetic information are rather commonly found in mitochondria. As a whole, these elements represent the mitochondrial genome, also termed the *chondriome* (Table 1). In all cases, a rather large molecule corresponds to the so-called high-molecular-weight mitochondrial DNA (hmw mtDNA). In the following part of this chapter, for reasons of simplicity, the DNA will be termed mtDNA. In many cases, different types of additional, autonomous elements of a lower molecular weight were identified. Among these, mitochondrial plasmids may be of either circular or linear structure. Some of these elements are derived from the mtDNA. In other, rare cases, their origin is not so clear. In addition, double-stranded RNA (dsRNA) species, elements that are related to mycoviruses, may also be part of fungal mitochondria (e.g., *Ophiostoma ulmi*) [5].

Total DNA from Mitochondria

After a genetic analysis revealed that a particular phenotype is encoded by genetic factors located in the cytoplasm, a subsequent molecular analysis may lead to the identification of the corresponding traits. In such an analysis DNA is isolated from mitochondria of the wild-type strain and the corresponding mutant, and a comparative molecular characterization is performed.

Isolation of total DNA from mitochondria of *P. anserina*

Experiments. Total DNA from mitochondria may be extracted in different ways. The DNA may be isolated from mycelial homogenates after concentration of mitochondria by differential centrifugation or by banding in sucrose gradients. It is important to include a digestion with proteinase K in the DNA preparation procedure. This step removes most of the proteins bound to the ends of linear plasmids. In many earlier protocols this step was not included, and linear plasmids became lost during later steps since the bound proteins did not allow the DNA to enter either CsCl gradients or agarose gels properly.

Method

1. Grow strains (wild-type and mutant, respectively) on 10 agar plates each, containing BMM. Grow the mycelium until it covers the plates.
2. Scrape off the top of the culture using a sterile scalpel and transfer hyphae derived from one agar plate to an Erlenmeyer flask containing 200 mL liquid medium each. Grow cultures in an incubator for about 2 days at 27°C under constant rotation (150 rounds per minute; rpm).
3. Transfer culture from the content of two small Erlenmeyer flasks to one Erlenmeyer flask containing 2 L of liquid medium each. Inoculate these cultures for 5 days at 27°C under constant rotation (130 rpm).
4. Recover mycelium by filtration through four layers of sterile cheesecloth and remove liquid carefully.
5. Determine the wet weight of the mycelium and add a few milliliters of sterile glycerol until the mycelium looks shiny.
6. Add the mycelium to a mortar that has been cooled with liquid nitrogen. Add some liquid nitrogen to the mycelium and disrupt it to a powder.
7. For each gram of mycelium add about 4 mL of mitochondria buffer to a beaker of a Waring blendor.
8. Transfer the disrupted mycelium to the beaker and homogenize it further with three bursts for about 15 sec each.
9. Transfer homogenate to centrifuge tubes (GSA rotor, Sorvall) which were preincubated on ice and centrifuge for 5 min at 4°C at 4.000 rpm.
10. Recover supernatant and repeat centrifugation in a new centrifuge tube for 10 min.
11. Recover supernatant and centrifuge in a new GSA tube at 12.500 rpm, 30 min, 4°C.

12. Remove the supernatant carefully and recover the mitochondria-containing pellet in TES/SDS. Add a volume of TES/SDS to equal the initial wet weight of the harvested mycelium.

13. Transfer redissolved mitochondria to a Teflon homogenizer and homogenize carefully.

14. Add 100 μg/mL proteinase K and incubate homogenate for at least 2 h at 60°C.

15. Add one-fourth of the volume of 5 M NaCl and mix carefully by inversion of the reaction tube. Incubate for 1 h (alternatively overnight) at 4°C.

16. Centrifuge at 5000 rpm in a SS34 rotor (Sorvall), 20 min, 4°C.

17. Transfer supernatant to a fresh centrifugation tube, add one-third volume of sterile 40% polyethylene glycol 6000, and incubate at 4°C for 1 to 12 h.

18. Centrifuge at 5000 rpm (SS34) for 20 min at 4°C.

19. Discard supernatant and redissolve pellet at 37°C for 1 h in 4 mL of TES/CsCl.

20. Add 50 μL bisbenzimide (Hoechst 33258: 2 mg in TES) and adjust the refraction index to 1.3985.

21. Centrifuge in an TV865 rotor (Sorvall) for 20 h at 20°C and 48,000 rpm.

22. Visualize DNA bands under UV light and recover the upper band, which represents mtDNA (the lower one is nuclear DNA).

23. Remove bisbenzimide by repeated extractions of the DNA fraction with isopropanol saturated with CsCl.

24. Remove CsCl from the DNA fraction by dialysis at least two times with 2 L of TE buffer at 4°C. Dialysis should be performed at least for 10 h.

Results. DNA isolated by this procedure is generally suitable for further molecular analysis. Since proteins are removed from the terminal ends, linear plasmids are able to enter the gels and will lead to distinctive bands. Also, circular plasmids are fractionated in this way but will lead to several bands according to their structure—i.e., covalently closed circular (ccc), open circular (oc), or even linearized molecules.

Fractionation of total mitochondrial DNA by gel electrophoresis. Isolated DNA may be tested by subjecting an undigested aliquot to agarose gel electrophoresis.

Method

1. Total mtDNA of the wild-type strain and of the extrachromosomal mutant are loaded into two adjacent wells of an 1% agarose gel (1% agarose in TAE buffer). Load 1 μg of a DNA standard into a third well of the same gel.

2. Electrophoresis is performed in 1× TAE buffer for several hours until the DNA bands of the DNA standard are well separated.

3. DNA is stained with ethidiumbromide (0.5 μg/mL TAE buffer) for approximately 20 min.

4. Destain the agarose gel in water for 15 min.

5. Visualize DNA under UV light (254 nm).

6. Take a picture of the gel and compare the lanes loaded with DNA from the different strains.

Results. Differences may be found in the lanes containing DNA from different strains (Fig. 3). mtDNA isolated by the method described above corresponds to a diffuse band migrating between about 18 and 40 kbp. This band represents DNA molecules which, due to shearing, were linearized during DNA preparation. In addition, one or more distinctive bands may migrate at lower weight. These bands represent autonomous DNA molecules (e.g., linear plasmids) present in mitochondria of certain fungal strains.

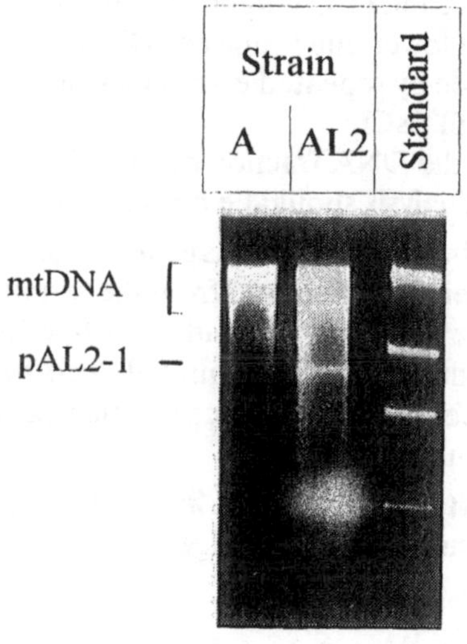

Figure 3 Comparison of total DNA isolated from mitochondria of wild-type strain *A* and long-lived mutant *AL2* of *Podospora anserina*. In addition to sheared molecules of mtDNA which are migrating as a diffuse band at about 18–40 kbp in size, a distinct DNA band is migrating at 8.5 kbp in mutant *AL2* but not in the wild-type strain. DNA of bacteriophage lambda, which was digested with restriction endonuclease *Hind*III, was used as a DNA standard in a separate lane of the 1% agarose gel.

High-Molecular-Weight Mitochondrial DNA (mtDNA)

Usually, fungal mtDNA is a covalently closed circular (ccc) DNA. Only a few linear mtDNA species were demonstrated unequivocally [6,7]. Although the coding capacity of the mtDNA is rather conserved, a remarkable size polymorphism is observed ranging from a minimal size of about 17 kbp in a strain of the yeast *Schizosaccharomyces pombe* [8] to about 176 kbp in the basidiomycete *Agaricus bisporus* [9]. One of the reasons for this size polymorphism is the presence or absence of sequences which interrupt the coding sequences of mitochondrial genes. These intron sequences may, as in the case of *Podospora*

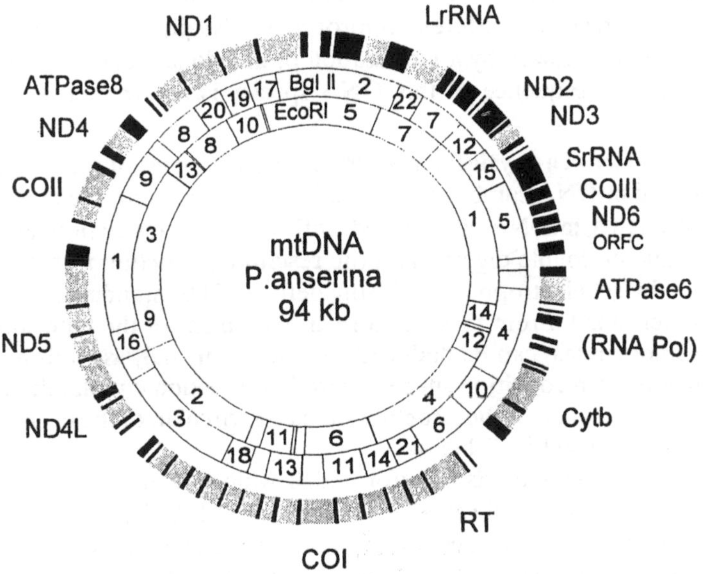

Figure 4 mtDNA of *Podospora anserina* wild-type strain s. The physical map based on a restriction enzyme analyses with *Bgl*II and *Eco*RI is shown in the inner two circles. On the outer circle, the genetic map is indicated as it became clear from hybridization studies and from a determination of the complete nucleotide sequence [55]. Black bars correspond to coding sequences for proteins, the large (*LrRNA*) and the small (*srRNA*) ribosomal RNA, and the whole set of tRNAs (not indicated in detail). Intron sequences are indicated by gray bars. In addition to the standard protein encoding genes a reverse transcriptase (*RT*) is encoded by the first intron of the cytochrome c oxidase gene subunit I (*COI*), which gives rise to the formation of plDNA, a circular plasmid in senescent cultures. Remnants of a viral-type RNA polymerase gene (*RNA Pol*) were identified between the apocytochrome b gene (*Cytb*) and the ATPase subunit 6 gene (*ATPase6*).

anserina (Fig. 4), represent a rather substantial fraction of the mtDNA. Some introns code for polypeptides (e.g., RNA maturases, reverse transcriptases) and thus represent coding sequences for products additional to the "well-defined" set of gene products which are typically encoded by mtDNAs. Finally, in species with mtDNAs of an intermediate or large size, some unassigned open reading frames (ORFs) of substantial length were identified. In some cases, good evidence exists that these sequences are derived from the integration of mitochondrial plasmids and represent nonfunctional, rearranged genes [10].

In contrast to linear or circular plasmids, which are not found in mitochondria of all fungi, the mtDNA isolated by the procedure introduced above is a standard component of these organelles. The structure of this molecule can be characterized by different approaches. Usually, a physical map is constructed first, followed by a localization of individual genes on this map. Subsequently, the sequence of this DNA species is determined in part or completely.

Construction of a physical map—Restriction enzyme analysis. A physical map of the mtDNA can be constructed by restriction enzyme analysis. In these experiments mtDNA is digested with different restriction enzymes, either in single or double digests, and the resulting restriction fragments are fractionated by gel electrophoresis. Using defined DNA standards, fragment sizes are determined. From these data restriction maps of the type shown in Figure 4 (inner circles) can be deduced. A restriction analysis leads not only to the detection of recognition sites for certain restriction endonucleases but also to the identification of the structure, linear vs. circular, of the corresponding molecule (see also Fig. 7, below).

If a comparative analysis is performed, in which the mtDNA of different strains (e.g., *P. anserina* wild-type *A* and mutant *AL2*; Fig. 5) is analyzed, differences in the mtDNA can be directly identified on agarose gels. This type of approach can also be used to discriminate between different wild-type isolates (races) of the same species. It is known that the mtDNAs of rather closely related strains may clearly differ from each other.

Construction of a genetical map—Southern blot analysis. The localization of certain genes on a particular physical map is usually done by Southern blot hybridization using cloned mitochondrial genes of related species as specific probes.

Method

1. Digest mtDNA with restriction enzymes (parallel digestions with different enzymes are performed).
2. Fractionate the resulting DNA fragments by electrophoresis.

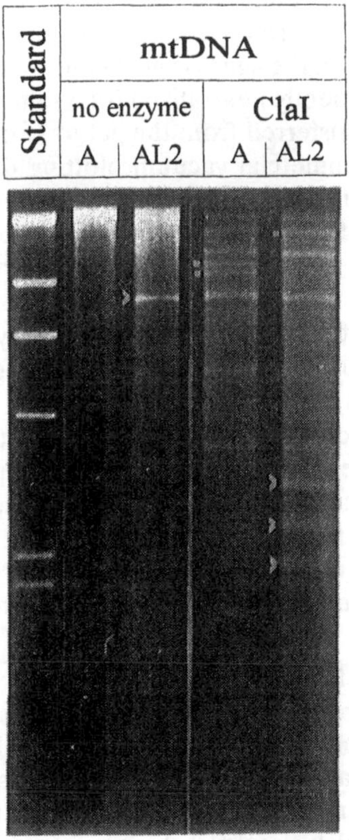

Figure 5 *Podospora anserina*, comparative restriction enzyme analyses of long-lived strain *AL2* and wild-type strain *A*. Undigested, total mitochondrial DNA or DNA digested with restriction enzyme *Cla*I was fractionated by gel electrophoresis. In the undigested DNA fraction of mutant *AL2* linear plasmid *pAL2-1* (white arrow) can be detected in addition to a diffuse band corresponding to sheared mtDNA. In both digested DNA fractions further differences may be clearly seen. Two DNA fragments in the DNA of wild-type *A* disappear, whereas a novel fragment migrates at about 17.8 kbp (white dots). Since the sum of the two fragments in the wild-type mtDNA is larger than 17.8 kbp, it can be concluded that a deletion of parts of the mtDNA led to the differences in the *Cla*I digestion patterns. Finally, a hybridization analyses revealed that the three *Cla*I fragments in the mtDNA fraction of mutant *AL2* which are indicated by white arrows arose from the digestion of the autonomous linear plasmid *pAL2-1*.

3. After staining with ethidium bromide (be careful, this is a powerful mutagen), the fragment pattern is photographed under UV light.

4. DNA fragments in the agarose gel are incubated for 5 min in denaturation buffer and subsequently neutralized for 5 min in neutralization buffer.

5. The DNA is transferred from the gel to a solid medium (e.g., nylon membranes) using a commercial vacuum blotting device (e.g., VacuGene, Pharmacia). SSC buffer is used to transfer the DNA.

6. After removal of excess buffer, the DNA is fixed to the membrane by UV crosslinking using a commercial available UV crosslinker (e.g., Stratalinker, Stratagene).

7. Hybridization of the DNA is performed in hybridization buffer containing a radiolabeled probe of a typical mitochondrial gene of another species. Hybridization conditions can be modified, depending on the conservation of the DNA probe and the corresponding gene of the mtDNA under analyses.

8. After removing excess amounts of the labeled probe and a brief drying of the filter, the moist filter is covered with Saran wrap followed by exposure to an x-ray film for a few hours to several days.

9. Finally, on the autoradiograph restriction DNA fragments hybridizing to a specific gene probe can be identified and located on the physical map (Fig. 4).

Cloning of mtDNA fragments. After digestion of mtDNA with a suitable restriction enzyme, the resulting DNA fragments can be ligated to a vector molecule digested with the same restriction endonuclease. Using an *E. coli* plasmid vector (e.g., a plasmid of the pUC series), a suitable host strain can be transformed with the resulting hybrid plasmids. Subsequently, individual *E. coli* transformants, which should each contain a different hybrid plasmid with a different mtDNA fragment, can be isolated. This procedure usually leads to the cloning of most mtDNA fragments. However, not all fragments may be cloned in this way. In fact, fragments of a larger size are usually not included in the different *E. coli* transformants. In this case, a different restriction enzyme may be used to digest the mtDNA. This procedure, which can be repeated with other enzymes, will most certainly lead to the cloning of overlapping mtDNA fragments which cover the complete mtDNA molecule. Alternatively, individual mtDNA fragments may be isolated from agarose gels, reisolated from the gel, and finally be ligated to a compatible vector.

Sequence analysis of mtDNA. After cloning of overlapping DNA fragments the nucleotide sequence of these fragments can be determined following one or different strategies by which DNA sequences can be elucidated. A subsequent analysis of the sequence results in the identification of the coding potential of the corresponding genome fraction. In addition, the sequence data can be used to identify mtDNA polymorphisms in different

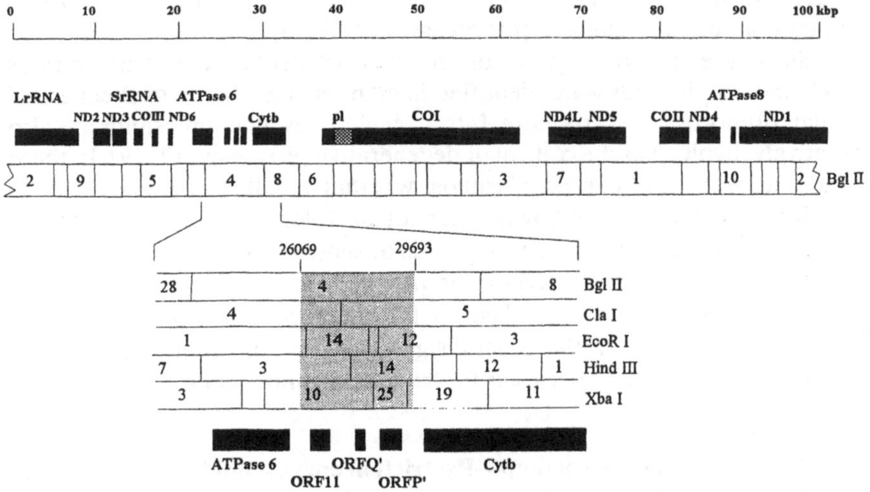

Figure 6 Localization of a deletion in long-lived mutant *AL2* of *P. anserina.* The exact position of the deletion between positions 26069 and in the mtDNA of the wild-type 29693 [55] is derived from a sequence analyses of the cloned, rearranged DNA fragment from the mtDNA of mutant *AL2.* In the upper part a linearized physical map of the mtDNA of *P. anserina* is indicated (*Bgl*II restriction map). Above this map the localization of individual genes and open reading frames (ORFs) is shown. The hatched region in the *COI* gene corresponds to the first intron of this gene, which gives rise to the formation of a circular plasmid (plDNA). In the lower part, a detailed map of the mtDNA, which is rearranged in mutant *AL2,* is shown for several restriction endonucleases. The sequence that is deleted in the mutant is indicated by the gray area.

strains of the same species or of different species. Using this approach it was possible to identify exactly the deleted DNA region in long-lived mutant *AL2* of *P. anserina* (Fig. 6).

Low-Molecular-Weight Mitochondrial DNA (Mitochondrial Plasmids)

Soon after the discovery of the 2-μm plasmid of the yeast *Saccharomyces cerevisiae*, a circular eukaryotic plasmid associated with the nucleus (see below), the first circular plasmid in mitochondria was identified in *Podospora anserina* [11–13]. This plasmid, which was termed plDNA (or α-SEN DNA), is one of a few plasmids that are associated with a particular phenotype—that is, senescence. In juvenile cultures of *P. anserina* it is the first intron of the mitochondrial gene coding for cytochrome oxidase subunit I (*COI*) (Fig. 4).

During senescence this intron becomes liberated and can be isolated as an autonomous, circular element [for reviews see 14–16].

Since the first description of this type of genetic elements, various mitochondrial plasmids were identified in many species. Some of them are of circular, others of linear structure. Interestingly, a few linear plasmids are also causatively involved in the control of degenerative processes [17–19]. In these cases it appears that certain plasmids, whether circular or linear, lead to instabilities of that mtDNA which cannot be tolerated since they lead to enzyme deficiencies and as a consequence to senescence.

Recent, systematic searches for plasmids revealed that this type of genetic element is a more general part of the fungal genome than was believed before [20–22]. However, the general significance of fungal plasmids, except for a few strains of a few species, is far from being understood.

Experiments

Structure and physical map—Restriction enzyme analysis. The DNA migrating at a certain position in a gel can be recovered using different methods. Among these, the binding of electrophoretically fractionated DNA molecules to different matrices and a subsequent recovery from melted gel pieces are currently the most frequently used procedure. The bound DNA is finally released from the corresponding matrix by incubation in low-salt buffers. This DNA can be used for further analyses—e.g., restriction analyses or ligation.

Method

1. 0.5 μg of DNA recovered from an agarose gel is incubated for 1–2 h in the appropriate buffer together with 1 unit of a particular restriction endonuclease. Parallel digests are performed using different endonucleases. Since restriction enzymes are needed for this analysis that cut the DNA only at one or a few positions, enzymes that recognize six base pairs (six cutters), e.g., *Eco*RI, *BamH*I, *Hind*III, are tested first.
2. After digestion, the incubation mixtures are loaded on individual lanes of an agarose gel, electrophoresed, stained, and visualized under UV light (see above).
3. From the restriction patterns the endonucleases are selected that give rise to only two DNA bands.
4. In a next set of experiments, DNA is incubated simultaneously with two of the selected enzymes.
5. After electrophoresis the restriction pattern is analyzed.

Results. As indicated in Figure 7, these simple experiments can discriminate between circular and linear DNA species. If the analyzed DNA is of linear structure, restriction enzymes which, in a single digest, give rise to

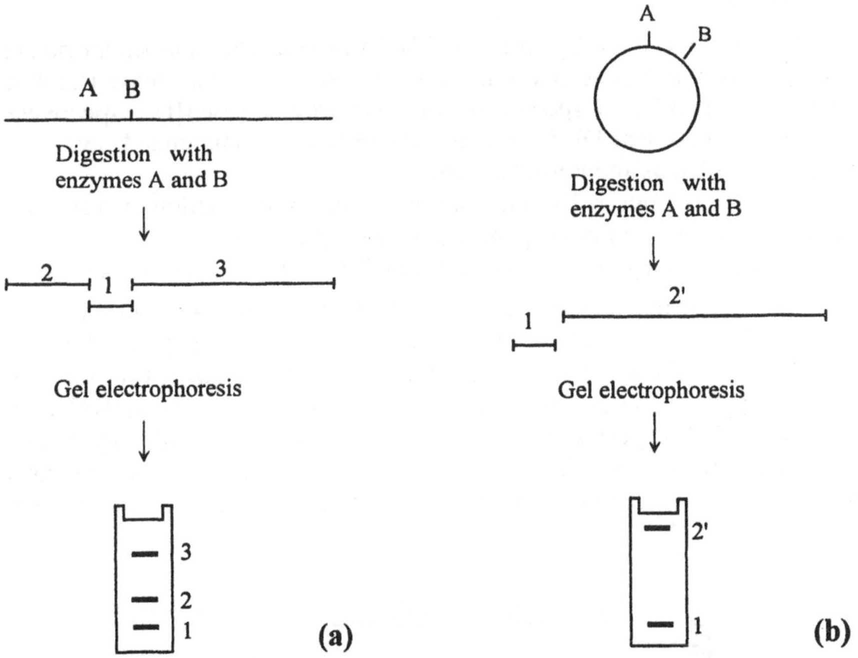

Figure 7 Discrimination of linear and circular DNA species by restriction analyses. An aliquot of the DNA under question is digested with two restriction enzymes which in initial experiments gave rise to a single DNA fragment. After gel electrophoresis of the double-digested DNA, three DNA fragments can be located if the uncut DNA species has a linear structure (a). Only two DNA fragments arise from a circular DNA species (b).

two fragments, will lead to three DNA fragments when a double digestion is performed. Ideally, two fragments will be the result of the double digestion of a circular molecule. It is important to verify the results of an experiment with a certain combination of enzymes with at least another combination, since the experiments may not always lead to clear results. This may be due to co-migrating bands or, more importantly, to small fragments that may run out of the gel.

Analyses of the ends of linear elements—Exonuclease digestion. The structure of a DNA can be verified by digestion of the isolated element with enzymes that degrade linear molecules from their 5' and 3' termini, respectively. In addition, protected DNA ends can be identified by this method.

Method

1. Four individual digests of total DNA isolated from mitochondria are performed with different amounts of 1.) a 5'-specific exonuclease (lambda exonuclease), and 2.) a 3'-specific exonuclease (exonuclease III), respectively. To each digest a linear DNA of a known size (e.g., a commercially available plasmid) is added as an internal control.

2. After an appropriate incubation, the different reaction mixtures are fractionated by conventional agarose gel electrophoresis.

3. After staining with ethidium bromide, the corresponding gel is analyzed.

Results. If the analyzed DNA molecule is of linear structure and not protected at its 5' and 3' ends, it will be completely digested by both types of exonucleases. In the example shown in Figure 8 this is the case for the internal control, that is, linearized plasmid pBR322, and the sheared mtDNA. In contrast, the DNA species migrating at 8.5 kbp (*pAL2-1*) is only digested by exonuclease III but not by lambda exonuclease. This DNA shares the typical characteristics of a number of known linear plasmids and of certain viruses

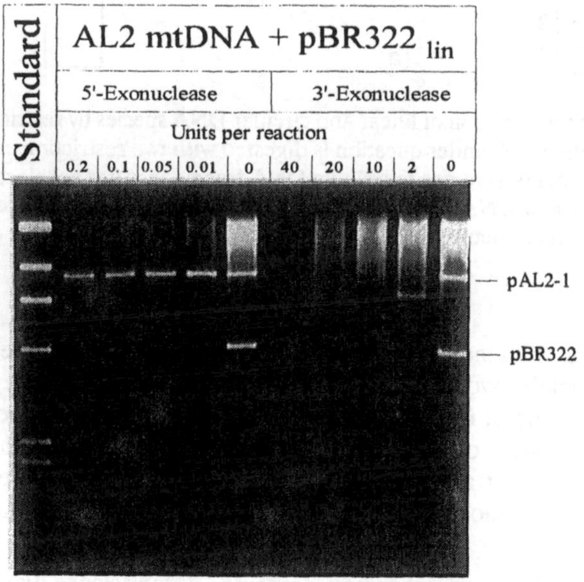

Figure 8 Exonuclease digestion of total mitochondrial DNA from *P. anserina* mutant *AL2* with a 5'-specific exonuclease (lambda exonuclease) and a 3'-specific exonuclease (exonuclease III). As an internal control linearized pBR322 DNA was added to the reaction mixture.

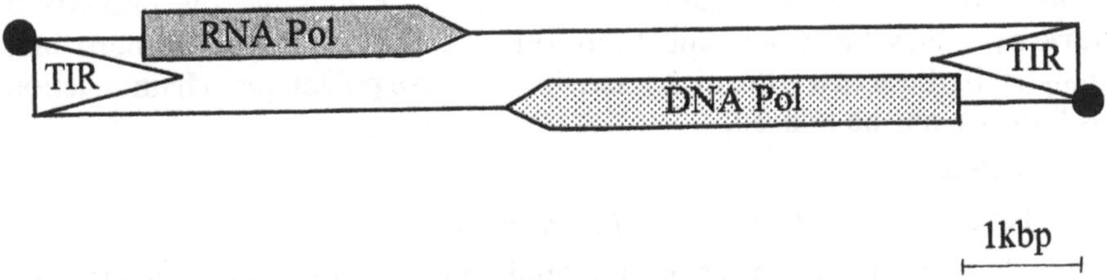

Figure 9 Structure of linear plasmid *pAL2-1* of *P. anserina* mutant *AL2*. This is an example indicating some characteristics of a linear plasmid which encodes for both an RNA polymerase (*RNA Pol*) and a DNA polymerase (*DNA Pol*). Terminal inverted repeats (TIR) are indicated by arrows on the ends of the plasmid. Black circles indicate 5'-bound proteins.

that contain proteins bound to their 5' ends. These proteins are important for the replication of these types of genetic elements and are responsible for the protection against 5'-specific exonucleases.

Sequence analysis of linear plasmids. The last step in the structural characterization of a linear plasmid is the determination of its nucleotide sequence. This can be performed after cloning overlapping plasmid fragments in the same way as explained for the mtDNA. However, the cloning of the terminal fragments of a linear plasmid with the protected 5' end is usually not easy. Although the 5'-bound protein can be digested with proteinase K, at least one or a few amino acids appear to remain linked to the ends. In a few cases it was possible to clone the terminal fragment of a linear plasmid after digestion of this fragment with calf intestinal phosphatase [e.g., 23].

After sequencing the linear plasmid, the nucleotide sequence is analyzed using different packages of computer software. This type of analysis leads to a fine structure of the genetic element under question and to the identification of regions with a potential function. Generally, linear plasmids are characterized by long terminal inverted repeated, open reading frames able to code for proteins with homology to viral type RNA and/or DNA polymerases (Fig. 9). In cases where a linear plasmid codes for only one type of polymerase, usually a second plasmid is present in the mitochondrium which codes for the second polymerase. However, although the sequence data suggest a coding function, this function needs a molecular confirmation (e.g., by transcript analyses).

Coding functions of linear plasmids—Transcript analyses. Specific transcripts can either be identified by conventional Northern blot analyses in which total RNA is fractionated by gelelectrophoresis and RNA blots are hybridized against specific DNA probes. The sensitivity of this method is limited. In cases like the linear plasmid *pAL2-1* of *P. anserina*, the amounts of transcripts may be to low and therefore not detectable by Northern blot analysis. In these cases, the highly sensitive reverse polymerase chain reaction (RT-PCR) may be utilized.

Method

Isolation of total RNA from *P. anserina*:

1. 4–6 g mycelium is frozen in liquid nitrogen and ground in a motor.
2. After transfer of the ground mycelium to a sterile beaker 40 mL of GTC/β-ME solution is added. This solution was preincubated to 60°C.
3. After incubation for 10 min at 60°C the mixture is homogenized in a Waring Blendor.
4. Cellular debris is separated by centrifugation of the homogenate in an SS35 rotor (Sorvall) at 10,000 rpm for 10 min.
5. The supernatant is centrifuged through a 5.7 M CsCl cushion (5.7 M CsCl, 0.1 M EDTA, pH 7.4) using a TST41.14 (Kontron) rotor. Centrifugation is performed for 18 h at 34,000 rpm at 20°C.
6. After careful removal of the supernatant, the RNA pellet is washed three times with 200 μL 70% ethanol.
7. The pellet is redissolved in 400 μL DEPC-treated water.
8. The RNA is precipitated with ethanol by adding 1/10 volumes of 3 M sodium acetate, pH 5.0, and 3 volumes ethanol.
9. After 30 min of incubation at –70°C the RNA is recovered by centrifugation for 15 min in a microcentrifuge.
10. The RNA pellet is dried and redissolved in DEPC-treated water.

Reverse polymerase chain reaction:

1. 20 μg of RNA is incubated for 1 h at 37°C with 100 units of RNase-free DNase I in 0.1 M sodium acetate, 5 mM $MgSO_4$, pH 5.0.
2. Extraction of the mixture with 1 volume phenol is followed by a phenol/chloroform and 1 chloroform extraction.
3. Precipitate RNA with ethanol.
4. Redissolve the RNA pellet in DEPC-treated water.
5. Anneal a suitable primer combination to 2 μg RNA for 10 min.
6. Perform first-strand cDNA synthesis using 200 units of MMLV reverse transcriptase for 60 min at 37°C in 50 mM Tris-Cl, pH 8.3, 75 mM KCl, 3 mM $MgCl_2$, 10 mM DTT, 0.5 mM dNTP.
7. Amplify cDNA using a specific primer combination and *Taq* DNA polymerase. The reaction is carried out in a thermocycler under the following conditions: 35 cycles (1 min, 58°C; 1–3 min, 72°C; 45 sec, 93°C).
8. Analysis of amplification products on a conventional agarose gel.

Results. Amplification products obtained by this method indicate that regions of the genetic element (e.g., open reading frames) are transcribed at a low level. However, since this method is highly sensitive and, due to contaminations, may easily produce artifacts, it is important to work extremely carefully and to use appropriate controls.

B. Plasmids in the Nuclear/Cytoplasmic Compartment

In this section only two plasmid systems will be introduced: one plasmid because it was the first eukaryotic plasmid to be discovered, and the other system because it is one of the few systems clearly associated with a phenotype. Note the remarks in the introduction which deal with the localization of plasmids, either in the nucleus or in the cytoplasm. For further reference on plasmids, their occurrence, their characteristics and their functions the interested reader is referred to [24,25].

The 2 μm Plasmid of Saccharomyces cerevisiae

The first eukaryotic plasmid was identified in the yeast *Saccharomyces cerevisiae* by electron microscopy [26]. This plasmid has a circular structure. Due to its contour length it was termed 2 μm plasmid. The 2 μm plasmid is found in most laboratory strains and in about 80% of all wild isolates of *S. cerevisiae*. In most strains the copy number is about 60 molecules per cell [27]. Although the plasmid shows a non-Mendelian segregation pattern, it is located in the nucleus [28]. More interestingly, the 2 μm plasmid is packaged into nucleosomes, thus, like eukaryotic chromosomes, showing a chromatin like structure. The replication of the 2 μm DNA is similar to the replication of the chromosomal DNA. It takes place during the S phase of a cell cycle. Replication starts at one specific origin of replication.

Although the plasmid and different yeast strains were thoroughly analyzed, no particular phenotype could be attributed to the 2 μm DNA. The different coding regions identified in the sequence of the plasmid appear to be involved in the control of the copy number and in the maintenance of the plasmid [29,30]. However, the identification of the 2 μm plasmid in S. *cerevisiae* was a very important step forward to establish a molecular engineering technology in general and, in particular, for this eukaryotic microorganism. Today *S. cerevisiae* is the best-analyzed eukaryote.

Killer Plasmids in Kluyveromyces lactis

In *Kluyveromyces lactis* plasmids were identified that are linear elements [31]. Like those linear plasmids that have been introduced as part of the mitochondrial genome of many fungi, the *K. lactis* plasmids contain different reading frames, two of which code for a RNA polymerase and a DNA polymerase.

They contain long inverted terminal repeats and proteins bound to their 5' ends. Again, this basic structure resembles the structure of certain DNA viruses (e.g., adenovirus).

In contrast to the 2 μm plasmid of *S. cerevisiae*, the *K. lactis* plasmids, termed pGKL1 and pGKL2, are located in the cytoplasm and not in the nucleus [32]. The two plasmids are associated with a clear phenotype. Plasmid-containing strains are able to kill a number of different yeast strains. This phenotype, and the resistance of the corresponding killer strains to their own toxin, is encoded by the smaller plasmids, pGKL1. The toxin is a polypeptide consisting of three subunits. It becomes secreted from the producer cell and has a cytostatic effect on sensitive cells that become arrested in the G1 phase of the cell cycle [33]. The larger plasmid, pGKL2, may be viewed as a helper plasmid.Without the presence of this plasmid, pGKL1 is not stabily maintained in a cell.

C. Viruslike Elements

In the preceding part of this chapter the structural similarity of linear plasmids with the genome of certain DNA viruses has been stressed. Moreover, in fungi a large number of viruses were reported [for a review, see 34]. Generally these elements were identified as protein particles, termed viruslike particles (VLPs). They mainly contain a double-stranded RNA as a genetic material. Although in most cases the infectiousness of these elements is not proven, most VIPs are generally viewed as mycoviruses. However, many of these elements may in fact represent defective viruses, and therefore may be termed viruslike elements. Moreover, in addition to VLPs dsRNA molecules which were never found to be encapsidated by proteins were identified in certain fungi. These elements may represent a further step of reduction. Interestingly, some of the encapsidated as well as the naked dsRNA species appear to be correlated with a particular phenotype. As one example a killer phenotype was demonstrated to be encoded by dsRNAs in several yeasts and in the basidiomycete *Ustilago maydis* [for reference see 35]. Cells containing the dsRNA virus synthesize a secreted protein that can kill other cells of the same or different species that lack the corresponding genetic element. The toxin-producing cells themselves are immune to the toxin [for review see 34,36,37]. Remarkably, in *Kluyveromyces lactis* a killer activity was demonstrated to be encoded by linear plasmids (see above).

Another interesting phenotype associated with the presence of a mycovirus is hypovirulence in *Cryphonectria parasitica*. This ascomycete is the causative agent of chestnut blight, a severe disease that nearly eliminated American chestnut trees. From a chestnut tree that was recovering from this disease a *Cryphonectria* strain with a reduced virulence was isolated. Interestingly, this strain contains a mycovirus with dsRNA which appears to be responsible for a hypovirulent phenotype. Moreover, it was shown that hypovirulent strains are able to convert resident virulent strains to hypovirulence and result in healing of a *Cryphonectria*-induced plant cancers. Consequently, it was suggested to use the corresponding virus in the biological control of the disease [for a recent review see 38].

In the following section only selected techniques are introduced which are used to demonstrate the presence of VLPs in a particular strain and to determine the chemical structure of the nucleic acid in these strains. Techniques for extraction of total nucleic acids—DNA and RNA—are not mentioned in detail, since examples of the many different procedures have been introduced in the preceding part of this chapter.

Experiments

After total nucleic acid is isolated from a particular strain, this DNA may be subjected to agarose or polyacrylamide electrophoresis. In cases in which distinctive bands can be localized in a gel, the chemical nature of the nucleic acid, that is RNA or DNA, can be identified by incubation of aliquots with different RNases or DNase I. If the nucleic acid that gives rise to a distinctive band is an RNA, the corresponding band will no longer appear in a gel after digestion with RNase. In this case, in which good evidence about the RNA structure of a given extrachromosomal nucleic acids was obtained, the following procedure may result in the purification of viruslike particles.

Isolation of viruslike particles (procedure according to [39])

1. Disruption of cells (about 10 g wet weight) in 50 mL of phosphate buffer (100 mM sodium phosphate pH 7.5, 10 mM $MgCl_2$) by vortexting with class beads.
2. Removal of cellular debris by centrifugation at 10,000 g for 10 min at 4°C.
3. Recover the supernatant in 9% polyethylene glycol 8000, 2.5% NaCl, and incubate over night at 4°C after gentle agitation.
4. Centrifugation for 15 min at 10,000 g.
5. Recovery of the pellet in phosphate buffer.
6. Centrifugation for 3 h at 10,000 g.
7. Recovery of the pellet in a minimal amount of phosphate buffer.
8. The redissolved material is layered onto a 10–50% linear sucrose gradient and centrifuged at 30,000 rpm for 90 min in a swinging bucket rotor (RPS 40, Hitachi).
9. Collection of different pools from the gradient after measuring the UV absorbance at 260 nm.
10. Pools of gradient fractions with UV absorbance peaks are dialyzed against phosphate buffer.

11. Viruslike particles are pelleted by centrifugation at 35,000 rpm for 4 h in a representative 40 rotor (Hitachi).

Electron microscopy (procedure according to [39])

1. Staining of a suspension of presumptive VLPs with 2% potassium phosphotungstate, pH 7.0, on Formvar and carbon-coated 200-mesh copper grids.

2. Microscopic observation of the sample in an electron microscope.

Results. Following the outlined strategy, VLPs and dsRNA molecules may be identified in different fungi. Since not all elements are known to be related to a particular phenotype, VLPs containing dsRNAs or naked dsRNA species may be more uniquely found in fungi than can be expected now.

4. GENETIC ELEMENTS IN THE NUCLEAR GENOME

A. Transposable Elements

Genetic elements which, due to their mobility, cause unstable mutations have been known for a long time and have been studied in detail mainly in bacteria, plants, nematodes, and in *Drosophila*. In fungi, a number of transposable elements were identified in *Saccharomyces cerevisiae* (*Ty*: transposon yeast), and these are the elements that were studied most thoroughly in this group of eukaryotic microorganisms (Table 2). Interestingly, the *Ty* elements are members of the so-called retrotransposons, a group of mobile elements that transpose via an RNA intermediate. *Ty* elements were identified in different

Table 2 Selection of fungal transposable elements

Species	Element	References
Myxomycetes:		
Dictyostelium discoideum	*DIRS-1*	[42]
	Tdd-3	[43]
Physarum polycephalum	*Tp2*	[44]
Yeasts:		
Saccharomyces cerevisiae	*Ty*	[45]
Filamentous fungi:		
Cladosporium fulvum	*CfT-1*	[46]
Fusarium oxysporum	*Fot1*	[47]
	Foret1	[48]
Neurospora crassa	*Tad*	[49,50]
Podospora anserina	*repa*	[51]

yeast strains and are present in 35–50 copies. The length of these elements is about 5.9 kbp and consists of a unique region flanked by direct repeats of about 340 bp. At their integration site *Ty* elements are flanked by short, 5-bp direct repeats. *Ty* elements lead to the synthesis of two abundant *Ty*-specific polyadenylated RNAs. From these RNAs different proteins are translated. These proteins are assembled to viruslike particles (VLPs) that contain a single-stranded RNA (ssRNA) and a reverse transcriptase. Most likely these VLPs represent units that are transposition intermediates. According to these characteristics, the *Ty* retrotransposons of yeast resemble a group of animal viruses, the retroviruses, which also propagate via an ssRNA intermediate [for review see 40,41].

In yeast and other systems it was demonstrated that transposable elements are involved in genome reorganization and, in many cases, in gene expression. Rather recently the analysis of transposable elements gained further interest since additional elements were identified in different fungal species. Some of these species are of high significance since they are known to cause specific diseases in animals and plants. It may well be that transposable elements are determinants of strain specificity of the various fungal strains with a pathogenic potential. In addition, other phenotypes, such as unstable spore color mutations in *Ascobolus immersus* or degenerative processes like senescence, are considered to be, at least in part, the consequence of the presence and/or the activity of transposable sequences.

Even the presence of inactive sequences, sequences that no longer move from one position to another, may lead to genomic instabilities due to their presence in multiple copies. These characteristics of transposons led to the development of different strategies which, in a few cases, were successfully used to identify this type of genetic elements. In the following section, two approaches are introduced that appear to be applicable to a wide variety of different fungi. These approaches are based on 1.) the ability of transposons to inactivate specific genes due to their integration (transposon trapping), and 2.) the presence of multiple copies of active and inactive transposon sequences in the genome of fungi. Since many fungi are characterized by rather low amounts of repetitive DNA and, in addition, most of the repetitive DNA is represented by the rDNA fraction, the isolation of non-rDNA repetitive DNA will most likely result in the identification of transposon-associated sequences.

Experiments

Transposon trapping. This approach relies on the inactivation of a selectable marker gene by the integration of an active transposon into the gene sequence. The selectable marker gene can be either a cloned endogenous gene or a cloned heterologous gene introduced into the fungal genome. Both approaches were successfully used to isolate fungal transposons [e.g., 47,49]. Since in many cases no suitable, cloned genes are available for a given fungus, the second approach is outlined below.

Method

1. Introduce a specific marker gene into the fungal strain of interest (e.g., the *niaD* gene coding for nitrate reductase, mutation of which leads to chlorate resistance; the hygromycin phosphotransferase gene (*hph*), which leads to hygromycin B resistance). This can be achieved via integrative transformation of protoplasts of specific strains. Using the *niaD* gene, it is important to first select a stable *niaD* mutant in which the endogenous gene is inactivated. A procedure for this has been described [52]. Choosing the *hph* gene it is important to use a transforming plasmid which contains a fungal promoter sequence which allows for expression of this prokaryotic gene in the used fungus. In addition, it is a prerequisite that the strain under analysis is naturally sensitive to hygromycin B.
2. Selection of mutants of strains (e.g., strains sensitive to chlorate resistance).
3. Isolation of DNA from individual mutants.
4. Southern blot analysis using the cloned selection gene (e.g., *niaD*) as a probe.
5. Selection of mutants that contain an insertion sequence in the selectable marker gene.
6. Cloning of the corresponding insertion sequence and sequence analysis.

Identification of transposons in repetitive sequences. Sequences that may be remnants of transposons have been identified in a few filamentous fungi [48,51,53]. This was either by accident or by following a specific strategy.

1. After digestion of total fungal DNA and fractionation of the resulting fragments by agarose gel electrophoresis, DNA fragments are identified that correspond to the rDNA repetitive DNA fraction. In many fungi this is the predominant repetitive DNA. The identification of the rDNA-containing fragments is performed by Southern blot hybridization using the cloned homologous or heterologous rDNA as a probe.
2. DNA fragments which, in contrast to single fragments giving rise to a background smear, occur as distinctive bands in an ethidium bromide–stained gel, are recovered from the gel and cloned in a suitable vector. These fragments should contain repetitive DNA sequences which, due to their abundance, give rise to distinctive bands.
3. The cloned DNA fragments are recovered and used to verify whether they correspond to repetitive sequences in the genome of the analyzed fungus.

4. Finally, DNA fragments that give rise to hybridization of multiple fragments are sequenced and the sequence is compared to known transposable element.

Results. Using these strategies it is possible to isolate sequences that are active as transposons or are derived from this type of elements. These sequences may now be used to analyze the number of elements present in the strain of interest or in other closely related strains. In addition, it may be investigated whether the corresponding element is still active (e.g., by the selection and analysis of revertants) and whether it is transposing at particular stages during development. At this point it should be stressed that the first procedure—that is, transposon trapping—is a procedure that results in the isolation of active transposons. The second procedure may result in the identification of active and inactive transposons or even in the identification of other types of repetitive sequences. Among the group of inactive transposons it may be possible that, as in other systems (e.g., plants), elements may be identified that are only transposing if a transacting factor (transposase) is provided by a second active element. On the other hand, inactive transposable sequences may be remnants of formerly active elements.

5. EXTENSIONS

1. The identification of low-molecular-weight DNA species may result from the analysis of total DNA, DNA that contain the nuclear–cytoplasmic DNA fraction as well as DNA from mitochondria. The analysis of this DNA may lead to the identification of plasmids associated with the nucleus or the cytoplasm.

2. Sometimes distinctive extrachromosomal genetic traits may be identified as additional DNA bands in CsCl density gradients which contain ethidium bromide to visualize DNA bands under UV light. In this way the first circular mitochondrial plasmid was identified in DNA preparations from senescent cultures of the ascomycte *Podospora anserina* [11].

3. The structure of a genetic element may be analyzed by electron microscopy. In cases in which the element is small enough and does not become broken artificially during the DNA preparation procedure, this may indeed be a very sensitive and efficient method [11].

4. Genetic traits that are inherited according to Mendelian rules and lead to the "killing" of ascospores under certain genetically defined conditions were described for certain *Neurospora* and *Podospora* strains. This "spore killer" phenotype resembles the "killer" phenotype specified by certain linear plasmids or dsRNAs in different yeasts and basidiomycetes. However, spore killing

appears not to be encoded by extrachromosomal genetic traits but clearly by different, genetically localized chromosomal genes [for a reference see 54].

APPENDIX

In this appendix media and buffers needed for the procedures introduced in the preceding parts of this chapter are listed. Only material not directly mentioned in the corresponding procedure is indicated.

Media

BMM. 1.5 g malt extract and 20 g agar in 1 L cornmeal extract. Cornmeal extract is obtained from 250 g cornmeal incubated in 10 L water at 60°C overnight. After this time the supernatant is filtered through several layers of cheesecloth, and the cornmeal is discarded.

CM medium. 0.15% KH_2PO_4, 0.05% KCl, 0.05% $MgSO_4$, 1% D-glucose, 0.37% NH_4Cl, 0.2% Pepton, 0.2% yeast extract, 1 mg/L $ZnSO_4$, 1 mg/L $FeCl_2$.

Buffers

Denaturation buffer: 1.5 M NaCl, 0.5 M NaOH

Hybridization buffer: 50% Formamide (stringent hybridization), 5× SSPE, 0.5% sodium dodecyl sulfate (SDS), 0.1 mg/mL salmon sperm DNA. (The stringency of hybridization may be reduced by adding 30–50% of formamide).

Mitochondria buffer: 0.05 M Tris/Cl, 0.01 M EDTA, 0.5 M sucrose, pH 8.3

Mitochondria lysis buffer: 1% SDS, 0.05 M EDTA, 0.02 M sodium acetate, pH 5.0; autoclaved

Neutralization buffer: 2 M NaCl, 1 M Tris/Cl, pH 5.5

20X SSC: 1 L contains 175.3 g NaCl, 88.2 g sodium citrate, pH 7.0 (adjusted with 10 N NaOH)

20X SSPE: 1 L contains 174 g NaCl, 27.6 g $NaH_2PO_4 \times H_2O$, 7.4 g EDTA, pH adjusted to 7.4 with 10 N NaCl

TE: 10 mM Tris/Cl, 1 mM EDTA, pH 8.0

TES: 30 mM Tris/Cl, 5 mM EDTA, 50 mM NaCl, pH 8.0

TES/SDS: 30 mM Tris/Cl, 5 mM EDTA, 50 mM NaCl, 4% SDS, pH 8.0

TES/CsCl: Add 1.1 g CsCl per mL TES and adjust refraction index to 1.3985

TAE *buffer (20×)*: 400 mM sodium acetate, 800 mM Tris/Cl, 40 mM EDTA, pH 8.3 adjusted with acetic acid

GTC/βME *buffer*: 5.5 M Guanidium isothiocyanate, 0.5% sarcosyl, 25 mM sodium citrate, 0.1 M β-mercaptoethanol, pH 7.0

RNA CsCl: 5.7 M CsCl, 0.1 M EDTA, pH 7.4

REFERENCES

1. Lederberg, J. (1952). Cell genetics and hereditary symbiosis. *Physiol. Rev. 32*:403.
2. Esser, K. (1982). *Cryptogames*. University Press, Cambridge.
3. Slonimski, P. P., B. Ephrussi (1949). Action de l'acriflavine sur les levures. V. Le systeme de cytochromes des mutants 'petite colonie'. *Ann. Inst. Pasteur Paris 77*: 419.
4. Osiewacz, H. D., J. Hermanns, D. Marcou, M. Triffi, K. Esser (1989). Mitochondrial DNA rearrangements are correlated with a delayed amplification of the mobile intron (plDNA) in a long-lived mutant of *Podospora anserina. Mutat. Res. 279*:9.
5. Rogers, H. J., K. W. Buck, C. M. Brasier (1987). A mitochondrial target for double-stranded RNA in diseased isolates of the fungus that causes Dutch elm disease. *Nature 129*:558.
6. Wesolowski, M., H. Fukuhara (1981). Linear mitochondrial desoxyribonucleic acid from the yeast *Hansenula mrakii. Mol. Cell. Biol. 1*:387.
7. Kovacs, L., J. Lazowska, P. P. Slonimski (1984). A yeast with linear molecules of mitochondrial DNA. *Mol. Gen. Genet. 197*:420.
8. Zimmer, M., G. Lückemann, B. F. Lang, K. Wolf (1984). The mitochondrial genome of fission yeast *Schizosaccharomyces pombe.* 3. Gene mapping in strain *EF1* (CBS 356) and analysis of hybrids between strains *EF1* and ade 7–50 h$^-$. *Mol. Gen. Genet. 196*:473.
9. Hintz, W. E., M. Mohan, J. B. Anderson, P. A. Horgen (1985). The mitochondrial DNA of *Agaricus*: heterogeneity in A. bitorquis and homogeneity in *A. brunnescens. Curr. Genet. 9*:127.
10. Hermanns, J., H. D. Osiewacz (1994). Three mitochondrial unassigned open reading frames of *Podospora anserina* represent remnants of a viral-type RNA polymerase gene. *Curr. Genet. 25*:150.
11. Stahl, U., P. A. Lemke, P. Tudzynski, U. Kück, K. Esser (1978). Evidence for plasmid like DNA in a filamentous fungi, the ascomycete *Podospora anserina. Mol. Gen. Genet. 162*:341.
12. Stahl, U., U. Kück, P. Tudzynski, K. Esser (1980). Characterization and cloning of plasmid like DNA of the ascomycete *Podospora anserina. Mol. Gen. Genet. 178*: 369.
13. Cummings, D. J., L. Belcour, C. Grandchamps (1979). Mitochondrial DNA from *Podospora anserina*. II. Properties of mutant DNA and multimeric circular DNA from senescent cultures. *Mol. Gen. Genet. 171*:239.

14. Esser, K. (1985). Genetic control of aging. The mobile intron model. In: Bergener, M., M. Ermini, H. B. Stähelin (eds.): *The 1984 Sandoz Lectures in Gerontology, Thresholds in Aging.* Academic Press, London, p. 3.
15. Osiewacz, H. D. (1990). Molecular analysis of ageing processes in fungi. *Mutat. Res. 237*:1.
16. Osiewacz, H. D. (1992). The genetic control of aging in the ascomycete *Podospora anserina.* In: Zwilling, R., C. Balduini (eds.): *Biology of Aging.* Springer-Verlag, Berlin, Heidelberg, p. 153.
17. Chan, B. S.-S., D. A. Court, P. J. Vierula, H. Bertrand (1991). The *kalilo* linear senescence-inducing plasmid of *Neurospora* is an invertron and encodes DNA and RNA polymerases. *Curr. Genet. 20*:225.
18. Court, D. A., A. J. K. Griffiths, S. R. Krauss, P. J. Russel, H. Bertrand (1991). A new senescence-inducing mitochondrial linear plasmid in field-isolated *Neurospora crassa* strains from India. *Curr. Genet. 19*:129.
19. Hermanns, J., A. Asseburg, H. D. Osiewacz (1994). Evidence for a life span–prolonging effect of a linear plasmid in a longevity mutant of *Podospora anserina. Mol. Gen. Genet. 243*:297.
20. Yang, X., A. Griffiths (1993). Plasmid diversity in senescent and nonsenescent strains of *Neurospora. Mol. Gen. Genet. 237*:177.
21. Arganoza, M. T., J. Min, Z. Hu, R. A. Akins (1994). Distribution of seven homology groups of mitochondrial plasmids in *Neurospora*: evidence for widespread mobility between species in nature. *Curr. Genet. 26*:62.
22. Hermanns, J., A. J. M. Debets, R. F. Hoekstra, H. D. Osiewacz (1995). A novel family of linear plasmids with homology to pAL2-1 of *Podospora anserina*. *Mol. Gen. Genet. 247*:638.
23. Hermanns, J., H. D. Osiewacz (1992). The linear mitochondrial plasmid pAL2-1 of a long-lived *Podospora anserina* mutant is an invertron encoding a DNA and RNA polymerase. *Curr. Genet. 22*:491.
24. Esser, K., U. Kück, C. Lang-Hinrichs, et al. (1986). *Plasmids of Eukaryotes.* Springer-Verlag, Berlin, Heidelberg.
25. Meinhardt, F., F. Kempken, J. Kämper, K. Esser (1990). Linear plasmids among eukaryotes: fundamentals and applications. *Curr. Genet. 17*:89.
26. Sinclair, J. J., B. J. Stevens, O. Sanghavi, M. Rabinowitz (1967). Mitochondrial, satellite and circular DNA filaments in yeast. *Science 156*:1234.
27. Clark-Walker, G. D., L. G. Miklos (1974). Localization and quantification of circular DNA in yeast. *Eur. J. Biochem. 41*:359.
28. Livingston, D. M. (1977). Inheritance of 2 μm DNA plasmid from *Saccharomyces. Genetics 86*:73.
29. Mead, D. J., D. C. J. Gardner, S. G. Oliver (1986). The yeast 2 μ plasmid: strategies for the survival of selfish DNA. *Mol. Gen. Genet. 205*:417.
30. Futcher, A. B., B. Reid, D. A. Hichey (1988). Maintenance of the 2 μm circle plas- mid of *S. cerevisiae* by sexual transmission: an example of selfish DNA. *Genetics 118*:411.
31. Gunge, N., A. Tamaru, F. Ozawa, K. Sakaguchi (1981). Isolation and characterization of linear deoxyribonucleic acid plasmids from *Kluyveromyces lactis* and the killer[associated killer character. *J. Bacteriol. 145*:382.

32. Gunge, N., K. Murata, K. Sakaguchi (1982). Transformation of *Saccharomyces cerevisiae* with linear DNA killer plasmids from *Kluyveromyces lactis*. *J. Bacteriol.* *151*:462.
33. Sugisaki, Y., N. Gunge, K. Sakaguchi, M. Yamasaki, G. Tamura (1983). *Kluyveromyces lactis* killer toxin inhibits adenylate cyclase sensitive yeast cells. *Nature 304*:464.
34. Buck, K. W. (1986). Fungal virology—an overview. In Buck, K. W. (ed.): *Fungal Virology.* CRC Press, Boca Raton, Fla., p. 1.
35. Brown, G. G., P. M. Finnegan (1989). RNA plasmids. *Int. Rev. Cytol. 117*:1.
36. Bruenn, J. (1986). The killer systems of *Saccharomyces cerevisiae* and other yeasts. In Buck, K. W. (ed.): *Fungal Virology.* CRC Press, Boca Raton, Fla., p. 85.
37. Koltin, Y. (1986). The killer systems of *Ustilago maydis*. In Buck, K. W. (ed.): *Fungal Virology*. CRC Press, Boca Raton, Fla., p. 109.
38. Nuss, D. (1992). Biological control of chestnut blight: an example of virus-mediated attenuation of fungal pathogenesis. *Microbiol. Rev. 56*:561.
39. Castillo, A., V. Cifuentes (1994). Presence of double-stranded RNA and virus-like particles in *Phaffia rhodozyma. Curr. Genet. 26*:364.
40. Finnegan, D. J. (1985). Transposable elements in eukaryotes. *Int. Rev. Cytol. 93*: 281.
41. Kingsman, A. J., J. Mellor, S. Adams, et al. (1987). The genetic organization of the yeast *Ty* element. *J. Cell Sci. Suppl. 7*:153.
42. Cappello, J., K. Handelmann, H. F. Lodish (1985). Sequence of *Dictyostelium DIRS-1*: an apparent retrotransposon with inverted terminal repeats and an internal circle junction sequence. *Cell 43*:105.
43. Marschalek, R., G. Borschet, T. Dingermann (1990). Genomic organization of the transposable element *Tdd-3* from *Dictyostelium discoideum. Nucl. Acids Res. 18*: 5753.
44. McCurrach, K. J., H. M. Rothinie, N. Hardtman, L. A. Glover (1990). Identification of a second retrotransposon-related element in the genome of *Physarum polycephalum. Curr. Genet. 17*:403.
45. Cameron, J. R., E. Y. Loh, R. W. Davis (1979). Evidence for transposition of dispersed repetitive DNA families in yeast. *Cell 16*:739.
46. McHale, M. T., I. N. Roberts, S. M. Noble, et al. (1992). *CfT-1*: an LTR-retroposon in *Cladosporium fulvum*, a fungal pathogen of tomato. *Mol. Gen. Genet. 233*:337.
47. Daboussi, M.-J., T. Langin, Y. Brygoo (1992). *Fot1*, a new family of fungal transposable elements. *Mol. Gen. Genet. 232*:1233.
48. Julien, J., S. Poirier-Hamon, Y. Brygoo (1992). *Foret1*, a reverse transcriptase-like sequence in the filamentous fungus *Fusarium oxysporum. Nucl. Acids Res. 20*: 3933–3937.
49. Kinsey, J. A., J. Helber (1989). Isolation of a transposable element from *Neurospora crassa. Proc. Natl. Acad. Sci. USA 86*:1929.
50. Cambareri, E. B., J. Helber, J. A. Kinsey (1994). *Tad1-1*, an active *LINE*-like element of *Neurospora crassa. Mol. Gen. Genet. 242*:658.
51. Deleu, C., B. Turcq, J. Begueret (1990). *repa*, a repetitive and dispersed DNA sequence of the filamentous fungus *Podospora anserina. Nucl. Acids Res. 18*:4901.

52. Cove, D. J. (1976). Chlorate toxicity in *Aspergillus nidulans*: the selection and characterization of chlorate resistant mutants. *Heredity 36*:191.
53. Romao, J., J. E. Hamer (1992). Genetic organization of a repeated DNA sequence family in the rice blast fungus. *Proc. Natl. Acad. Sci. USA 89*:5316.
54. Raju, N. B., D. D. Perkins (1991). Expression of meiotic drive elements spore *killer-2* and spore *killer-3* in asci of *Neurospora tetrasperma. Genetics 129*:25.
55. Cummings, D. J., K. L. McNally, J. M. Domenico, E. T. Matsuura (1990). The complete DNA sequence of the mitochondrial genome of *Podospora anserina. Curr. Genet. 17*:375.

8

Genetic Transformation and Vector Developments in Filamentous Fungi

Maureen B. R. Riach* and James Robertson Kinghorn
University of St. Andrews, St. Andrews, Fife, Scotland

1. INTRODUCTION

Recombinant DNA technology, which has provided techniques for the routine isolation and analysis of genes from almost any organism, was originally successfully employed during the 1970s in prokaryotes, most notably *Escherichia coli*, and then in eukaryotes with the baker's yeast *Saccharomyces cerevisiae*. Genetic engineering procedures for the filamentous fungi followed thereafter in the 1980s and were originally developed in the genetically intensively studied ascomycetous fungi, first in *Neurospora crassa* and then a little later in *Aspergillus nidulans*, before being extended to less-tractable fungal species including plant and animal pathogens as well as industrial fungi that produce a variety of molecules of interest to man.

Essential for the advancement of molecular genetic research in filamentous fungi was the development of gene transfer (genetic transformation) systems for the introduction of exogenous DNA into fungal cells, usually achieved by means of a vector system which contains a selectable marker and permits the selection of cells that have been successfully transformed. Since this research first began in the 1980s, genetic transformation systems have been developed

**Current affiliation*: Blackwell Science Ltd., Edinburgh, Scotland

for a number of fungi, as have numerous transformation/cloning vectors containing a variety of selection markers. As a number of detailed reviews have been published on fungal transformation and fungal recombinant DNA procedures [for a selection see 1–9], this review will attempt to avoid reiterating much of the subject matter covered elsewhere and will instead attempt to emphasize recent transformation and vector developments in the filamentous fungi.

Some key concepts in genetic transformation and vector developments in filamentous fungi are as follows:

1. A number of methods have been used for the transformation of filamentous fungi, with varying degrees of success.

2. Many transformation selection systems are now available, and the strategy selected is largely strain dependent.

3. The successful development of efficient transformation systems for fungi has led to the development of attractive alternatives to the standard methods for the cloning of fungal genes.

4. In filamentous fungi, the transforming DNA most commonly becomes integrated into the host genome by recombination; however, autonomously replicating vectors have also been developed.

5. An *Aspergillis nidulans* sequence (AMA1), responsible for autonomous replication, has been isolated and used not only to successfully increase transformation frequencies in several filamentous fungi but also to synthesize "instant gene banks" when cotransformed with genomic DNA into an organism.

6. Fungal telomeric sequences have been isolated and used to create linear autonomously replicating vectors for fungal transformation.

7. Gene replacement and gene disruption events, obtained by homologous transformation, can be used to introduce desired or null mutations, respectively, into the recipient fungal cell.

8. A number of specialized vectors have been developed for the analysis of fungal genes and their products, including those designed for gene promoter analysis, fungal protein expression, and secretion of heterologous proteins from fungal cells.

9. Fungal transposons, which have only relatively recently been shown to exist, may be useful for the genetic manipulation of fungal species.

10. Once obtained, fungal transformants can be genetically characterized in a number of ways.

11. With both the increase in availability of transformation and selection systems for the genetically well characterized fungi and the use of molecular genetic techniques to study genetically unexplored fungi, numerous fungal genes will be isolated and characterized, thereby providing detailed information on the organization of a fungal genome at the molecular level.

2. TRANSFORMATION PROCEDURES

Genetic transformation of filamentous fungi, as with any organism, requires that the cells be made competent to take up the incoming (vector) DNA by rendering the normally impervious fungal cell wall permeable to the DNA. "Competent" cells are then treated with the transforming DNA, and selective pressure is subsequently applied to detect only those cells (transformants) that have successfully incorporated and expressed this DNA and are thus capable of growth under selective conditions.

By far the most common method for transformation of fungi involves the preparation of protoplasts in the presence of an osmotic stabilizer—e.g., high salt or high sugar concentration, to prevent cell lysis—followed by their regeneration to give transformed colonies. Protoplasts are obtained by digesting the cell walls of young mycelia, spores, or germlings with lytic enzymes such as the complex mixture of enzymes from snail gut [10] or the most commonly used, Novozym 234, which is a commercial preparation of enzymes from *Trichoderma viride* containing principally glucanases and chitinases. In order to produce sufficient quantities of regenerable protoplasts, some species require mixtures of several enzymes from several sources. Protoplasts in osmoticum are exposed to DNA in the presence of calcium ions ($CaCl_2$) and polyethylene glycol (PEG), which promote DNA uptake, the mechanism of which is unclear but is thought perhaps to be because PEG causes membrane fusions trapping the $CaCl_2$ precipitated DNA in the process. The protoplasts are then regenerated on osmotically buffered selection medium, which allows the growth only of transformed cells. Comprehensive descriptions of $CaCl_2$/PEG transformation procedures, which are normally modified and optimized by individual laboratories, can be found in earlier primary publications and review articles [2,3,5, 8,11–15].

A more recent development was the method of transforming fungal protoplasts by electroporation, which involves use of a high-voltage electric pulse to allow reversible permeabilization of the cell membrane and uptake of DNA. Electroporation has been successfully employed for the transformation of *Fusarium solani* and *A. nidulans* [16], *Gliocladium* [17], *Aspergillus awamori* and *Aspergillus niger* [18], *Trichoderma harzianum* [19], and *Neurospora crassa* and *Penicillium urticae* [20].

The preparation of protoplasts for transformation is laborious. To reduce this tedium, if needed regularly, protoplasts of certain fungi can be aliquoted, frozen, and used successfully at a later date [8,21]. However, to circumvent this step completely, researchers have attempted to identify alternative transformation methods using intact fungal cells. One such procedure involves exposing intact fungal spores to transforming DNA in the presence of lithium acetate to induce DNA uptake, and has been used successfully in

N. crassa [22], *Coprinus cinereus* [23,24] and *Ustilago violaceae* [25,26]. Another transformation method involving intact fungal cells is particle bombardment (biolistics; biological ballistics), whereby tungsten particles (microprojectiles) coated with DNA are accelerated at high velocity directly into fungal spores or hyphae. Developed primarily as a means to transform plant cells in situ [27], biolistics has also been successfully applied to a number of filamentous fungi such as *N. crassa* [28], *Magnoportha griseia* [29], *Phytophthora capsici*, *P. citricola*, *P. cinnamomi* and *P. citrophthora* [30], and *Trichoderma harzianum* and *Gliocladium virens* [31]. Biolistics has advantages over conventional transformation procedures in that it is technically simple, it can be used to transform any fungal species whether it be amenable or less tractable to existing transformation procedures, and it can be reliably used to transform fungal mitochondria [reviewed in 29]. Additionally, biolistic transformation in the fungi *Trichoderma harzianum* and *Gliocladium virens* was shown to result in increased transformation frequency and genetic stability of transformants when compared with protoplast-mediated transformation of these organisms [31], illustrating the potential value of this technique for use in future transformation experiments. However, in addition to safety implications, a major drawback with biolistics is the requirement for expensive equipment not readily available in many laboratories, and the necessity in some countries, for example the U.K., to obtain special (firearms) licenses to use such equipment. Therefore, as yet, this technique has not been in widespread use.

Similarly, although the lithium acetate and electroporation procedures may provide alternative means for transforming fungi that cannot readily be transformed by more conventional techniques, the fact that the transformation frequencies reported for these methods show no significant improvements over those of the general $CaCl_2$/PEG technique, in addition to the fact that electroporation requires the use of expensive equipment, has meant that these methods have not been widely used.

3. TRANSFORMATION SELECTION MARKERS

Many selection systems are now available for fungi (reviewed in [1,3–5]). Those selectable markers in routine use fall into two broad classes: nutritional selective markers and dominant selective markers.

The most common method of selection of transformants involves nutritional selection markers whereby a cloned wild-type gene is used to complement an auxotrophic mutation in the recipient strain [for a recent review see 8]. Both homologous and heterologous markers have been used for this purpose. However, one drawback of nutritional markers is the requirement for the presence of the corresponding mutation in the recipient strain for transformation to be successful. This requires the isolation of auxotrophic mutant strains, which is time-consuming and laborious and may be difficult or undesirable, particularly in industrial strains, because undefined mutations affecting, for instance, metabolites of industrial interest, may be introduced along with the desired mutation. Nevertheless, the *pyr-4* and *niaD* systems, for example, have proved to be very useful, as these markers are functional in several species and the respective uridine auxotrophic mutants and nitrate nonutilizing mutants required can be isolated directly by positive selection on the basis of resistance to 5-fluoroorotic acid [11] or chlorate [32,33], respectively.

It is possible to both select and counterselect for the mutant (resistant) and wild-type (sensitive) phenotypes of these markers, which makes them particularly useful for genetic manipulation purposes. Other such two-way or bidirectional selection systems include those that utilize the markers *acuA* [34] and *sC* [35]. The *acuA* gene encodes acetyl CoA synthase, and the required mutants, selected by their resistance to fluoroacetate, exhibit poor growth on acetate as the sole carbon source. This system has been used for several fungal species, including *Ustilago maydis* [34]. The *sC* gene encodes ATP sulfurylase, and the required mutants, selected on the basis of resistance to selenate, are unable to utilize inorganic sulfur as the sole sulfur source [35].

Dominant selectable markers provide an alternative means of selection as they can be used to transform both wild-type and mutant strains, the only requirement being that the recipient organism is sensitive to the selective pressure applied. These dominant selection markers are mainly encoded by drug-resistance genes, most of which are of bacterial origin, such as the phleomycin, G418, or hygromycin resistance genes [36], and are fused to fungal promoter and terminator regions to allow expression of the marker gene and to permit growth of the transformant cells in the presence of the appropriate antibiotic. Such markers have not been used extensively for *A. nidulans* and *N. crassa* because in certain cases they have a high natural resistance to the antibiotic, and also there are many auxotrophic mutants available. Nevertheless, they have found wide application in a number of filamentous fungi. Oligomycin [37] and benomyl [38,39] resistance genes are examples of dominant selection markers of fungal origin, which have the advantage of already possessing, of course, fungal transcriptional signals. Many such antibiotic-resistance marker genes have a broad host range. However, one disadvantage is that an allele for resistance must be isolated in order that the wild-type strain can be transformed to resistance. Another drawback is that the resistance allele may not show significant dominance over the wild-type allele, resulting in selection difficulties. The *A. nidulans amdS* gene, which is often regarded as a dominant selection system, is actually a nutritional marker encoding the enzyme acetamidase. Where the recipient strain lacks acetamidase, *amdS* permits growth of transformants on acetamide as the sole source of carbon or nitrogen. Alternatively, even if the host strain naturally contains a copy of the *amdS* gene, transformants can still be distinguished, because they exhibit enhanced growth on acetamide as the nitrogen source due to an increase in the *amdS* gene copy number [5].

Finally, although not a transformation selection system per se, conidial color markers can be used as a screening approach. For instance, in *A. nidulans*, if a white *wA2* mutant is transformed with a wild-type allele, then wA^+ transformants that are green in color can be observed [40]. Obviously, this approach can only be applied to fungi in which mutants affected in conidial color can be generated.

In recent years the use of existing, previously reported transformation selection systems has spread to a whole battery of hitherto untransformed fungal species, and, more excitingly, new selection markers have also been described. As examples of the former, the *N. crassa pyr4* gene was used to transform *Trichoderma viride* [41]; *Gibberella fujikuroi* was transformed using the *A. niger niaD* gene [42]; the *trp-2* gene of *Coprinus cinereus* was used to transform *C. bilanatus* [43]; the hygromycin B phosphotransferase gene was used in the transformation of *Pleurotus ostreatus* [44], *Hebeloma cylindrosporum* [45], and *Gibberella pulicaris* [46]. The oomycetous fungi *Phytophthora infestans* and *P. megasperma* f. sp. *glycinea*, in which gene transfer procedures have only recently become available, have been transformed using the dominant selectable marker hygromycin phosphotransferase for resistance to hygromycin [47–49], and *P. infestans* has also been transformed using the gene encoding the enzyme neomycin phosphotransferase, which gives resistance to G418 [47,48]. With regard to the new selection systems, the aspartase gene (*aspA*) from *E. coli*, for example, has recently been transformed into *A. nidulans*. When expressed in *A. nidulans*, which naturally lacks aspartase, the *E. coli aspA* gene has the ability to alter the metabolic patterns of the recipient strain. *aspA* can therefore be used as a dominant selectable marker for transformation of *A. nidulans* and has been demonstrated to be useful for obtaining multicopy transformants [50]. Moreover, as aspartase has never been reported to be present in fungi [50], *aspA* could potentially be useful as a dominant marker for a broad range of filamentous fungi.

Gene copy number in transformants can often be important, and a high copy number is frequently desirable. The *amdS* gene [51] is routinely employed to achieve this goal, and the *aspA* system is also of potential use in this regard [50]. If a low copy number is desired, the selection markers *trpC* [14], *argB* [52], and *niaD* [33] are often employed.

If a transforming gene cannot be selected for directly, a common tactic is to cotransform a vector containing this gene with one containing a selectable marker. There is a high probability that cells that are competent to take up

DNA will take up both types of vector molecule; thus, one can select for cotransformants by applying selective pressure to detect transformants containing the selectable marker and subsequently testing these cells for expression of the nonselectable gene. However, cotransformation rates vary and, at least in *A. nidulans*, they appear to be dependent on the selection marker employed as, for example, rates of 15% and 95% were observed when *niaD* [53] and *amdS* [51], respectively, were used for selection in *A. nidulans*.

4. ISOLATION OF FUNGAL GENES

Fungal genes have been cloned by several routinely used methods including "reverse genetics," DNA homology to heterologous probes, differential hybridization, antibody recognition, and functional expression in *E. coli* or *S. cerevisiae* [reviewed in 3,5,9,54]. However, such approaches have their limitations, and subsequent to the successful development of efficient transformation systems for fungi, self-cloning of fungal genes has provided an alternative and attractive cloning method, particularly for *A. nidulans* and *N. crassa*, which have many well-characterized mutant strains. This technique involves the use of fungal genomic libraries constructed in plasmid or cosmid vectors to "complement" the corresponding fungal mutant strain by transformation, and is thus limited to organisms carrying appropriate, well-defined mutations. The clone containing the complementing gene can then be isolated either by marker rescue in *E. coli* or by subselection from pools of clones from plasmid- or cosmid-based genomic libraries. Both of these procedures for recovery of cloned genes are laborious and can be facilitated by employing a cosmid-based gene library, as these vectors can harbor relatively large fragments of chromosomal DNA, thereby reducing the number of clones that may need to be screened before the gene of interest is detected [5,8,9].

In organisms that have a well-characterized genetic map and a number of existing cloned genes, one can employ the more laborious procedure of chromosome "walking" from a known, previously cloned gene to the desired gene, identified only by its genetic map position, using a genomic library. Clones containing overlapping DNA segments located progressively distal to the previously identified gene are tested for complementation of a mutation in the gene of interest. Chromosome "walking" can be achieved using bacteriophage or cosmid vectors harboring relatively large DNA insert fragments [55]. Overlapping clones can be organized into large contiguous chromosomal regions designated "contigs," and used to construct physical genomic maps (contig maps). In *A. nidulans*, for instance, cosmid libraries have been divided into chromosome-specific subcollections containing a sufficient number of clones to permit the development of a contig map [56]. Additionally, these chromosome-specific libraries should greatly facilitate chromosome-walking experiments by reducing the numbers of clones that require screening. Genes cloned by this method can be positively identified by transformation and complementation of an appropriate mutant strain or by gene disruption (see below). This procedure could also be applied to other fungi.

5. INTEGRATIVE VERSUS AUTONOMOUSLY REPLICATING TRANSFORMATION VECTORS

In filamentous fungal species, the transforming DNA most commonly becomes integrated into the host genome by recombination. Three types of integration events have been demonstrated [14]: type I, homologous integration at a resident site; type II, nonhomologous integration at an ectopic site; and type III, gene replacement. Multiple integration events are common, both at homologous and nonhomologous sites, and the relative frequencies of the number and types of integration events appear to depend on the selection system or the strain of organism used [for a review see 3].

Numerous efforts have been made to develop autonomously replicating vectors for the filamentous fungi which normally exhibit integrative transformation, such as the ascomycetes *Aspergillus* and *Neurospora.* Autonomously replicating vectors are advantageous in that they substantially increase the efficiency of transformation and their subsequent recovery in *E. coli* is easier than with integrative vectors. Recovery of transforming DNA from the fungus without its excision from the genome, and meiotic and mitotic instability of transformants grown under nonselective conditions, are among the main diagnostic criteria for the identification of autonomously replicating vectors. However, vectors that have unquestionably been integrated into the fungal chromosome have been recovered by transformation of *E. coli* with the undigested genomic DNA from the fungal transformant, due to reversal of the integration event [57]. Therefore, caution must be exercised when interpreting data for evidence of autonomously replicating vectors.

A number of claims have been made for the evidence of autonomous replication of vectors in *Neurospora crassa* [reviewed 1,58,59], but these have been treated with skepticism by some [57] and remained unacknowledged by others [4,5]. Nevertheless, bona fide autonomously replicating vectors were successfully developed, first for *Mucor circinelloides* [60] and later for *Phycomyces blakesleeanus* [61], *Absidia glauca* [62], *Podospora anserina* [63], *Ustilago maydis* [64,65] and *Ustilago violacae* [25,26]. More recently, a vector containing an *A. nidulans* DNA fragment that exhibits replicon activity has been developed for *A. nidulans* by Clutterbuck and colleagues [66] (see below for further details). Not only does this vector replicate autonomously in *A. oryzae* and *A.*

niger [66], but a derivative has also been shown to have replicon activity in *Gibberella fujikuroi* [67]. As anticipated, using these autonomously replicating fectors, transformation frequencies were improved significantly in *A. nidulans*, *A. oryzae*, and *A. niger* [66] but, contrary to expectations, were not greatly enhanced in *Gibberella fujijuroi* [67]. Other autonomously replicating vectors have been obtained for the dimorphic human-pathogenic fungus *Histoplasma capsulatum* [68], the mushroom-forming fungus *Pleurotus ostreatus* [44], and the plant pathogens *Ashbya gossypii* [69], *Fusarium oxysporum* [70], *Cryphonectria parasitica* [70], *Nectria haematococca* [70,71], and *Phanaerochaete chrysosporium* [72–74].

In the filamentous hemiascomycete *Ashbya gossypii*, plasmid replication was successfully achieved using autonomously replicating sequences (ARS) derived from *Saccharomyces cerevisiae*, a feat not previously accomplished with filamentous fungi [69].

The 6.1-kb *A. nidulans* sequence (*AMA1*) responsible for autonomous replication [66] not only dramatically increases transformation frequency in *A. nidulans* but also can be used to synthesize "instant gene banks" whereby an *A. nidulans* mutant is cotransformed with the plasmid ARp1 containing *AMA1*, or its derivative (pHELP), and *A. nidulans* genomic DNA, which can be either endonuclease cleaved, sheared, or even uncut. Additionally, ARp1 or pHELP may be used to cotransform a genomic library constructed in a conventional integrative vector. Here, cotransformation efficiency is markedly improved and results in the formation of a replicating hybrid plasmid derived from the two transforming plasmids, allowing one to screen for the presence of a gene of interest by complementation of mutant alleles. This method has been used to clone the *A. nidulans adD* and *adC* genes [75]. Further experiments in *A. nidulans* employing this method used genomic DNA from other fungal species to clone heterologous genes. Such applications yielded clones containing the *P. canescens trpC* gene [76] and the *pyrG* gene from the plant pathogen *Gaeumannomyces graminis* [77], supporting the notion that this is an exciting and useful new procedure which obviates the need for conventional gene bank construction. It appears that joining of the chromosome fragments and the "helper" plasmid may occur either by ligation or recombination depending on whether the plasmid is in linear or covalently closed circular form [76].

In the basidiomycete *P. chrysosporium*, transforming plasmids are apparently maintained extrachromosomally as a result of their recombination with the plasmid pME which is endogenous to *P. chrysosporium* [73]. In the oomycete *Phytophthora infestans*, it was demonstrated that cotransformed linearized plasmids were ligated together in vivo but, instead of being extrachromosomally maintained, they cointegrated into the genome [48].

Telomeric DNA sequences have been characterized from the filamentous fungi *Neurospora crassa* [78,79] and *Fusarium oxysporum* [70]. Linear transformation vectors containing telomere consensus sequences were created in *Fusarium oxysporum* by fungal rearrangement of an integrating vector, and functioned with high efficiency as autonomously replicating vectors in *N. haematococca* and *C. parasitica* as well as *F. oxysporum* [70]. A rather similar situation has been found in *Histoplasma capsulatum* in which the transforming plasmid undergoes in vivo modification including duplication and addition of telomeric sequences at the termini of linear DNA to produce multicopy linear plasmids which replicate autonomously [68]. The isolation and characterization of telomeres from other fungi could lead to the development of further linear autonomously replicating vectors for fungal transformation. If fungal centromere sequences could be similarly isolated, they could be incorporated into such telomeric plasmids, thus forming artificial filamentous fungal chromosomes for use as vectors. In this regard, yeast artificial chromosome (YAC) vectors have been developed that can harbor extremely large DNA fragments and may be exploited for the construction of filamentous fungal genomic libraries, facilitating the screening for cloned genes. A YAC library has, for example, been generated for *Erysiphe graminis*, an obligate fungal pathogen of barley [80].

In zygomycetous fungi such as *Mucor circinelloides*, *Phycomyces blakesleeanus*, and *Absidia glauca*, transforming vectors predominantly exhibit autonomous replication [reviewed by 1,4,5,58] and, at least in the case of *Mucor*, difficulties were experienced in integrating vector DNA into the host genome. Indeed, in these fungi effort has been made to obtain transformation vectors that integrate into the host genome in order that gene disruption and gene replacement techniques (see below) can be developed to gain an insight into gene function and regulation and to allow gene manipulations. Integrative transformation has been demonstrated for *Absidia glauca* using a normally autonomously replicating vector into which had been inserted repetitive DNA elements [81], indicating that the development of vectors for targeted integration should therefore be feasible.

6. GENE REPLACEMENT AND GENE DISRUPTION

Gene replacement and gene disruption events can be obtained only by integration of the transforming DNA into the homologous site on the host chromosome.

Gene replacement (known also as gene conversion) is used to introduce desired mutations into the recipient cell and occurs by two methods, either direct or indirect. The direct (one-step) replacement method involves double crossover of a linear transforming molecule at the homologous locus and the in vitro-created mutations must be bordered by normal chromosomal sequences on the linear fragment for homologous integration to occur. The indirect (two-step) replacement method involves circular DNA molecules integrated at the homologous locus by a single crossover event which creates tandem duplications of the target sequence separated by vector sequences. Such integration can be reversed by allowing self-fertilization of transformants under nonselective conditions, and the ensuing plasmid loss due to unequal crossing over can result in the retention of either the mutant or wild-type gene sequence, depending on the position of the recombination event [82]. By using selective markers for which there is both forward and reverse selection, for example *niaD* or *pyrG*, one can circumvent the need for a sexual cycle by selecting for sensitivity to chlorate and 5-fluoroorotic acid, respectively, and hence rare mitotic plasmid loss in transformants [reviewed in 8].

Gene disruption occurs by homologous integration of either a circular vector containing a defective gene or a linear DNA molecule containing the target gene interrupted by a selectable marker. Both methods can be used to create null mutations: the former method results in a duplication of the gene in which neither copy contains the entire coding region; the latter method results in replacement of the gene with the defective gene containing the marker (insertional inactivation by direct gene replacement). Such techniques can be used, for example, to confirm that a gene has indeed been cloned, to study the physical role of the products of such genes (e.g., in fungal development processes or pathogenesis), or to eliminate genes with undesirable properties in pathogens or industrial organisms. Nevertheless, caution must be exercised with the interpretation of results of gene disruption experiments as problems can potentially arise [reviewed by 8]. It is usually necessary to ensure that the resultant strain does, in fact, carry the desired mutation. This can be conveniently achieved by restriction endonuclease, by DNA hybridization, or by PCR amplification and analysis (see Chapter 5).

7. OTHER FUNGAL VECTORS

A number of other specialized vectors are available that were originally developed for the analysis of genes, and their products were isolated mainly from the aspergilli. Nevertheless, their use will most probably be extended to many other groups of fungi.

Vectors have been developed for promoter analysis that identify sequences in the 5′ control regions of fungal genes, which are important for transcriptional regulation. These vectors contain bacterial "reporter" genes *lacZ* and *uidA* encoding β-galactosidase [83] or β-glucuronidase [84,85], respectively, and DNA fragments harboring the 5´ promoter sequences of the gene of interest are inserted into a convenient restriction site at the N-terminus of the reporter gene. Such constructs are often made in a fusion vector carrying a homologous marker to direct the integration of the recombinant plasmid to a defined genomic site in the fungal transformant, such as the *argB* locus. An embellishment of this approach has been the development of a so-called "twin reporter" vector in which the genes encoding β-galactosidase and β-glucuronidase are both employed. This construct is useful for studying functionally related but divergently transcribed genes and has been used to investigate the contiguous *niiA–niaD* genes for nitrate assimilation in *A. nidulans* [85].

Often, it is desirable to obtain high levels of a particular fungal protein, and preferential to produce it in a fungus rather than in *E. coli*. Several fungal expression vectors have been developed for this purpose. Such vectors contain strong promoters, including those of the alcohol dehydrogenase (*alcA* [86]), glucoamylase (*glaA* [87]), and glyceraldehyde-3-phosphate dehydrogenase (*gpdA* [88,97]) genes, and carry convenient restriction sites and selectable markers.

Finally, a number of vectors are available that allow secretion of a heterologous protein. Such vectors contain a strong promoter (as discussed above) followed by either a fungal signal sequence, such as that of the *glaA* gene [87,89] or the amylase (*amy*) genes [90], or one that has been designed and synthesized artificially. A further refinement to this system is where the gene of interest is fused to the C-terminus of the glucoamylase gene, which appears to increase the amount of gene product expressed. Separation of the glucoamylase protein from the protein of interest is effected by incorporating a KEX-2 (proteolytic) site between these protein sequences [91], which the natural protease of the host filamentous fungus is apparently able to recognize and cleave.

8. TRANSPOSABLE ELEMENTS

Transposons have been used routinely for genetic manipulation in a wide range of organisms. Fungal transposons have only relatively recently been shown to exist, and have been observed in strains of *N. crassa* [92], *Fusarium oxysporum* [93], and *A. niger* [94]. The transposons in both *F. oxysporum* and *A. niger* were obtained by "trapping" them in the *niaD* gene, which resulted in the formation of chlorate-resistant strains mutated in their *niaD* gene. This approach failed to isolate transposons in *A. nidulans* where over 100 *niaD* mutants were examined without success (J.R.K., unpublished). It would appear that these *A. nidulans* laboratory strains, chosen originally for their genetic stability, lack transposons. It therefore appears that for trapping transposons it may be beneficial to use a less stable fungal strain, if available, as the vehicle of choice.

9. GENETIC ANALYSIS OF TRANSFORMANTS

Transformants can be genetically characterized in a number of ways. In the imperfect fungi, which have no sexual cycle, the only course open for genetic analysis of transformants is that of Southern blot analysis. This can be used to determine the copy number of transforming DNA sequences and establish whether the vector is replicating autonomously or has integrated into the chromosome and, indeed, ascertain the type of integration that has occurred. Transformants of fungi that do have a sexual cycle can be characterized not only by Southern analysis, but also by the classical (formal) genetic technique of sexual crosses (described, for example, for *A. nidulans* by [95,96]) which can be used to determine the site of integration of cloned genes and establish their linkage to other characterized loci. Classical genetics can thus provide a powerful means of confirming results obtained by Southern analysis. Parasexual and haploidization analysis of transformants can also be carried out in organisms with sexual or parasexual cycles to assign genes to linkage groups. In recent years, the advent of pulse-field gel electrophoresis (PFGE), which permits the size separation of linear chromosomes, has resulted in the production of electrophoretic karyotypes for a number of filamentous fungal species, including *Ustilago maydis, U. hordei, Cephalosporium acremonium, A. nidulans, A. niger, A. niger* var *awamori, A. oryzae,* and *N. crassa* [reviewed by 8,9]. Moreover, PFGE has provided a means of allocating cloned genes to specific chromosomes by Southern hybridization in species where this would previously have been impossible due to their lack of a classical genetic system (see Chapter 5).

10. FUTURE PROSPECTS

The number of filamentous fungi that have been genetically transformed has increased quite considerably over the past few years, thereby opening the door to more in-depth molecular analysis. Nevertheless, there still remain well over 1 million fungal species that could potentially benefit from such manipulations. Many of these as yet genetically unexplored fungi are of interest to man as they are pathogenic, produce molecules of biotechnological importance, or are interesting from an environmental standpoint. Therefore, there is little doubt that even more fungi will gain the attention of molecular biologists. As many transformation and selection systems are available to achieve this purpose, one must consider individual strains independently and choose and optimize appropriate systems for each organism. Although there are already many to choose from, further systems will no doubt be developed with other inherent characteristics and virtues. It seems likely that the technique of $CaCl_2$/PEG treatment of protoplasts will continue to be the most commonly used fungal transformation procedure, but, given that the technique is tedious and often unreliable, one might expect that it may be refined or even supplanted by other methods.

The "instant gene bank" methodology recently described by Clutterbuck and colleagues [75,76] certainly opens the door for the isolation, in *A. nidulans*, of genes from a myriad of fungi [77]. Furthermore, cosmids, YAC vectors, and techniques such as contig mapping will probably facilitate both the isolation of interesting traits in intractable fungi, and the molecular analysis of more tractable strains in greater depth. To this end, vector systems per se may also continue to be improved, as will those useful for expression or secretion studies. More fungal species will benefit from YAC technology, especially fungi with very large genomes, such as the oomycetes, and other intractable fungi. Even more advantageous than YAC-based chromosome libraries would be the development of artificial filamentous fungal chromosomes (FACs?), which could be designed to contain genes of interest, such as those for antibiotic or enzyme products, as required.

A few hundred fungal genes have so far been isolated and sequenced, including household genes such as those involved in glycolysis and amino acid biosynthesis, as well as various specialized genes encoding enzymes involved, for example, in cell cycle, mitosis, or cellular differentiation. In addition, a number of DNA:protein-binding regulatory genes have been isolated. Over the next few years, many more fungal genes will be sequenced and characterized, giving a clearer picture of the molecular mechanisms of life in these lower eukaryotes. Given that with *A. nidulans* and *N. crassa* 1.) self-cloning has become routine and 2.) mature genetic maps are available from classical genetic experiments performed over 50 years, it would not be surprising if most of the genes from these fungi, which are already characterized by formal genetics, were cloned and sequenced within the next decade. Indeed, it is a realistic possibility that the complete genome of a filamentous fungus will be determined at the nucleotide level, with *Aspergillus nidulans* as the strongest candidate, followed closely by *Neurospora crassa*.

11. EXPERIMENTS

A. Protoplast Preparation and Transformation*†

1. Inoculate complete medium plates each with a single inoculum of the *Aspergillus* strain to be transformed and incubate for 4–7 days at 37°C until the

---

*Please note that there are several variations of these methods currently in use.
†Media are described in the Appendix.

entire surface of the plate is covered with conidia. On the day prior to transformation, suspend *Aspergillus* conidia from two plates of complete medium in 2 × 10 mL saline Tween solution, vortex vigorously, and seed into 2 × 400 mL minimal medium containing a nitrogen source and supplements appropriate to the selection system being employed. Incubate overnight with orbital shaking. Please note that the incubation times and temperatures employed should be optimized for each strain to give very young mycelial cells as the starting material for protoplast preparation. Different workers incubate strains at, e.g., 25°C, 30°C, or 37°C for various lengths of time depending on the strain being used.

2. Harvest the mycelium by filtration through two layers of muslin and rinse with 500 mL cold (4°C) 0.6 M $MgSO_4$. Resuspend the mycelium in 5 mL cold osmotic medium in a precooled 150 mL conical flask; then add 50 mg Novozym 234 (suspended in 2 mL ice-cold osmotic medium) and incubate on ice for 5 min. To this add 1.25 mL BSA solution (12 mg/mL BSA in ice-cold osmotic medium). Incubate at 30°C with slow shaking for 60–90 min until the protoplasts are released (protoplasts can be viewed under a microscope using the low-power objective lens), and then place the mixture on ice to stop the reaction.

3. Vigorously swirl the flask to liberate the protoplasts from the mycelial debris, then gently carry out the remaining manipulations. Using a 10-mL pipette, divide the above mixture equally between two 30-mL Corex tubes that have been precooled on ice, avoiding touching the sides of the tubes. Rinse the flask with 4–5 mL cooled osmotic medium, and add an equal volume to each tube. Overlay each protoplast mixture with an equal volume of ice-cold trapping buffer by allowing it to run very slowly down the side of the tubes.

4. Balance the tubes with ice-cold trapping buffer and centrifuge for 20 min at 5000 rpm, 4°C, in a Sorvall HB-4 swing-out rotor. This will pellet the mycelial debris, and a bushy band of protoplasts will form at the interface. Using a Pasteur pipette, pool the complete protoplast bands from the two tubes into another precooled 30-mL Corex tube, and then add an equal volume of ice-cold 1 × STC and centrifuge at 7000 rpm, 4°C, for 5 min the Sorvall HB-4 rotor. Discard the supernatant, resuspend the pellet in 10 mL ice-cold 1 × STC, then spin again for 5 min at 7000 rpm, 4°C, in the Sorvall HB-4 rotor and discard the supernatant.

5. Resuspend the protoplasts in just sufficient 1 × STC to provide enough protoplast aliquots for all treatments, using 50 μL protoplasts per treatment. Add DNA to the protoplast aliquots as follows:

In a typical experiment, in 10-mL plastic tubes, one adds 10 μg/μL DNA to 50 μL protoplasts; the volume of 2 × STC added must equal the volume of DNA and the volume is made up to a total of 100 μL with 1 × STC, e.g.:

DNA (1 μg/μL)	2 × STC	1 × STC	Protoplasts	Total volume
10 μL	10 μL	30 μL	50 μL	100 μL

To each tube, add 25 μL 60% polyethylene glycol (PEG) 6000, mix gently using the tip, and incubate on ice for 30 min. Add 1 mL 60% PEG 6000, mix gently by rolling the tube for a few minutes, then incubate at room temperature for 30 min. Add 5 mL ice-cold 1 × STC, then spin at 5000 rpm for 5 min in a swing-out rotor at room temperature.

6. Remove the supernatant without disturbing the pellet and gently resuspend the pellet in 300 μL ice-cold 1 × STC. Spread 3 × 100 μL aliquots of each batch of transformed protoplasts on three selection plates of protoplast medium containing the appropriate nitrogen source and supplements except for that required for growth of any untransformed organisms. Incubate plates at 37°C for up to 4 days.

As a control system, treat protoplasts as above but without the addition of DNA and plate out as follows:

1. Dilute control protoplasts 10^{-2} in sterile distilled H_2O and spread 100 μL on nonselective plates of protoplast medium supplemented with all the requirements for growth for untransformed organisms. Incubate plates at 37°C for up to 4 days. This gives the count of colonies derived from nonprotoplast (intact) cells, because protoplasts lyse in hypotonic conditions.
2. Dilute control protoplasts 10^{-2}, 10^{-3}, and 10^{-4} in 1 × STC and spread 100 μL of each dilution on plates of nonselective protoplast medium supplemented with all the requirements for growth of untransformed organisms. Incubate plates at 37°C for up to 4 days. This gives the total count of colonies derived from both intact cells and protoplasts.

Subtract the colony count of step 1 from step 2 to obtain the true count of viable protoplasts, which can then be used to calculate the frequency of transformation.

B. Large-Scale Fungal Genomic DNA Preparation Using the Nucleon II Kit*

1. Grind to a fine powder 300–400 mg pressed wet-weight mycelium in liquid N_2 (a roughly equivalent amount of freeze-dried mycelium can alternatively be used).
2. Suspend the powder in 2 mL Nucleon reagent B in a 15-mL screw-capped polypropylene tube with 15 mm internal diameter.

*Adapted for filamentous fungi by Shiela Unkles.

3. Add 1 μL 10 mg/mL RNase A and incubate at 37°C for 30 min.

4. Add 1.5 mL 5M sodium perchlorate and rotary mix (at approx. 100 rpm) at room temperture for 15 min.

5. Incubate at 65°C for 25 min, inverting once or twice during incubation.

6. Add 5.5 mL chloroform (stored at –20°C). Rotary mix at room temperature for 10 min.

7. Centrifuge at 800 × *g* for 1 min.

8. Add 800 μL, Nucleon Silica suspension (shaken vigorously to resuspend) without remixing, and centrifuge at 1400 × *g* for 3 min.

9. Remove upper aqueous layer, avoiding the interface, and add 0.8–1 volume of ethanol.

10. Gently invert. The threadlike DNA precipitate can be rinsed out using a sterile Pasteur pipette.

11. Wash the DNA in 70% ethanol by swirling the pipette.

12. Remove the DNA from the pipette into a fresh tube, dry the pellet, and resuspend in TE. This may take several hours.

For *Aspergillus nidulans* the yield should be around 400–500 μg. For *Phytophthora* the yield should be around 200 μg (Shiela Unkles, unpublished).

Nucleon II Kit can be obtained from Scotlab.

APPENDIX

A. Media and Buffers for *Aspergillus* Transformation

Unless otherwise indicated, solid media are prepared by the addition of 1.2% agar to the appropriate liquid media, and all media and buffers are sterilized by autoclaving at 15 lb/inch^2 for 15 min.

Fungal Media

Complete and minimal medium for *Aspergillus* are based on the recipes described by Cove [98] and Pontecorvo et al. [95].

Complete medium

10 g glucose
50 M salts solution (see below)
1 mL trace elements solution (see below)
1 mL vitamin solution (see below)
2 g peptone
1 g yeast extract
1 g casein hydrolysate

Make up to 1 L with distilled H_2O and pH 6.5 with NaOH.

Minimal Medium (nitrogenless)

10 g glucose
50 M salts solution (see below)
1 mL trace elements solution (see below)

Make up to 1 L with distilled H_2O and pH 6.5 with NaOH.

Nitrogen sources The various nitrogen sources either are incorporated directly into the medium prior to autoclaving or are kept as sterile 1 M stock solutions and added to nitrogenless minimal medium precooled to 55°C.

Trace elements solution

1.1 g $(NH_4)_6Mo_7O_{24}\cdot 4H_2O$
11.1 g H_3BO_4
1.6 g $CoCl\cdot 6H_2O$
1.6 g $CuSO_4\cdot 5H_2O$
50.0 g EDTA (disodium salt)
5.0 g $FeSO_4\cdot 7H_2O$
5.0 g $MnCl_2\cdot 7H_2O$
22.0 g $ZnSO_4\cdot 7H_2O$

Make up to 1L with distilled H_2O and boil with stirring. Cool the solution to 60°C, adjust to pH 6.5–6.8 with KOH, and store in the dark at 4°C.

Vitamin solution

25.0 mg biotin
2.5 g nicotinic acid
0.8 g para-amino benzoic acid
1.0 g pyridoxine HCl
2.0 g pantothenic acid
2.5 g riboflavin
1.5 g aneuric acid
20.0 g choline chloride

Make up to 1 L with distilled H_2O.

Supplements The following supplements are sterilized by filtration and stored as concentrated aqueous solutions at 4°C. The appropriate amounts of supplements are then added, as required, to media precooled to 55°C.

Supplement	Concentration of stock solution	Amount added to 100 mL medium
Arginine-HCl	4.2 g/100 mL	1 mL
Biotin	10.0 mg/100 mL	1 mL
Choline-HCl	2.0 g/100 mL	1 mL

Supplement	Concentration of stock solution	Amount added to 100 mL medium
1 M glutamate	18.7 g/100 mL	1 mL
Methionine	0.5 g/100 mL	1 mL
Nicotinic acid	10.0 mg/100 mL	1 mL
PABA	0.14 g/100 mL	0.5 mL
1 M proline	11.15 g/100 mL	1 mL
Putrescine	0.2 g/100 mL	1 mL
Pyridoxine-HCl	0.5 g/100 mL	1 mL
Riboflavin	0.8 g/100 mL	1 mL
1 M uridine	24.42 g/100 mL	1 mL

Salts solution

10.4 g KCl
10.4 g $MgSO_4 \cdot 7H_2O$
30.4 g KH_2PO_4

Make up to 1 L with distilled H_2O.

Saline Tween solution

0.01% Tween 80
0.9% NaCl

Osmotic medium

1.2 M $MgSO_4$
10 mM sodium phosphate pH 7.0

Adjust to pH 5.8 with 0.2 M Na_2HPO_4, filter sterilize, and dispense in 100-mL aliquots.

Protoplast medium

10 g glucose
1.2 M sorbitol
50 mL salts solution
1 mL trace elements solution

Make up to 1 L with distilled H_2O and pH 6.5 with NaOH. Add agar to 1.2%.

Buffers

1 × STC	2 × STC
1.2 M sorbitol	2.4 M sorbitaol
10 mM Tris-HCl pH 7.5	20 mM Tris-HCl pH 7.5
10 mM $CaCl_2$	20 mM $CaCl_2$

Trapping buffer
0.6 M sorbitol
100 mM Tris-HCl pH 7.0

ACKNOWLEDGMENTS

Research in the St. Andrews laboratory has been supported by funds from the Science and Engineering Research Council (SERC), the Agricultural and Food Research Council (AFRC), and the Commission of the European Communities (CEC).

REFERENCES

1. Rambosek, J., J. Leach (1987). Recombinant DNA in filamentous fungi: progress and prospects. *CRC Crit. Rev. Biotechnol. 6*:357.
2. Turner, G., D. J. Ballance (1985). Cloning and transformation in *Aspergillus*. In: *Gene Manipulations in Fungi* (J. W. Bennett, L. L. Lasure, eds.). Academic Press, London, p. 259.
3. Fincham, J. R. S. (1989). Transformation in fungi. *Microbiol. Rev. 53*:148.
4. Van den Hondel, C. A. M. J. J., P. J. Punt (1990). Gene transfer systems and vector development for filamentous fungi. In: *Applied Molecular Genetics of Fungi* (J. F. Peberdy, C. E. Caten, J. E. Odgden, J. W. Bennett, eds.). Cambridge University Press, Cambridge, p. 1.
5. Ballance, D. J. (1991). Transformation systems for filamentous fungi and an overview of fungal gene structure. In: *Molecular Industrial Mycology: Systems and Applications for Filamentous Fungi* (S. A. Leong, R. M. Berka, eds.). Marcel Dekker, New York, p. 1.
6. Goosen, T., C. J. Bos, H. W. J. Van den Broek (1991). Transformation and gene manipulation in filamentous fungi: an overview. In: *Handbook of Applied Mycology (Fungal Biotechnology* Vol. 4) (D. K. Arora, K. G. Mukerji, R. P. Elander, eds.). Marcel Dekker, New York.
7. Timberlake, W. E. (1991). Cloning and analysis of fungal genes. In: *More Gene Manipulations in Fungi* (J. W. Bennett, L. L. Lasure, eds.). Academic Press, London, p. 51.
8. May, G. (1992). Fungal technology. In: *Applied Molecular Genetics of Filamentous Fungi* (J. R. Kinghorn, G. Turner, eds.). Blackie Press, Glasgow, p. 1.
9. Kinghorn, J. R., S. E. Unkles (1994). Molecular genetics and expression of foreign proteins in the genus *Aspergillus*. In: *Biotechnology Handbooks* (Vol. 7). *Aspergillus* (J. E. Smith, ed.). Plenum Press, London, p. 65.
10. Mink, M., H.-J. Holtke, C. Kessler, L. Ferenczy (1990). Endonuclease-free, protoplast-forming enzyme preparation and its application in fungal transformation. *Enzyme Microb. Technol. 12*:612.
11. Ballance, D. J., F. P. Buxton, G. Turner (1983). Transformation of *Aspergillus nidulans* by the orotidine-5′-phosphate decarboxylase gene of *Neurospora crassa*. *Biochem. Biophys. Res. Commun. 112*:284.

12. Tilburn, J., C. Scazzocchio, G. G. Taylor, J. H. Zabicky-Zissman, R. A. Lockington, R. W. Davies (1983). Transformation by integration in *Aspergillus nidulans. Gene. 26*:205.
13. John, M. A., J. F. Peberdy (1984). Transformation of *Aspergillus nidulans* using the *argB* gene. *Enzyme Microb. Technol. 6*:386.
14. Yelton, M. M., J. E. Hamer, W. E. Timberlake (1984). Transformation of *Aspergillus nidulans* by using a *trpC* plasmid. *Proc. Natl. Acad. Sci. USA 81*:1470.
15. Peberdy, J. F. (1989). Fungi without coats—protoplasts as tools for mycological research. *Mycol. Res. 93*:1.
16. Richey, M. G., E. T. Marek, C. L. Schardl, D. A. Smith (1989). Transformation of filamentous fungi with plasmid DNA by electroporation. *Phytopathology 79*:844.
17. Thomas, M. D., C. M. Kenerly (1989). Transformation of the mycoparasite *Gliocladium. Curr. Genet. 15*:415.
18. Ward, M., K. H. Kodama, L. J. Wilson (1989). Transformation of *Aspergillus awamori* and *Aspergillus niger* by electroporation. *Exp. Mycol. 13*:289.
19. Goldman, G. H., M. Van Montagu, A. Herrera-Estrella (1990). Transformation of *Trichoderma harzianum* by high-voltage electric pulse. *Curr. Genet. 17*:169.
20. Chakraborty, B. N., M. Kapoor (1990). Transformation of filamentous fungi by electroporation. *Nucl. Acids Res. 18*:6737.
21. Vollmer, S. J., C. Yanofsky (1986). Efficient cloning of genes of *Neurospora crassa. Proc. Natl. Acad. Sci. USA 83*:4869.
22. Dhawale, S. S., J. V. Paietta, G. A. Marzluf (1984). A new, rapid and efficient transformation procedure for *Neurospora. Curr. Genet. 8*:77.
23. Binninger, D. M., C. Skrzynia, P. J. Pukkila, L. A. Casselton (1987). DNA-mediated transformation of the basidiomycete *Coprinus cinereus. EMBO J. 6*:835.
24. Binninger, D. M., L. Le Chevanton, C. Skrzynia, C. D. Shubkin, P. J. Pukkila (1991). Targeted transformation in *Coprinus cinereus. Mol. Gen. Genet. 227*:245.
25. Bej, A. K., M. H. Perlin (1989). A high efficiency transformation system for the basidiomycete *Ustilago violaceae* employing hygromycin resistance and lithium acetate treatment. *Gene 80*:171.
26. Bej, A. K., M. H. Perlin (1991). Acquisition of mitochondrial DNA by a transformation vector for *Ustilago violacea. Gene 98*:135.
27. Sanford, J. C., T. M. Klein, E. D. Wolf, N. Allen (1987). Delivery of substances into cells and tissues using a particle bombardment process. *J. Part. Sci. Technol. 5*:27.
28. Armaleo, D., G.-N. Ye, T. M. Klein, K. B. Shark, J. C. Sanford, S. A. Johnston (1990). Biolistic nuclear transformation of *Saccharomyces cerevisiae* and other fungi. *Curr. Genet. 17*:97.
29. Klein, T. M., R. Arentzen, P. A. Lewis, S. Fitzpatrick-McElligott (1992). Transformation of microbes, plants and animals by particle bombardment. *Bio/Technology 10*:286.
30. Bailey, A. M., G. L. Mena, L. Herrera-Estrella (1993). Transformation of four pathogenic *Phytophthora* spp by microprojectile bombardment on intact mycelia. *Curr. Genet. 23*:42.

31. Lorito, M., C. K. Hayes, A. Di Pietro, G. E. Harman (1993). Biolistic transformation of *Trichoderma harzianum* and *Gliocladium virens* using plasmid and genomic DNA. *Curr. Genet. 24*:349.
32. Unkles, S. E., E. I. Campbell, D. Carrez, et al. (1989). Transformation of *Aspergillus niger* with the homologous nitrate reductase gene. *Gene 78*:157.
33. Malardier, L., M. J. Daboussi, J. Julien, F. Roussel, C. Scazzocchio, Y. Brygoo (1989). Cloning of the nitrate reductase gene (*niaD*) of *Aspergillus nidulans* and its use for transformation of *Fusarium oxysporum. Gene 78*:147.
34. Hargreaves, J. A., G. Turner (1989). Isolation of the acetyl CoA synthase gene from the corn smut pathogen, *Ustilago maydis. J. Gen. Microbiol. 135*:2675.
35. Buxton, F. P., D. I. Gwynne, R. W. Davies (1989). Cloning of a new bidirectionally selectable marker for *Aspergillus* strains. *Gene 84*:329.
36. Austin, B., R. M. Hall, B. M. Tyler (1990). Optimized vectors and selection for transformation of *Neurospora crassa* and *Aspergillus nidulans* to bleomycin and phleomycin resistance. Gene 93:157.
37. Ward, M., L. J. Wilson, C. L. Carmona, G. Turner (1988). The *oliC3* gene of *Aspergillus niger*: isolation, sequence and use as a selectable marker for transformation. *Curr. Genet. 14*:37.
38. May, G. S., J. Gambino, J. A. Weatherbee, N. R. Morris (1985). Identification and functional analysis of beta-tubulin genes by site specific integrative transformation in *Aspergillus nidulans. J. Cell Biol. 101*:712.
39. Orbach, M. J., E. B. Porro, C. Yanofsky (1986). Cloning and characterization of the gene for β-tubulin from a benomyl-resistant mutant of *Neurospora crassa* and its use as a dominant selectable marker. *Mol. Cell. Biol. 6*:2452.
40. Tilburn, J., F. Roussel, C. Scazzocchio (1990). Insertional inactivation and cloning of the *wA* gene of *Aspergillus nidulans. Genetics 126*:81.
41. Cheng, C., N. Tsukagoshi, S. Udaka (1990). Transformation of *Trichoderma viride* using the *Neurospora crassa pyr4* gene and its use in the expression of a Taka-amylase A gene from *Aspergillus oryzae. Curr. Genet. 18*:453.
42. Sanchez-Fernandez, R., S. E. Unkles, E. I. Campbell, J. A. Macro, E. Cerda-Olmedo, J. R. Kinghorn (1991). Transformation of the filamentous fungus *Gibberella fujikuroi* using the *Aspergillus niger niaD* gene encoding nitrate reductase. *Mol. Gen. Genet. 225*:231.
43. Burrows, D. M., T. J. Elliott, L. A. Casselton (1990). DNA-mediated transformation of the secondarily homothallic basidiomycete *Coprinus bilanatus. Curr. Genet. 17*:175.
44. Peng, M., N. K. Singh, P. A. Lemke (1992). Recovery of recombinant plasmids from *Pleurotus ostreatus* transformants. *Curr. Genet. 22*:53.
45. Marmeisse, R., G. Gay, J.-C. Debaud, L. A. Casselton (1992). Genetic transformation of the symbiotic basidiomycete fungus *Hebeloma cylindrosporum. Curr. Genet. 22*:41.
46. Salch, Y. P., M. N. Beremand (1993). *Gibberella pulicaris* transformants: state of transforming DNA during asexual and sexual growth. *Curr. Genet. 23*:343.
47. Judelson, H. S., B. M. Tyler, R. W. Michelmore (1991). Transformation of the oomycete pathogen, *Phytophthora infestans. Mol. Plant-Microbe Interact. 4*:602.

48. Judelson, H. S. (1993). Intermolecular ligation mediates efficient cotransformation in *Phytophthora infestans. Mol. Gen. Genet. 239*:241.
49. Judelson, H. S., M. D. Coffey, F. R. Arredondo, B. M. Tyler (1993). Transformation of the oomycete pathogen *Phytophthora megasperma* f. sp. *glycinea* occurs by DNA integration into single or multiple chromosomes. *Curr. Genet. 23*:211.
50. Hunter, G. D., C. R. Bailey, H. N. Arst (1992). Expression of a bacterial aspartase gene in *Aspergillus nidulans*: an efficient system for selecting multicopy transformants. *Curr. Genet. 22*:377.
51. Wernars, N., T. Goosen, L. M. J. Wennekes, et al. (1985). Gene amplification in *Aspergillus nidulans* by transformation with vectors containing the *amdS* gene. *Curr. Genet. 9*:361.
52. Upshall, A. (1986). Genetic and molecular characterization of *argB*$^+$ transformants of *Aspergillus nidulans. Curr. Genet. 10*:593.
53. Campbell, E. I., S. E. Unkles, J. Macro, C. A. M. J. J. Van den Hondel, J. R. Kinghorn (1989). An improved transformation system for *Aspergillus niger. Curr. Genet. 16*:53.
54. Turner, G. (1990). Strategies for cloning genes from filamentous fungi. In: *Applied Molecular Genetics of Fungi* (J. F. Peberdy, C. E. Caten, J. E. Ogden, J. W. Bennett, eds.). Cambridge University Press, Cambridge, p. 29.
55. Wahl, G. M., K. A. Lewis, J. C. Ruiz, B. Rothenberg, J. Zhao, G. A. Evans (1987). Cosmid vectors for rapid genomic walking, restriction mapping, and gene transfer. *Proc. Natl. Acad. Sci. USA 84*:2160.
56. Brody, H., J. Griffith, A. J. Cuticchia, J. Arnold, W. E. Timberlake (1991). Chromosome-specific recombinant DNA libraries from the fungus *Aspergillus nidulans. Nucl. Acids Res. 19*:3105.
57. Ballance, D. J., G. Turner (1985). Development of a high-frequency transforming vector for *Aspergillus nidulans. Gene 36*:321.
58. Ballance, D. J. (1986). Sequences important for gene expression in filamentous fungi. *Yeast 2*:229.
59. Kinsey, J. A. (1985). Neurospora plasmids. In: *Gene Manipulations in Fungi* (J. W. Bennett, L. L. Lasure, eds.). Academic Press, London, p. 245.
60. Van Heeswijck, R. (1986). Autonomous replication of plasmids in *Mucor* transformants. *Carlsberg Res. Commun. 51*:433.
61. Revuelta, J. L., M. Jayaram (1986). Transformation of *Phycomyces blakesleeanus* to G-418 resistance by an autonomously replicating plasmid. *Proc. Natl. Acad. Sci. USA 83*:7344.
62. Wostemeyer, J., A. Burmester, C. Weigel (1987). Neomycin resistance as a dominantly selectable marker for transformation of the zygomycete *Absidia glauca. Curr. Genet. 12*:625.
63. Perrot, M., C. Barreau, J. Begueret (1987). Nonintegrative transformation in the filamentous fungus *Podospora anserina*: stabilization of a linear vector by the chromosomal ends of *Tetrahymena thermophila. Mol. Cell. Biol. 7*:1725.
64. Tsukuda, T., S. Carleton, S. Fotheringham, W. K. Holloman (1988). Isolation and characterization of an autonomously replicating sequence from *Ustilago maydis. Mol. Cell. Biol. 8*:3703.

65. Samac, D. A., S. A. Leong (1989). Characterization of the termini of linear plasmids from *Nectria haematococca* and their use in construction of an autonomously replicating transformation vector. *Curr. Genet. 16*:187.
66. Gems, D., I. L. Johnstone, A. J. Clutterbuck (1991). An autonomously replicating plasmid transforms *Aspergillus nidulans* at high frequency. *Gene 98*:61.
67. Bruckner, B., S. E. Unkles, K. Weltring, J. R. Kinghorn (1992). Transformation of *Gibberella fujikuroi*: effect of the *Aspergillus nidulans* AMA1 sequence on frequency and integration. *Curr. Genet. 22*:313.
68. Woods, J. P., W. E. Goldman (1992). *In vivo* generation of linear plasmids with addition of telomeric sequences by *Histoplasma capsulatum. Mol. Microbiol. 6*:3603.
69. Wright, M. C., P. Philippsen (1991). Replicative transformation of the filamentous fungus *Ashbya gossypii* with plasmids containing *Saccharomyces cerevisiae* ARS elements. *Gene 109*:99.
70. Powell, W. A., H. C. Kistler (1990). In vivo rearrangement of foreign DNA by *Fusarium oxysporum* produces linear self-replicating plasmids. *J. Bacteriol. 172*: 3163.
71. Kistler, H. C., U. Benny (1992). Autonomously replicating plasmids and chromosome rearrangement during transformation of *Nectria haematococca. Gene 117*:81.
72. Randall, T., C. A. Reddy (1991). An improved transformation vector for the lignin-degrading white rot basidiomycete *Phanerochaete chrysosporium*. *Gene 103*:125.
73. Randall, T. A., C. A. Reddy (1992). The nature of extra-chromosomal maintenance of transforming plasmids in the filamentous basidiomycete *Phanerochaete chrysosporium. Curr. Genet. 21*:255.
74. Randall, T., C. A. Reddy, K. Boominathan (1991). A novel extrachromosomally maintained transformation vector for the lignin-degrading basidiomycete *Phanerochaete chrysosporium. J. Bacteriol. 173*:776.
75. Gems, D. H., A. J. Clutterbuck (1993). Co-transformation with autonomously-replicating helper plasmids facilitates gene cloning from an *Aspergillus nidulans* gene library. *Curr. Genet. 24*:520.
76. Gems, D., A. Aleksenko, L. Belenky, et al. (1994). An "instant gene bank" method for gene cloning by mutant complementation. *Mol. Gen. Genet. 242*:467.
77. Bowyer, P., A. E. Osbourn, M. J. Daniels (1994). An "instant gene bank" method for heterologous gene cloning: complementation of two *Aspergillus nidulans* mutants with *Gauemannomlyces graminis* DNA. *Mol. Gen. Genet. 242*:448.
78. Schechtman, M. G. (1987). Isolation of telomere DNA from *Neurospora crassa. Mol. Cell. Biol. 7*:3168.
79. Schechtman, M. G. (1990). Characterization of telomere DNA from *Neurospora crassa. Gene 88*:159.
80. Borbye, L., H. Giese (1994). Genome manipulation in recalcitrant species: construction and characterization of a yeast artificial chromosome (YAC) library from *Erysiphe graminis*, an obligate fungal pathogen of barley, f. sp. *hordei. Gene, 144*: 107–111.
81. Burmester, A., A. Wostemeyer, J. Wostemeyer (1990). Integrative transformation of a zygomycete, *Absidia glauca*, with vectors containing repetitive DNA. *Curr. Genet. 17*:155.

82. Miller, B. L., K. Y. Miller, W. E. Timberlake (1985). Direct and indirect gene replacements in *Aspergillus nidulans. Mol. Cell. Biol. 5*:1714.
83. Van Gorcom, R. F. M., P. J. Punt, P. H. Pouwels, C. A. M. J. J. Van den Hondel (1986). A system for the analysis of expression signals in *Aspergillus. Gene 48*:211.
84. Roberts, I. N., R. P. Oliver, P. J. Punt, C. A. M. J. J. Van den Hondel (1989). Expression of the *Escherichia coli* β-glucuronidase gene in industrial and phytopathogenic filamentous fungi. *Curr. Genet. 15*:177.
85. Punt, P. J., P. A. Greaves, A. Kuyvenhoven, et al. (1991). A twin-reporter vector for simultaneous analysis of expression signals of divergently transcribed, contiguous genes in flamentous fungi. *Gene 104*:119.
86. Waring, R. B., G. S. May, N. R. Morris (1989). Characterization of an inducible expression system in *Aspergillus nidulans* using *alcA* and tubulin-coding genes. *Genetics 79*:119.
87. Berka, R. M., K. H. Kodama, M. W. Rey, L. J. Wilson, M. Ward (1991). The development of *Aspergillus niger* var. *awamori* as a host for the expression and secretion of heterologous gene products. *Biochem. Soc. Transact. 19*:681.
88. Cullen, D., G. L. Gray, L. J. Wilson, et al. (1987). Controlled expression and secretion of bovine chymosin in *Aspergillus nidulans. Bio/Technology 5*:369.
89. Ward, M., L. J. Wilson, K. H. Kodama, M. W. Rey, R. M. Berka (1990). Improved production of chymosin in *Aspergillus* by expression as a glucoamylase-chymosin fusion. *Bio/Technology 8*:435.
90. Christensen, T., H. Woeldike, E. Boel, S. B. Mortensen, K. Hjortshoej, L. Thim, M. T. Hansen (1988). High level expression of recombinant genes in *Aspergillus oryzae. Bio/Technology 6*:1419.
91. Contreras, R., D. Carrez, J. R. Kinghorn, C. A. M. J. J. Van den Hondel, W. Fiers (1991). Efficient KEX2-like processing of a glucoamylase-interleukin-6-fusion protein in *Aspergillus nidulans* and secretion of mature interleukin-6. *Bio/Technology 9*:378.
92. Kinsey, J. A., J. Helber (1989). Isolation of a transposable element from *Neurospora crassa. Proc. Natl. Acad. Sci. USA 86*:1929.
93. Daboussi, M.-J., T. Langin, Y. Brygoo (1992). Fot1, a new family of fungal transposable elements. *Mol. Cell. Biol. 232*:12.
94. Glayzer, D. C., I. N. Roberts, D. B. Archer, R. P. Oliver (in press). *Mol. Gen. Genet.*
95. Pontecorvo, G., J. A. Roper, L. H. Hemmons, K. D. Macdonald, A. W. J. Bufton (1953). The genetics of *Aspergillus nidulans. Adv. Genet. 5*:141.
96. Clutterbuck, A. J. (1974). *Aspergillus nidulans.* In: *Handbook of Genetics*, Vol. 1 (R. C. King, ed.). Plenum Press, London, p. 447.
97. Punt, P. J., M. A. Dingemarse, A. Kuyvenhoven, R. D. M. Soede, P. H. Pouwels, C. A. M. J. J. Van der Hondel (1990). Functional elements in the promoter region of the *Aspergillus nidulans gpdA* gene encoding glyceraldehyde-3-phosphate dehydrogenase. *Gene 93*:101.
98. Cove, D. J. (1966). The induction and repression of nitrate reductase in the fungus *Aspergillus nidulans. Biochem. Biophys. Acta 113*:51.

9

Mapping and Breeding Strategies

Klaas Swart
Wageningen Agricultural University, Wageningen, The Netherlands

1. INTRODUCTION

Strategies in genetic mapping vary among different fungal species because fungi differ in the availability of genetic markers and in basic processes that allow application of genetic analysis. *Aspergillus nidulans* and *A. niger* have been used in many genetic studies and will therefore serve as good examples to demonstrate genetic mapping strategies.

In *A. nidulans* a sexual reproduction cycle comprising sexual differentiation, dikaryon formation, karyogamy, meiosis, and ascospore formation is present, so meiotic mapping of genes is applicable. Next to this, haploid nuclei in the multinuclear vegetative cells of the fungus may occasionally fuse to form a somatic diploid that subsequently can segregate into haploids in the absence of meiosis. A heterozygous diploid arises upon fusion of genetically different nuclei, as they are present in heterokaryotic cells. Such a diploid can give rise to recombinant haploids which allow genetic analysis in this so-called parasexual cycle.

A. niger is an imperfect fungus, thus lacking meiotic mapping possibilities. The alternative, parasexual cycle is operative, however, and genetic analyses are feasible. More recently, molecular techniques have been introduced as new strategies in genetic mapping.

The application of breeding strategies with the aim to improve fungal strain performance has not been widely practiced. Industrial strain improvement traditionally relies on random mutagenesis and subsequent screening procedures. Although this approach has been successful in many cases, the whole procedure is time-consuming, and the final outcome is not predictable. Repeated mutagenesis, e.g., may severely weaken a strain in characteristics that may not be directly involved in the formation of the desired product. General physiology, growth behavior, performance in large-scale processes, etc., may be impaired. Such disadvantages imply that many new mutants have to be tested to get one strain with improved performance. If available, breeding strategies can overcome several problems encountered by the traditional approach.

In this chapter a general outline of breeding strategies will be discussed and a few examples will be worked out.

2. COMPARISON OF GENETIC MAPPING TECHNIQUES

Genetic mapping by sexual crosses is a routine procedure in many organisms. This is seldom satisfactory in *A. nidulans*, however, because the meiotic recombination frequency is extremely high and linkage will not be observed in the vast majority of cases. Clutterbuck [1] calculated the total number of map units in each chromosome of *A. nidulans*: more than 4000 map units are distributed over eight linkage groups. The eight linkage groups vary in size; the map units per linkage group are depicted in Figure 1.

Figure 1 shows that several crossovers occur per meiosis in each chromosome. A second feature of meiotic recombination in *A. nidulans* concerns the absence of crossover interference, which means that genes are randomly distributed over the genetic map (n.b.: this is not necessarily identical to the physical map). So, genetic mapping by meiotic analysis is only feasible when short distances have to be determined.

Mitotic mapping was introduced by Roper [2] and Pontecorvo and coworkers [3] and was found to be extremely useful in the allocation of genes to chromosomes (see Chapter 4 for more details). The so-called parasexual cycle consists of the fusion of two genetically different haploid nuclei that may be present in a heterokaryon. Thus, a heterozygous diploid is formed and subsequent breakdown of this diploid can occur by nondisjunction. The process of nondisjunction involves random loss of one of each pair of chromosomes, and many different combinations of parental chromosomes are produced ($n = 8 \rightarrow 2^8 = 256$ combinations). In a somatic diploid crossovers may occur at a low frequency, producing intrachromosomal recombinants. The frequency of these crossovers is very low, and selective procedures have to be used to

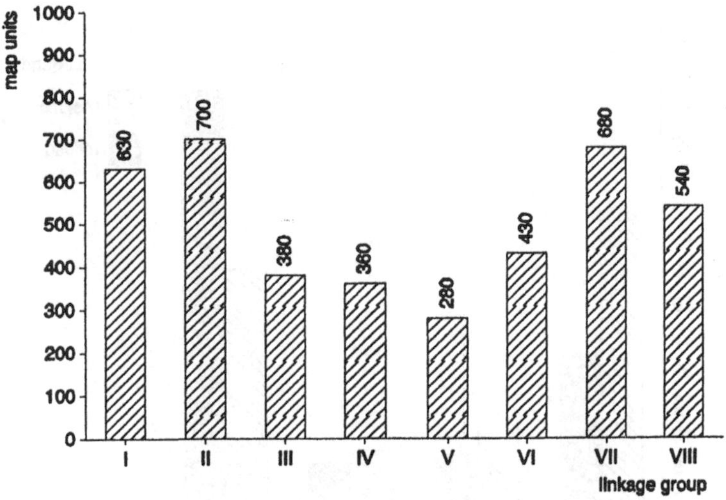

Figure 1 Genetic length of linkage groups of *Aspergillus nidulans*.

recover such recombinants. Due to the rarity of these intrachromosomal recombinants the parasexual cycle is useful to assign genes to chromosomes (Chapter 4). Mitotic crossing over in a diploid results in homozygosity of the chromosome arm distal to the site of crossing over. Depending on the markers available on the chromosome arm involved, such homozygotes can be recovered by selective procedures, and the site of crossing over can be deduced from the frequency of different genotypes among the homozygotes. Also the order of genes on a chromosome arm can be determined (Chapter 4). Crossing over in one arm will not affect the other arm, and thus information is obtained about the location of the centromere. The site of the centromere cannot be determined by standard meiotic analysis, unless tetrad analysis can be applied.

Pontecorvo and Käfer [4] and Käfer [5,6] compared the relative frequencies of mitotic and meiotic recombination in several chromosomal regions of *A. nidulans*. They concluded that mitotic crossing over occurred preferentially in the regions adjacent to the centromeres. As an example, the relative meiotic map units (from Clutterbuck [7], corrected by Haldane's mapping function [8]) and the mitotic recombination frequencies of a part of the left arm of chromosome I are depicted in Figure 2.

Pulsed-field gel electrophoresis (PFGE) has recently been applied to map genes on a chromosome. PFGE separates chromosomes into distinct bands. In cases where a probe of a gene is available, blotting and hybridizations

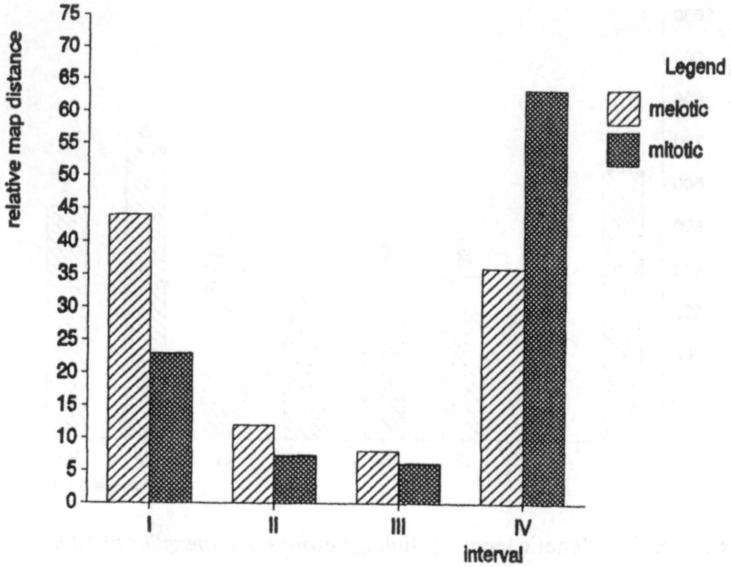

Figure 2 Comparison of relative meiotic and mitotic map distances in part of the left arm of chromosome I of *A. nidulans*. The map comprises the following genes and intervals:

Genes:	*suAadE*——*riboA*——*anA*——*adG*——*centromere*
Interval:	I II III IV

can be used to find the chromosome of the resident gene. This way of genetic analysis is the only way if no phenotype of a particular gene is known—e.g., the rRNA genes [9] (see also Chapter 5 and Swart et al. [10] for an overview).

3. PITFALLS IN GENETIC MAPPING

Standard genetic mapping, according to the general descriptions mentioned in Chapters 3 and 4, looks more or less straightforward. In practice, however, unexpected problems can be met due to specific (genetic) characteristics of the strains that are examined. We will focus on two important characteristics–translocations and clonal segregation of mitotic recombinants.

Translocations in a fungal strain, in general, cannot be detected cytologically but are easily found when linkage of marker genes can be studied in the parasexual cycle. Genes in one linkage group segregate together if both parents are nontranslocation strains. In the case in which one strain harbors a translocation, haploidization will not segregate all individual chromosomes because some segregants will contain a duplication and others will be missing an essential chromosome part (Fig. 3). The duplication-type segregant will in almost all cases not be picked up from the haploidization medium due to poor growth of this unbalanced genotype. The other one, deficiency type, will be inviable because in the haploid fase essential genes are missing.

Thus, markers of the two chromosomes involved in the translocation segregate together in the viable haploid segregants, and genetic linkage is concluded for genes that are on different chromosomes. Detection of translocations in meiotic genetic analysis is more difficult, although it is possible in several cases. First, linkage between previously unlinked genes can be found if marker genes are present close enough on both sides of the translocation breakpoint. Secondly, the duplication-type colony is often viable and will occur in 1 out of 3 progeny colonies of a cross, provided plating conditions allow the growth and detection of these deviant (crinkled) phenotypes.

In the course of growth of a heterozygous diploid, mitotic crossing over may occur, and the resultant recombinant can become a significant part of the diploid colony. Such a recombinant can become the main part of a diploid colony if it has some growth advantage over the original heterozygous diploid. Upon haploidization a fairly high recombinant frequency will then be detected

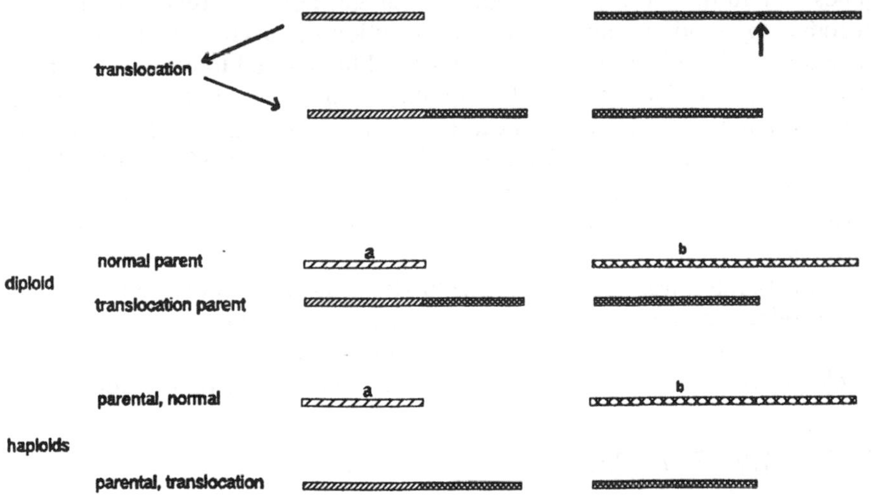

Figure 3 Mitotic mapping with a translocation in one of the parental strains results in apparent linkage of the markers a and b. The arrow indicates the translocation breakpoint.

between genes that are definitely in the same linkage group. In our work with *A. niger* we have repeatedly found high percentages of recombination between the *bioA* and *lysA* marker genes on linkage group III. At the time we clearly recognized this problem, we adjusted our experimental protocol: a diploid was propagated as little as possible, and, if available, spores of different original diploids were taken to be plated (separately) on the haploidization medium [11]. If by chance one sample of spores would contain a clonal segregant, this should be detected by the differences of linkage results obtained from the separate analyses of the different diploids. The data of a thus detected clonal segregant can be eliminated from the results, and again low percentages of recombination are found for marker genes on the same linkage group. This problem has not been reported for *A. nidulans*.

4. OUTLINE OF BREEDING STRATEGIES

Breeding strategies can be applied if genetically marked strains are available. Extensive genetics of the species involved is not directly necessary; however, genetic mechanisms should be accessible. These can be genetics based on meiotic recombination (see Chapter 3) and/or based on the parasexual processes (see Chapter 4).

The application of genetics requires good genetic markers: a strain collection of the species containing several auxotrophic and resistance marker genes will fulfill these requirements. Such markers are relatively easy to introduce by random mutagenesis and selection of specific phenotypes (see Chapter 2). Genetic mapping techniques will be applied to establish linkage groups and to construct multiple marked strains. In different species such strains are already available and can be obtained from the laboratory where they were collected and/or from a stock center (e.g., FGSC or ATCC). A prerequisite is, of course, that no incompatibility exists between the strains that have to be combined or that any incompatibility could be overcome by protoplast fusion.

In our laboratory *Aspergillus niger* has been used to establish an extensive strain collection, to elaborate on genetic techniques, and to construct master strains [12–15]. In different applied projects this knowledge was employed to reach specific goals that were related to some sort of production performance.

5. EXAMPLES OF GENETIC APPROACHES

A. Controlled Recombination of Desired Mutations

Many independent mutants affected in the production/excretion of the enzyme glucose oxidase (GOX) were isolated using specific screening techniques

[16,17]. In order to prevent undesired "genetic background" mutations, each mutant was obtained after low-dose UV irradiation in a parental strain that contained only one marker gene. Three parental strains with dissimilar auxotrophic markers were used to enable subsequent selection of forced heterokaryons between different mutants aimed at testing allelism. Thus seven complementation groups were distinguished among the GOX-overproducing mutants, and two different genetic loci gave nonproducing phenotypes. Whereas each of the overproducing types was affected in a distinct part of the overall route leading to GOX production/excretion, it was of interest to test the possibility of combining different mutations in one final strain. To reach this goal the different *gox* genes were allocated to linkage groups by parasexual genetic analysis using master strains with all but one linkage group marked (note: at that time no suitable marker on linkage group VIII was available in a qualified master strain). From these experiments, several haploid segregants containing different marker genes in addition to the specific *gox* mutation were saved for subsequent use in recombination experiments. Diploids were constructed from haploids that contained *gox*-complementing genes on different chromosomes and marker genes on other relevant chromosomes. Second-generation haploid segregants were obtained, and selection against the auxotrophic markers on the chromosomes homologous to those containing the *gox* genes resulted in a recombination strain harboring two different *gox* mutations. This procedure could be repeated another two times resulting in a strain containing four different *gox* mutations. Any auxotrophic marker can be removed in a subsequent recombination experiment using an appropriate partner strain. The possibility of testing and characterizing each of the mutations separately before it is taken as a candidate to be included in the ultimate strain is an important advantage of this procedure. An ideal setup is illustrated in Figure 4. Finally, phenotypic testing of the ultimate strain should be done to monitor the production performance.

B. Genetic Mapping in a "Wild-Type" Strain

Suppose genetic analysis is required in a strain with specific qualities (production performance or one obtained after genetic engineering) but lacking genetic markers. One option could be to introduce a new genetic marker by mutagenesis; however, this might not be desired because of the risk of the introduction of nonrecognizable additional mutation(s) or intrachromosomal recombination in, e.g., the gene constructs that were added to the host genome by genetic engineering. We have successfully applied a different approach in *A. niger*. Many marker genes are available in our strain collection of this fungus, and several master strains for mitotic mapping have been constructed, covering

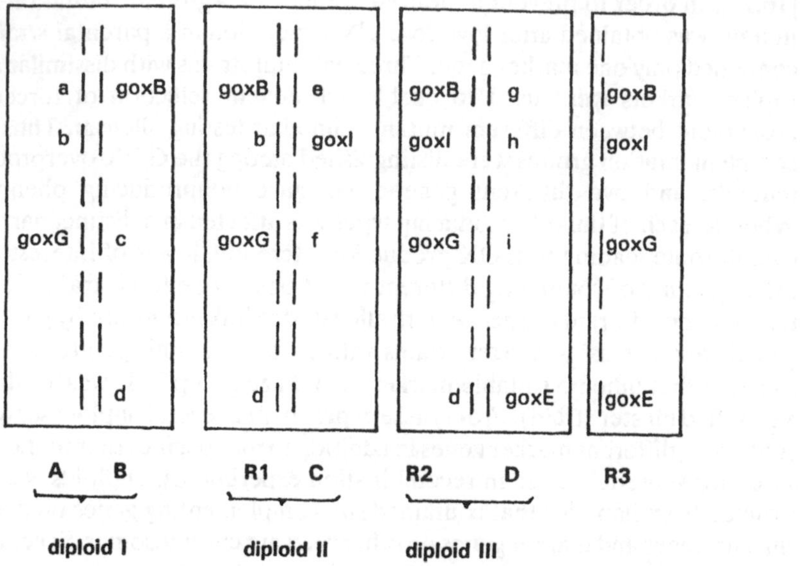

Figure 4 Diagram of a controlled strain construction schedule. A, B, C, and D = single gox-mutant strains with auxotrophic markers (a, b, c, d, e, f, g, h, i). R1, R2, and R3 = strains constructed by recombination.

all eight linkage groups. Most marker genes confer auxotrophies, but a few resistant genes are included in some master strains.

Especially, dominant resistance markers are useful to solve the above-mentioned genetic analysis problem. A dominant resistance mutation will give the resistance phenotype to a heterokaryon and to a constructed diploid as well. Thus, a combination of a "wild-type" strain with a master strain containing several auxotrophic markers and a dominant resistance marker enables the isolation of a heterokaryon and heterozygous diploids by selection upon prototrophy (characteristic of the wild-type strain) and resistance (characteristic of the master strain). The wild-type strains were transformants of *A. niger* N402 [18]. Different transformants harbored each independent multicopy inserts of the *A. niger* glucoamylase gene combined with the *A. nidulans amdS* gene. Such a transformant is at risk to recombine within the multicopy insert upon treatment with mutagens, so any manipulation to introduce a new marker should be avoided. Parasexual genetic analysis was performed on these strains by combining each transformant with a tester strain containing several auxot-

	I	II	III	IV	V	VI	VII	VIII	?
N915	fwnA1	argH12	bioA1	leuA1	pheA1	+	oliC2	crnB12	
	===	===	===	===	===	=	===	===	===
B1									glaA/amdS

Heterokaryons and heterozygous diploids were selected on MM + 3 μg/ml oligomycin. Haploid segregants of these diploids were tested for linkage of the amdS-gene to one of the tester strain markers.

Result:	amdS present	amdS absent
fwnA1$^+$	48	2
fwnA1	3	48

Conclusion: insert with glaA/amdS is on linkage group I

Figure 5 Strategy of gene mapping in a wild-type strain.

rophic markers and the dominant nuclear oligomycin resistance mutation oliC2 [19]. Whereas the phenotype of the glucoamylase multicopy insert was difficult to assess, the *amd*S genc(s), accompanying the *gla*A genes, were used to test presence or absence of the transformed sequences. This strategy is illustrated in Figure 5.

The inserts in seven different transformants could be assigned to linkage groups, and in each transformant the insert appeared to be in only one linkage group. This suggested the occurrence of the insertion of an array of transforming sequences at only one locus. Knowing the linkage group allocation of the insert in each transformant, a recombination procedure similar to the one discussed in the previous paragraph could be used to construct new strains containing *gla*A multicopy inserts on two different linkage groups. Hence, it could be tested whether the glucoamylase protein production of different transformants is additive or is reaching a limit.

C. A Mutation Strategy to Block the Degradation of an Intermediate Compound

Several (bio-)chemical compounds are not easily produced by chemical synthesis, because either one or more steps in the synthesis are extremely difficult to perform, or a monostereospecificity of the product is required, which often cannot be obtained by in vitro synthesis. In that case, a biotechnological production method can be the right alternative. Hydroxylated aromatic compounds belong to this category. We have recently investigated the para-hydroxylation of benzoate by *Aspergillus niger* as a model [20]. Here the desired compound, p-hydroxybenzoate, is an intermediate in a degradation pathway. Ideally, the first step(s) of the degradation route should be increased and subsequent breakdown of the intermediate should be blocked. A mutational analysis of the pathway will give information about steps and genes in the degradation route.

In the model study [20] strain ATCC 1015 was used. Mutations were induced by low-dose UV irradiation aimed at a survival rate of approximately 70%. In the first experiments we found that a much higher UV dose was required to get 70% survival compared to our standard laboratory strain N400 (= ATCC 9029). Additionally, hardly any mutants (e.g., simple amino acid auxotrophic ones) could be isolated after a filtration enrichment procedure. These two findings may indicate that the conidiospores that were used for mutagenesis were not mononucleate. A cytological investigation of the spores using the fluorescent DNA-specific dye DAPI indeed showed that most of the spores were binucleate. In subsequent mutation induction experiments a segregation step was included: irradiated spores were propagated on complete medium (CM), and the "second-generation" spores were harvested, split up into several portions, and used for mutant enrichment.

The filtration enrichment was done by incubation of 10^6 spores/mL liquid minimal medium (supplemented for the auxotrophy of the strain) with 0.1% benzoate or 4-hydroxybenzoate as carbon source. Germinating conidiospores and growing mycelium were removed by filtration over cotton–wool every 8 or 16 hours; ungerminated spores were collected by centrifugation and resuspended in fresh medium for the next round of enrichment. Whereas the substrates used in this procedure are poor carbon sources, the enrichment process had to be continued for 3 to 4 days. Finally, the ungerminated conidiospores were rescued by plating on CM, and mutants were identified by growth tests on supplemented medium (SM) + glucose and on SM + benzoate. This procedure yielded several mutants affected in the catabolism of benzoic acid—e.g., Bph mutants lacking benzoate-4-hydroxylase, and Phh mutants lacking benzoate-3-hydroxylase. Thus, the route of biodegradation of benzoate in *Aspergillus niger* was confirmed, and it became possible to clone the respective genes. Further analyses of the steps in the degradation route revealed new problems which hampered manipulation of this route [21,22]. These problems, however, are not in the scope of the present overview.

REFERENCES

1. Clutterbuck, A. J. (1992). Sexual and parasexual genetics of *Aspergillus* species. In: J. W. Bennett, M. A. Klich (eds.). Apsergillus: *Biology and Industrial Applications*. Butterworth-Heinemann, Boston, pp. 3–18.
2. Roper, J. A. (1952). Production of heterozygous diploids in *Aspergillus nidulans*. *Experientia 8*:14–15.
3. Pontecorvo, G., J. A. Roper, L. M. Hemmons, K. D. MacDonald, A. W. J. Bufton (1953). The genetics of *Aspergillus nidulans*. *Adv. Genet. 5*:141–238.
4. Pontecorvo, G., E. Käfer (1958). Genetic analysis based on mitotic recombination. *Adv. Genet. 9*:71–104.
5. Käfer, E. (1958). An 8-chromosome map of *Aspergillus nidulans*. *Adv. Genet. 9*:105–145.
6. Käfer, E. (1977). Meiotic and mitotic recombination in *Aspergillus* and its chromosomal aberrations. *Adv. Genet. 19*:33–131.
7. Clutterbuck, J. (1993). *Aspergillus nidulans*. In S. J. O'Brien (ed.). *Genetic Maps, Locus Maps of Complex Genomes.* Cold Spring Harbor Laboratory Press, Cold Spring Harbor, NY, pp. 3.87–3.90.
8. Haldane, J. B. S. (1919). The combination of linkage values, and the calculation of distance between the loci of linked factors. *J. Genet. 8*:299–309.
9. Debets, A. J. M., E. F. Holub, K. Swart, H. W. J. van den Broek, C. J. Bos (1990). An electrophoretic karyotype of *Aspergillus niger. Mol. Gen. Genet. 224*:264–268.
10. Swart, K., A. J. M. Debets, E. F. Holub, C. J. Bos, R. F. Hoekstra (1994). Physical karyotyping: genetic and taxonomic applications in Aspergilli. In K. A. Powell, A. Renwick, J. F. Peberdy (eds.). *The genus* Aspergillus. Plenum Press, New York and London, pp. 233–240.
11. Debets, A. J. M., K. Swart, C. J. Bos (1990). Genetic analysis of *Aspergillus niger*: Isolation of chlorate resistance mutants, their use in mitotic mapping and evidence for an eighth linkage group. *Mol. Gen. Genet. 221*:453–458.
12. Bos, C. J., A. J. M. Debets, K. Swart, A. Huybers, G. Kobus, S. M. Slakhorst (1988). Genetic analysis and the construction of master strains for assignment of genes to linkage groups in *Aspergillus niger. Curr. Genet. 14*:437–443.
13. Bos, C. J., S. M. Slakhorst, A. J. M. Debets, K. Swart (1993). Linkage group analysis in *Aspergillus niger. Appl. Microbiol Biotechnol. 38*:742–745.
14. Bos, C. J., F. Debets, K. Swart (1993). Aspergillus niger genetic loci (n = 8). In S. J. O'Brien (ed.). *Genetic Maps, Locus Maps of Complex Genomes.* Cold Spring Harbor Laboratory Press, Cold Spring Harbor, NY, pp. 3.87–3.90.
15. Debets, F., K. Swart, R. F. Hoekstra, C. J. Bos (1993). Genetic maps of eight linkage groups of Aspergillus niger based on mitotic mapping. Curr. Genet. 23: 47–53.
16. Witteveen, C. F. B., P. van de Vondervoort, K., Swart, J. Visser (1990). Glucose oxidase overproducing and negative mutants of *Aspergillus niger. Appl. Microbiol. Biotechnol. 33*:683–686.
17. Swart, K., P. J. I. van de Vondervoort, C. F. B. Witteveen, J. Visser (1990). Genetic localization of a series of genes affecting glucose oxidase levels in *Aspergillus niger. Curr. Genet. 18*:435–439.

18. Verdoes, J. C., P. J. Punt, J. M. Schrickx, H. W. Van Verseveld, A. H. Stouthamer, C. A. J. M. M. Van den Hondel (1993). Glucoamylase overexpression in *Aspergillus niger*: molecular genetic analysis of strains containing multiple copies of the *gla*A gene. *Transgenic Res. 2*:84–92.
19. Verdoes, J. C., A. D. Van Diepeningen, P. J. Punt, A. J. M. Debets, A. H. Stouthamer, C. A. M. J. J. Van den Hondel (1994). Evaluation of molecular and genetic approaches to generate glucoamylase overproducing strains of *Aspergillus niger. J. Biotechnol. 36*:165–175.
20. Boschloo, J. G., A. Paffen, T. Koot, et al. (1990). Genetic analysis of benzoate metabolism in *Apsergillus niger. Appl. Microbiol. Biotechnol. 34*:225–228.
21. Boschloo, J. G., E. Moonen, R. F. M. Van Gorcom, H. F. M. Hermes, C. J. Bos (1991). Genetic analysis of *Aspergillus niger* mutants defective in benzoate-4-hydroxylase function. *Curr. Genet. 19*:261–264.
22. Gorcom, R. F. M., van, J. G. Boschloo, A. Kuyvenhoven, et al. (1990). Isolation and molecular characterization of the benzoate-*para*-hydroxylase gene (*bph*A) of *Aspergillus niger*: a member of a new gene family of the cytochrome P450 superfamily. *Mol. Gen. Genet. 223*:192–197.

10

Mitochondrial Genetics of *Saccharomyces cerevisiae*

The Way from Genetic Crosses to Transposable Elements

Klaus Wolf
Rheinisch-Westfälische Technische Hochschule, Aachen, Germany

1. INTRODUCTION

Mitochondria are semiautonomous organelles that are responsible for respiration and oxidative phosphorylation in all eukaryotic cells. The mitochondrial genome carries the genetic information for only a few essential functions. Thus formation of mitochondria depends on a complex interplay between the nucleus and the mitochondria. The majority of mitochondrial proteins (more than 95%) are encoded by nuclear genes and synthesized on cytoplasmic ribosomes. Considerable effort, beginning in the early 1970s, has resulted in a very detailed understanding of mitochondrial genomes and genes, their expression, and their dependence on nuclearly encoded functions, so the mitochondrial genome of baker's yeast is one of the best characterized eukaryotic genomes. Genetics of mitochondria was initiated by the discovery of the respiratory-deficient petite colony mutant [for reviews see 1,2]. Later on, the discovery of mutants resistant to antibiotics and with defined lesions in genes for respiratory chain components [2,3] has provided the basis for a genetic and molecular analysis of transmission, segregation, and recombination of mitochondrial genomes. The discovery of mitochondrial introns and their functions has initiated a large series of studies on autocatalytic RNA splicing and mobile genetic elements.

Two key concepts of yeast as a model organism to study mitochondrial functions are the following:

1. The yeast *Saccharomyces cerevisiae* is an ideal eukaryotic organism because of its ease of cultivation, the availability of thousands of well-defined mutants, and the applicability of powerful molecular techniques. When the sequencing of the yeast genome is finished (probably in 1996), a solid basis for functional analyses will have been provided.

2. Since *Saccharomyces cerevisiae* can provide its energy by both fermentation and respiration, it is able to survive on fermentable carbon sources with grossly altered or even without mitochondrial DNA. This property of the so-called petite positive yeasts allows the introduction of any kind of mutational alterations in the mitochondrial DNA.

2. PRINCIPLES

A. The Mitochondrial Gene Map of *Saccharomyces cerevisiae*

Before entering the practical approach, the mitochondrial gene map with the relevant mutational sites will be presented (Fig. 1) and described briefly.

B. The Mitochondrial Standard Cross

Zygotes issued from the cross of two nonisomitochondrial cells are heteroplasmic, since they contain two distinct mitochondrial DNA populations and, later, the recombinants formed between them. The vegetative multiplication of these zygotes is accompanied by a rapid segregation of mitochondrial genomes, so that virtually all diploid progeny is homoplasmic after 20–25 generations. In the so-called standard cross, a large population of zygotes is formed by random mass mating between the parents and allowed to grow on nonselective medium. The entire progeny is then analyzed using a representative sample. Sporulation of a homoplasmic diploid clone will produce four spores with identical mitochondrial genotype, whereas the nuclear genes will segregate in a Mendelian fashion (2:2; see Chapter 3 for meiotic recombination). The mitochondrial standard cross is depicted schematically in Figure 2.

C. A Mobile Intron in the Mitochondrial Gene Encoding the Large Ribosomal RNA

Introns belonging to group I [for a review see 4] have been observed in a variety of locations in mitochondrial, chloroplast, cyanellar, and nuclear genetic systems in eukaryotes as well as in bacteriophages. Several group I introns are able to splice autocatalytically. The intron in the gene encoding the large ribosomal RNA (*rnl*) in mitochondria is able to splice autocatalytically. The

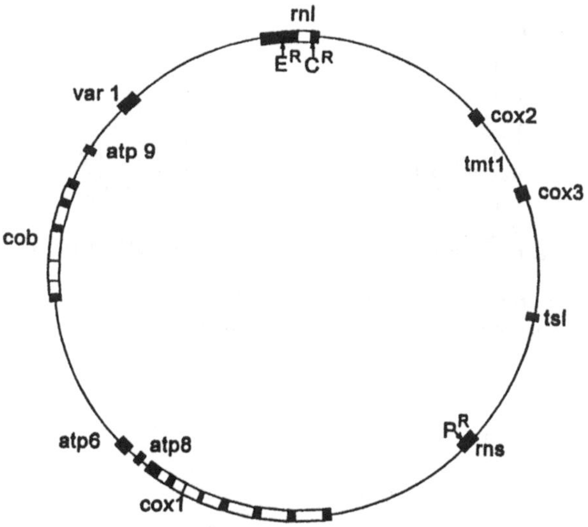

Figure 1 Simplified gene map of the mitochondrial genome of *Saccharomyces cerevisiae* (long form; i.e., with the maximum number of introns = 85 kbp). Black bars indicate exons, and white bars indicate introns. Nomenclature: *cox1, cox2, cox3* = genes for subunits 1, 2, and 3 of cytochrome c oxidase; *cob* = gene for apocytochrome b; *atp6, atp8, atp9* = genes for subunits 6, 8, and 9 of ATPase; *var1* = gene for a small ribosomal subunit protein; *tsl* = tRNA synthesis locus; tRNA genes are not shown except for tmt1; *tmt1* = gene for tRNAthr, the only gene located on the complementary strand; *rns* = gene for the small ribosomal RNA; *rnl* = gene for the large ribosomal RNA; E^R = mutational site conferring resistance to erythromycin; C^R = mutational site conferring resistance to chloramphenicol; P^R = mutational site conferring resistance to paromomycin.

most interesting aspect of this (and several other) introns is their ability to propagate themselves by inserting at predetermined positions into intronless sites of genes. This kind of intron movement is called intron homing. The mobility of group I introns was described genetically [4] long before the detection of introns. The systematic evaluation of crosses with mitochondrial antibiotic resistance markers has revealed two natural allelic forms of a genetic determinant, called *omega*$^+$ (ω^+) and *omega*$^-$ (ω^-). After the discovery of introns it was realized that the omega determinant corresponds to an optional intron in the *rnl* gene, the intron being present in ω^+ strains, but absent from ω^- strains. Sequence determination revealed an internal open reading frame in the *rnl* gene, whose gene product is a double strand-specific endonu-

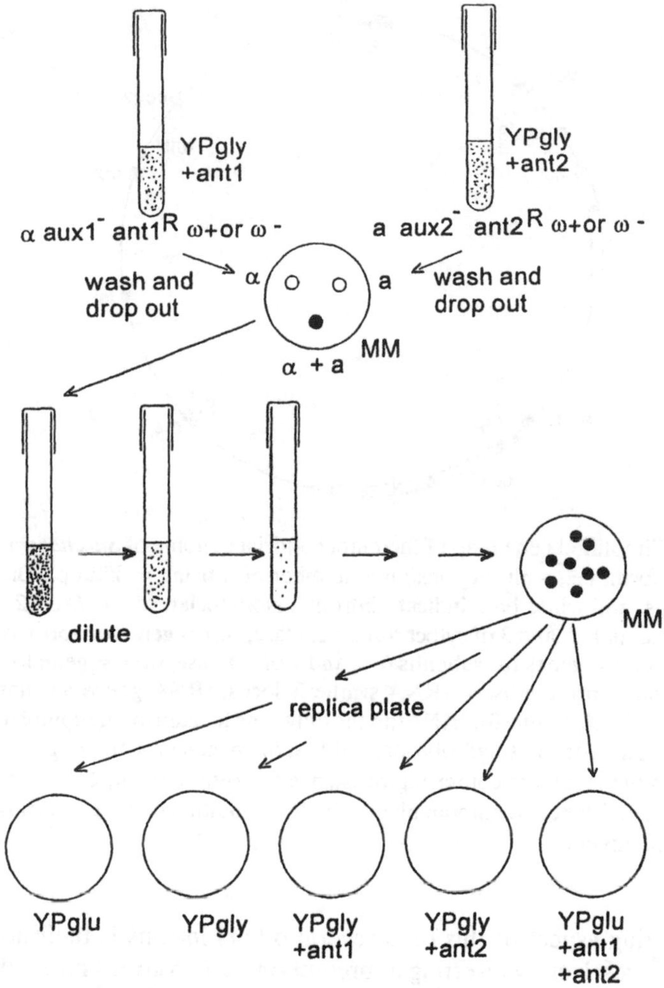

Figure 2 Schematic presentation of a mitochondrial standard cross. Further explanations and abbreviations of media are given in the text.

clease. In crosses between ω^+ and ω^- strains the intron was transmitted to more than 95% of the progeny. Molecular analysis of mitochondrial DNA in individual clones of the progeny shows that the intron conversion is associated with the co-conversion of flanking DNA sequences which carry the genetic markers conferring resistance to chloramphenicol (C) and erythromycin (E)

3. PRACTICE

A. Experiments

Isolation of Mitochondrial Mutants Resistant to Chloramphenicol, Erythromycin, and Paromomycin

Mitochondrial point mutations conferring resistance to antibiotics (ant^R) are rare and occur at a frequency of 10^{-7} to 10^{-8} per cell/division. They can easily be selected on glycerol complete medium (YPgly) supplemented with an appropriate concentration of the antibiotic. The following method should be used to select independent drug-resistant mutants. Alternatively, the desired mutants can be obtained from strain collections. Care should be taken, however, for purification of mutants and verification of the mutant phenotype. The procedure is the following:

1. Grow a haploid strain with at least one auxotrophic marker in YPgly.
2. Dilute in sterile water, plate 0.1 mL on YPgly plates, and incubate at 28°C for 3 days.
3. Pick single colonies (3 or more) and inoculate individually in 5 mL of YPgly medium. Incubate for 2 days at 28°C.
4. Plate 0.1 mL of each culture on YPgly plates containing the relevant antibiotics (see Materials).
5. Incubate antibiotic plates in a moist incubator at 28°C for at least 1 week and up to 3 weeks to allow appearance of resistant colonies.
6. Pick up a single colony per original subclone to obtain mutants of independent origin.
7. Inoculate 5 mL of YPgly and grow at 28°C for 2 days.
8. Purify mutant cultures by streaking on antibiotic plates to obtain single colonies.
9. Reisolate single clones from the antibiotic plate and store the mutants.

Standard "Homosexual" and "Heterosexual" Mitochondrial Crosses

The basic procedure of a mitochondrial standard cross is outlined below and summarized in a flow diagram (Fig. 2) for a given homosexual cross ($\omega^- \times \omega^+$) involving the two mitochondrial markers C^R and E^R. The same procedure has to be applied for a heterosexual cross ($\omega^+ \times \omega^-$).

1. Grow mutants in selective media with the relevant antibiotics.
2. Test on minimal medium (MM) for auxotrophic requirements.
3. Wash and mix the two parental strains.
4. Drop out on MM the individual haploid strains as control, together with a 1:1 mixture of the two parental strains.
5. Incubate until confluent growth of the mixture appears on MM (2–3 days, 28°C).

6. Suspend a loopful of the mixture in sterile water, count, and dilute.

7. Plate 50–100 cells on MM, and incubate at 28°C until colonies are visible (3–4 days).

8. Replica plate on YPglu, YPgly, YPgly+C, YPgly+E, and YPgly+C+E (more antibiotic plates are needed when more mitochondrial markers are involved).

9. Evaluate plates for the different phenotypic classes.

It is helpful first to mark all colonies on the lid of a Petri dish with black dots and then mark with color the individual resistant colonies. Then the lid is transferred to the next Petri dish, and the colonies resistant to the second antibiotic are marked with a different color. At the end, the final phenotypic composition of the colonies on one plate can be evaluated. Colonies that grow on both plates containing a single antibiotic but that fail to grow on plates supplemented with two antibiotics do not carry recombinant mitochondrial genomes, but represent a mixture of two mitochondrial genomes. These colonies should not be counted. Mutants that fail to grow on YPgly are respiratory-deficient mutants. These *rho*⁻ mutants should be counted, but should not be considered in the calculations.

Construction of Double and Triple Mitochondrial Mutants By Sporulation of Diploids

As outlined in section 2.B above, a homoplasmic diploid produces four mitochondrially identical spores. By sporulating individual homoplasmic diploid clones, haploid clones with the desired mitochondrial and nuclear phenotypes can be obtained. Due to the limited space, the reader is referred to Spencer et al. [5] and to the laboratory course manual *Methods in Yeast Genetics* by Rose et al. [6]. In this chapter we will perform and analyze one homosexual and one heterosexual cross with the two markers C^R and E^R and one homosexual cross involving the three markers C^R, E^R, and P^R.

B. Materials

Media

YPglu: 2% glucose, 1% yeast extract, 1% bactopeptone
YPgly: 2% glycerol, 1% yeast extract, 1% bactopeptone

YPgly+C: YPgly + 0.4% chloramphenicol, pH 6.5*
YPgly+E: YPgly + 0.5% erythromycin, pH 6.24
YPgly+P: YPgly + 0.2% paromomycin, pH 6.5

*pH is adjusted with 50 mM phosphate buffer.

MM: 2% glucose, 0. 67% yeast–nitrogen base without amino acids (for supplementations with bases or amino acids see below).

Supplements to MM. Prepare the following stock solutions, autoclave, and store in the refrigerator. Add 10 mL of the stock solution to the medium when temperature is below 60°C.

Stock Solutions

Supplement	Amount	Solute
Adenine	2 mg/mL	50 mM HCl
Uracil	1 mg/mL	water
Histidine	1 mg/mL	water
Isoleucine	6 mg/mL	water
Leucine	6 mg/mL	water
Lysine	1 mg/mL	water
Methionine	1 mg/mL	water
Phenylalanine	2 mg/mL	water
Tryptophan	2 mg/mL	water
Tyrosine	1 mg/mL	50 mM HCl
Valine	2 mg/mL	water

Other materials are standard equipments of a genetic/micro-biological laboratory.

Conservation of Strains

Yeast strains can be stored indefinitely in 50% (v/v) glycerol of –70% in vials containing 1 mL glycerol. Strains are grown on plates with YPglu. The cells are scraped off with a sterile toothpick and suspended in glycerol. The vials are shaken before freezing. Revival is obtained by dropping a small sample on a YPglu plate. Short-term storage, for less than 6 months, is possible on slants with YPglu, supplemented according to the auxotrophic requirements of strains.

C. Evaluation of Experiments

In this section the experimental data from crosses are summarized, which were carried out by the author [7]. This will help to compare your own results with published data. When this chapter is used in a theoretical course, these data can be used for evaluation.

In all crosses only respiratory-competent (RC) colonies are evaluated. Appropriate strains can be obtained from a stock center or from the Centre de Génétique Moléculaire du C.N.R.S., Gif-sur-Yvette, France.

A "Homosexual" Two-Factor Cross

a *aux*1^{-*} C^SE^S ω$^-$ × α *aux*2$^-$ C^RE^R ω$^-$

Result of a Typical Cross

C^SE^S	C^SE^R	C^RE^S	C^RE^R	Total
38.6	2.1	2.1	57.2	100

(Values in percent; at least 300 colonies should be analyzed.)

A "Heterosexual" Two-Factor Cross

a *aux*1$^-$ C^SE^S ω$^-$ × α *aux*2$^-$ C^RE^R ω$^+$

Result of a Typical Cross (values in percent)

C^SE^S	C^SE^R	C^RE^S	C^RE^R	Total
2.5	0	46.9	50.6	100

A "Homosexual" Three-Factor Cross

a *aux*1$^-$ $C^SE^Sp^R$ ω– × α *aux*2$^-$ $C^RE^Rp^S$ ω$^-$

In this case, the sum of four independent crosses is given (values in percent):

$C^SE^SP^R$	$C^SE^SP^S$	$C^SE^RP^R$	$C^SE^RP^S$	$C^RE^SP^R$	$C^RE^SP^S$	$C^RE^RP^R$	$C^RE^RP^S$
28.6	7.9	0.8	0.7	1.3	0.8	9.6	50.3

D. Conclusions from Different Crosses

1. In a homosexual cross both parents possess either an intronless *rnl* gene (ω$^-$) or an intron-containing *rnl* gene (ω$^+$). The transmission of the two alleles is 38.6 + 2.1 = 40.7% for E^R and 57.2 + 2.1 = 59.3% for C^R, which is the same order of magnitude. Both recombinant classes are equal (2.1% each). The absence of polarity among parental and recombinant types is an indication for a homosexual cross.

2. Unlinked markers show recombination frequencies of 10–15% for each reciprocal recombinant type. In the case of the $C \times E$ – cross, the

*At least one auxotrophic marker (*aux*$^-$) should be present; a and α are mating types.

frequency of both recombinant types is 2.1%, indicating linkage. Indeed, the E^R marker is located 671 bp from the intron, the C^R marker 54 from the intron (see Fig. 1).

3. In a heterosexual cross ($\omega^+ \times \omega^-$), the transmission of the C^R/C^S alleles is highly unequal. Transmission of C^R (from the ω^+ parent) is 46.5 + 50.6 = 97.1%. For the more distant E^R marker the transmission is 50.6%. Also, the two recombinant types are unequal—46.9% for C^RE^S, and 0% for C^SE^R.

4. In a three-point cross ($C^SE^Sp^R \times C^RE^Rp^S$), linkage or absence of linkage between markers can be determined:

*ant*1 (*C*)	*ant*2 (*E*)	*ant*3 (*P*)	Designation	Values (%)
S	*S*	*R*	*p1*	28.6
S	*S*	*S*	*r1*	7.9
S	*R*	*R*	*r2*	0.8
S	*R*	*S*	*r3*	0.7
R	*S*	*R*	*r4*	1.3
R	*S*	*S*	*r5*	0.8
R	*R*	*R*	*r6*	9.6
R	*R*	*S*	*p2*	50.3

Abbreviations: *ant* = antibiotic; *C* = chloramphenicol; *E* = erythromycin; *P* = paromomycin; *p* = parental type; *r* = recombinant; *R* = resistant; *S* = sensitive.

Calculations

Percent transmission of alleles from parent 1 ($C^SE^SP^R$):

% transmission ***ant1*** = % (***p1***+***r1***+***r2***+***r3***) = 38.0%
% transmission ***ant2*** = % (***p1***+***r1***+***r4***+***r5***) = 38.6%
% transmission ***ant3*** = % (***p1***+***r2***+***r4***+***r6***) = 40.3%

Percent recombinants between allelic pairs (two reciprocal recombinant classes):

% recombinants (***ant1-ant2***) = % (***r2***+***r3***+***r4***+***r5***) = 3.6%
% recombinants (***ant1-ant3***) = % (***r1***+***r3***+***r4***+***r6***) = 19.5%
% recombinants (***ant2-ant3***) = % (***r1***+***r5***+***r2***+***r6***) = 19.1%

The transmission of all three alleles from one parent is equal. The markers *ant1* and *ant2* (*C* and *E*) are linked, whereas markers *ant1* and *ant3* (*C* and *P*), as well as markers *ant2* and *ant3* (*E* and *P*), are unlinked.

Questions

1. The mitochondrial genome of *Saccharomyces cerevisiae* has a very active recombination system, since markers that are more distant than roughly 1000 bp appear as unlinked. The reader is confronted with the problem of recombination systems in other mitochondria (animals, plants) and in other organisms (prokaryotes, higher eukaryotes).

2. As a consequence of the high recombination rate, other strategies have been developed to localize markers genetically on the mitochondrial genome. The reader's attention should be drawn to the powerful deletion mapping by *rho*$^-$ clones.

3. Intron homing has consequences concerning the conservation of introns during evolution. The reader is encouraged to follow up the literature on the evolution of introns.

4. The homing endonuclease of the *Saccharomyces cerevisiae rnl* intron has (like other homing endonucleases) an extended recognition sequence; thus these enzymes can be used as rare cutters in genome projects. Experiments for rapidly analyzing the specificities of these endonucleases could be designed.

Extensions

1. Presence or absence of the intron, designated as ω^+ and ω^- on the basis of genetic crosses, can directly be assayed both by restriction enzyme analysis of the progeny of a given cross or by DNA–DNA hybridization with an intron probe. Methods for minipreparations of mitochondrial DNA are available.

2. The availability of thousands of respiratory-deficient mutants with defined lesions in mitochondrial genes (*mit*$^-$) allows the study of recombination between physically defined mutational sites.

3. A yeast cell contains in the order of 40–50 mitochondrial genomes; thus we are dealing with a multicopy genetic system. The transmission value in crosses is roughly the same for each marker, but is obviously dependent on the strains used and the physiological conditions. A number of experimental treatments of parental cells prior to crossing can be used to modify transmission [for literature see Dujon, 1]. It can be shown that the frequency of recombinants varies with the input ratio. The maximum value of recombinants is achieved when the two alleles in question are equally transmitted. Appropriate crosses could be performed. These results can be discussed in the light of the recombination model for phages by Visconti and Delbrück [8].

4. Until recently, the major drawback in mitochondrial molecular genetics came from the inavailability of a transformation system. The biolistic transformation [9] has opened the possibility of introducing DNA in mutants devoid of mitochondrial DNA (*rho°* mutants). Using this powerful technique, a wealth of experiments on the expression of mitochondrial genomes, including their interaction with nuclear genes, can be studied.

5. The yeast *Schizosaccharomyces pombe* is the second yeast with a comparable mitochondrial and nuclear genetic background. This yeast possesses a mitochondrial DNA of less than 19 kb, which makes this genome attractive for molecular studies [10].

REFERENCES

1. Dujon, B. (1981). Mitochondrial genetics and functions. In: *The Molecular Biology of the Yeast* Saccharomyces. *Life Cycle and Inheritance* (J. N. Strathern, E. W. Jones, J. R. Broach, eds.). Cold Spring Harbor Laboratory, Cold Spring Harbor, NY, pp. 505–635.
2. Wolf, K., L. De Giudice (1988). The variable mitochondrial genome of ascomycetes: organization, mutational alterations, and expression. *Adv. Genet. 25*:185–308.
3. Wolf, K. (1987). Mitochondrial genes of the budding yeast *Saccharomyces cerevisiae.* In: *Gene Structure in Eukaryotic Microbes* (J. R. Kinghorn, ed.). IRL Press, Oxford, pp. 41–68.
4. Dujon, B. (1989). Group I introns as mobile genetic elements; facts and mechanistic speculations—a review. *Gene 82*:91–114.
5. Spencer, J. F. T., D. M. Spencer, I. J. Bruce (1989). *Yeast Genetics.* Springer-Verlag, Berlin.
6. Rose, M. D., F. Winston, P. Hieter (1990). *Methods in Yeast Genetics: A Laboratory Course Manual.* Cold Spring Harbor Laboratory Press, Cold Spring Harbor, New York.
7. Wolf, K., B. Dujon, P. P. Slonimski (1973). Mitochondrial genetics. V. Multifactorial mitochondrial crosses involving a mutation conferring paromomycin-resistance in *Saccharomyces cerevisiae. Mol. Gen. Genet. 125*:53–90.
8. Visconti, N., M. Delbrück (1953). The mechanism of genetic recombination in phage. *Genetics 38*:5–33.
9. Fox, T. D., L. S. Folley, J. J. Mulero, et al. (1991). Analysis and manipulation of yeast mitochondrial genes. In: *Methods in Enzymology 194, Guide to Yeast Genetics and Molecular Biology* (C. Guthrie and G. R. Fink, eds.). Academic Press, San Diego, pp. 149–165.
10. Munz, P., K. Wolf, J. Kohli, U. Leupold (1989). Genetics overview. In: *Molecular Biology of the Fission Yeast* (A. Nasim, P. Young, B. F. Johnson, eds.). Academic Press, San Diego, pp. 1–25.

11

Construction of a Physical Map of the *Candida albicans* Genome

P. T. Magee
University of Minnesota, St. Paul, Minnesota

Stewart Scherer
University of Minnesota, Minneapolis, Minnesota

1. INTRODUCTION

Candida albicans is a pathogenic fungus that is becoming a serious medical problem. Although *C. albicans* causes thrush and vaginitis in otherwise healthy patients, its major medical importance is in infecting immunocompromised patients. Aggressive cytotoxic chemotherapy, bone marrow and other transplants, and AIDS all predispose patients to fungal infection, and *C. albicans* is the most common fungal pathogen. A major problem with fungal disease is that there are only a few effective antifungal drugs, and each of these has its drawbacks. Amphotericin B is toxic to the patient as well as the fungus and must be administered intravenously, while the imidazole and triazole drugs are fungistatic, not fungicidal. Fungal diseases are thus becoming a greater and greater health threat, and new treatments are not coming forth very rapidly.

Little is known about the pathogenesis of fungi in general and *Candida albicans* in particular. No virulence factors have been conclusively demonstrated to play a role in infection, although the extracellular proteinase [1,2], the yeast-to-hyphal dimorphic transition [3], and the phenotypic transition or colony switching [4] have all been postulated to be involved. A definitive demonstration of the involvement of any of these properties requires an unequivocal genetic test, comparing the virulence of null mutants in a potential virulence factor to their isogenic parents. But genetics in *C. albicans* is complicated by the fact that the fungus is diploid and has no known sexual cycle. Parasexual genetics has been used for genetic mapping of auxotrophies, but this sort of analysis is cumbersome and slow and suffers from the problem that it is impossible to be sure that the recombinants are euploid. Five years of effort in several laboratories was not sufficient to determine the chromosome number nor to locate a centromere.

Candida albicans molecular genetics, pioneered by Kurtz and co-workers at Squibb, has been developed to the point that one can clone genes, transform cells, and disrupt chromosomal genes [5–7]. An important step forward in molecular genetics was provided by the demonstration that *C. albicans* chromosomes can be separated by pulsed-field gel electrophoresis (PFGE) [8,9]. The use of Southern blots of karyotype separations demonstrated that the most common chromosome number is eight [10], and approximately 130 genes have now been assigned to bands on the electrophoretic karyotype (Table 1). Development of a macro restriction map of the *C. albicans* genome, using the 8-base-specific enzyme *Sfi* I, improved the resolution of the karyotypic mapping but did not allow fine structure mapping [11].

A complete physical map of the genome of *C. albicans* based on a set of overlapping clones would be extremely useful for a variety of reasons:

1. The organism has a highly varied karyotype, with as many as 25% of clinical isolates showing variation in the number or position of chromosome bands in PFGE. The mechanism by which the karyotype varies may be quite interesting.
2. The detailed genomic structure remains unknown. No centromeres have been isolated, and while several different middle-repeated DNA segments have been found, there is no information about their function.
3. There is evidence for aneuploidy in some strains, and it would be extremely interesting to know which chromosomal regions can be tolerated in the haploid state.
4. Transformation of some "switching" isolates (strains which undergo the phenotypic transition at a high rate) yields strains with autonomously replicating chromosomal fragments [12]. The structure of these fragments will tell us much about how Candida chromosomes are constructed.

A complete physical map would provide the tools to cast light on the mechanisms behind many of these observations. Furthermore, if sequence information is to accompany the map, many genes could be located by sequence homology, saving investigators the effort of attempting to clone inter-

Table 1 Cloned Genes of *Candida albicans* and Their Chromosome Assignments

Gene	Chromosome	Function
ACT1	*1*	actin
ADE1	*R*	CAIR:aspartate ligase
ADE2	*3*	AIR:carboxylase
ADH1		alcohol dehydrogenase
ALD1		aldehyde dehydrogenase
ALK8	*R*	cytochrome p450 enzyme
ALS1	*6*	—
APR1	*2*	proteinase A
ARD1	*6*	arabinitol dehydrogenase
ARF1	*R*	ADP ribosylation factor
ARG4	*7*	arginosuccinate lyase
ARH1		ATP-dependent RNA helicase
ARO3		3-deoxy-D-arabinoheptulosonate-7-P synthase
BEM1	*4*	budding
BGL2		beta-1,3 glucan transferase
BIN2		chaperonin
BKD1		branch-chain acid kinase dehydrogenase
CAG1	*5*	G protein alpha subunit
CPB1	*R*	corticosteroid binding protein
CCN1	*5*	G1 cyclin
CDC3	*1*	cell cycle
CDC10	*R*	cell cycle
CDC25	*3*	cell cycle
CDR1	*3*	multi-drug resistance
CHP1		transcription factor
CHS1	*7*	chitin synthase 1
CHS2	*R*	chitin synthase 2
CHS3	*R*	chitin synthase 3
CHT1	*R*	chitinase
CHT2	*5*	chitinase
CHT3	*R*	chitinase
CLN2		G1 cyclin
CMD1	*4*	calmodulin
COX2	*m*	cytochrome oxidase subunit 2
COX12	*6*	cytochrome oxidase subunit 6b
CPP1	*1*	protein tyrosine phosphatase
CPY1	*7*	carboxypeptidase Y
CRL1		GTP-binding protein
CRM1	*R*	—
CYP1		peptidyl proline isomerase

(continues)

Table 1 Continued

Gene	Chromosome	Function
CYP52		cytochrome P450 52A3
CZF1	*4*	zinc finger protein
DFR1	*7*	dihydrofolate reductase
DUT1	*1*	dUTPase
DYN1		dynein heavy chain
EBP1	*6*	estrogen binding protein
EBP2	*6*	estrogen binding protein
ECE1	*4*	cell elongation
EFG1		transcription factor
ENO1	*1*	enolase
ERG1	*1*	equalene epoxidase
ERG7	*2*	oxidosqualene cyclase
ERG16	*5*	lanosterol 14-alpha demethylase
ERK1	*4*	protein kinase
FAS1		fatty acid synthase beta subunit
FAS2	*3*	fatty acid synthase alpha subunit
GAL1	*1*	galactokinase
GLY1	*1*	glycine metabolism
HEM3	2	uroporphyrin I synthase
HEX1	*5*	beta-N-acetylglucosaminidase
HIS3	*2*	imidazole glycerol-P dehydratase
HIS4	*4*	histidine biosynthesis (3 enzymes)
HPT1	2	hypoxanthine phosphoribosyl transferase
HSP70		70K heat-shock protein
HST6	*3*	—
HST7	*R*	—
IMP1		IMP synthase
INO1	*R*	inositol-1-phosphate synthase
INT1	2	—
IPP1		inositol polyphosphate 5-phosphatase
ISP42	7	mitochondrial import protein
KGD1		alpha-ketoglutarate dehydrogenase
KRE1	*R*	beta-1,6-glucan biosynthesis
LEU2	*7*	isopropylmalate dehydrogenase
LYS1	*4*	saccharopine dehydrogenase
LYS2	*1*	2-aminoadipate reductase
MAL1	*R*	maltase
MDR1	*6*	multi-drug resistance
MGL1	*R*	alpha-methyl-glucoside utilization
MSM1		mitochondrial met-tRNA synthetase
MSW1		mitochondrial trp-tRNA synthetase
NAG1	*6*	glucosamine-6-P deaminase

Table 1 Continued

Gene	Chromosome	Function
NMT1	*4*	myristoyl-CoA:protein N-myristoyltransferase
NPS1	*3*	cell cycle
OPS4	*1*	opaque specific
PDE1	*5*	cyclic nucleotide phosphodiesterase
PEP1	*2*	—
PHR1	*4*	pH response
PKC1		protein kinase C
PLC1		ribosomal protein 10
PMA1	*3*	plasma membrane ATPase
PMI1		mannose phosphate isomerase
PMM1		phosphomannomutase
PMR1		Ca2+ transporting ATPase
PTR2	*R*	peptide transport
PYK1		pyruvate kinase
RAD5		DNA repair
RAD54	*R*	DNA repair
RBP1	*7*	rapamycin binding protein
RDN1	*R*	ribosomal RNA
RPL10		ribosomal protein L10e
RPL30		ribosomal protein L30
RPS8		ribosomal protein S8
RPS24		ribosomal protein S24
RPS25		ribosomal protein S25
SAP1	*6*	secreted aspartyl proteinasc
SAP2	*R*	secreted aspartyl proteinase
SAP3	*3*	secreted aspartyl proteinase
SAP4	*6*	secreted aspartyl proteinase
SAP5		secreted aspartyl proteinase
SAP6		secreted aspartyl proteinase
SAP7		secreted aspartyl proteinase
SEC18	*1*	ER-golgi transport
SER57	*1*	serine biosynthesis
SNR14		U4 snRNA
SOR2	*3*	sorbitol utilization
SOR9	*R*	sorbitol utilization
SPA2		morphogenesis and mating
SRV2		adenylate cyclase-associated protein
STE12	*1*	transcription factor
SUC1	*R*	sucrose utilization
SUG1		—

(continues)

Table 1 Continued

Gene	Chromosome	Function
TEF1		translation elongation factor
TEF2		translation elongation factor
TEF3		translation elongation factor
TIF1		translation factor eIF4a
TMP1	2	thymidylate synthase
TRL1		tRNA ligase
TRP1	*1*	phosphoribosyl anthranilate isomerase
TUB2	*1*	beta tubulin
UBC4		ubiquitin-conjugating enzyme
URA3	*3*	OMP decarboxylase
WHS11	2	white-specific
XOG1	*1*	exo-1,3-beta-glucanase
YBR162		—
YHR137		—
YIL124		—

esting genes by PCR, low-stringency hybridization, or complementation in *S. cerevisiae*, efforts that are not always successful.

We have therefore undertaken the preparation of a complete "contig" map of the *C. albicans* genome, using a library prepared in a fosmid vector [13]. Our mapping strain is a derivative of the one for which the *Sfi* I map was developed, 1006. Our first-order mapping uses probes that we have in hand. These are simultaneously assigned to the karyotype, the *Sfi* I map, and the fosmid library. Genes that have been sequenced but for which no clones are available can be assigned to the library by PCR; positive fosmids can then be assigned to the karyotype and the *Sfi* I map. In addition, a large new collection of probes is subjected to single-pass sequencing, and the resulting sequences are compared to the GenBank database to identify genes by homology.

A unique feature is the use of a World Wide Web server on which the latest mapping information is available. The server is updated on a daily basis. Thus, the latest mapping information is available to the community of scientists interested in *C. albicans* genetics.

2. PRINCIPLES

The principles of preparing a physical map of the genome of an organism are well known [14,15]. A physical map for these purposes is a library of clones which can be arranged along the chromosomes in order such that no part of the chromosome is uncovered (a contig map).

Since fungal chromosomes are much smaller than those of multicellular eukaryotes, the preparation of such a map is simplified. The chromosome ends serve as markers on the genomic map, and the chromosomal location of a gene, if it is known, provides a referral point for mapping. If the chromosomes of the organism to be mapped can be separated by PFGE, it is relatively easy to assign cloned genes to them. Alternatively, genetic assignments can be used. However, it is important to be sure that the cloned sequence corresponds to the mapped gene, since mutations in different genes may give a common phenotype. Thus, the first step in map preparation is to assign as many cloned genes as possible by hybridization to particular chromosomes separated by PFGE separations.

The second step is to prepare a library containing several genomes' worth of DNA sequences. The library is probed with cloned genes whose chromosomal location is known. If there is a relatively complete genetic map, the task is made much easier, since the location of clones from the library that hybridize to the probe will be known more exactly. Furthermore, a map with a large number of genes means that there is a reasonable chance that two genes will hybridize to one clone from the library. In such a case, a contig over the area adjacent to the two genes is easily assembled (Fig. 1).

Note that the ends of the clones are not precisely placed, and although the clones are drawn in a particular configuration relative to the genetic map, this configuration is arbitrary (e.g., clone 1 could have gene A at either end). At this stage of the map assembly, the clones are drawn as if they were all the average size of the inserts in the library. Needless to say, this is not accurate, and as they are characterized further, their size is corrected.

From such a contig grouping, the map can be expanded in either direction by walking. Walking can be accomplished by subcloning the ends of the sequences and probing the library anew. One doesn't know a priori in which direction a particular cloned end points in a grouping such as 4–7 (Fig. 1), but one can rapidly find out by determining whether the cloned end hybridizes to clone 2. If it does, either the other end points toward gene C or the cloned sequence is smaller than and completely contained within clone 2. The correct alternative can be determined by cloning the other end and using it to probe 2.

Once the appropriate end is determined, walking is accomplished by probing the library with sequences from the end. When a new grouping of clones is encountered, the process of assigning a contig is repeated. An appropriately positioned gene such as C or a previously established contig group can be used to verify that the walk is proceeding in the right direction.

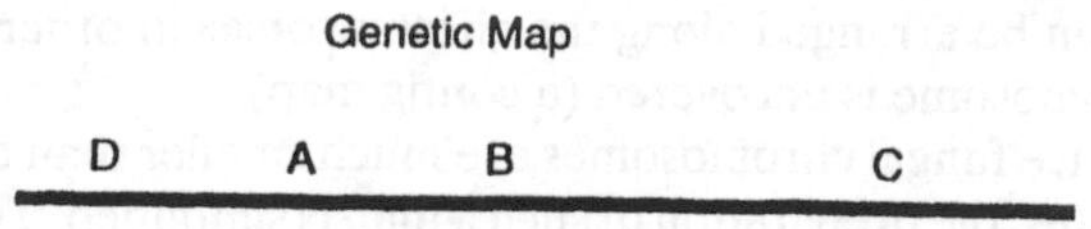

Fosmid	Hybridizing to gene A	B	C	D
1	+	-	-	-
2	+	+	-	-
3	+	-	-	+
4	-	+	-	-
5	-	+	-	-
6	-	+	-	-
7	-	+	-	-

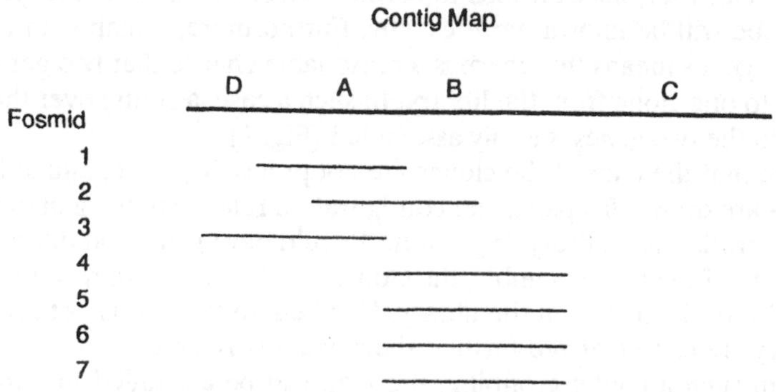

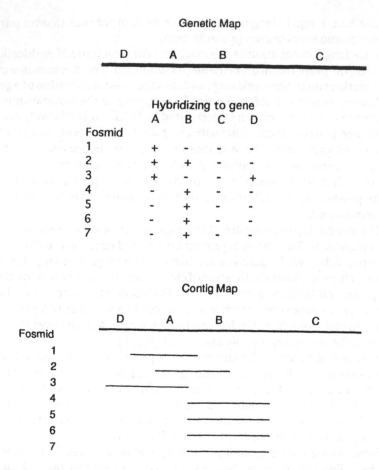

Figure 1 Establishing a contig. The letters A–D represent markers for which probes are available. The numbers indicate fosmids that have been shown to hybridize to one or more of the probes.

One should therefore assign all clones to the library before beginning extensive walking. If the density of cloned genes is great enough (and for almost any organism, the number of cloned genes is increasing rapidly), the map may largely self-assemble in the way that the contigs uniting A and B did. Walking may then be necessary only for a small number of gaps in which by chance no cloned genes occur. Walking may also be necessary if one wants early in the process to obtain a complete contig map of one region.

Repeated DNA constitutes a problem, since by definition it does not have a unique chromosomal location. Several classes of repeated DNA have been identified in *C. albicans*, including ribosomal DNA [16], a telomeric repeat [17], RPS1 [18], and several other sequences. Sequences small enough to be contained in a single fosmid will not affect the map, since we expect to find one or more fosmids with unique DNA on both sides of the repeat. Sequences larger than a fosmid will form an interruption in the contig map; we will be able to map the beginning and the end of the repeat, but we will not be able to determine its size.

A. Choice of Vectors

The choice of vectors for library preparation is determined by a variety of issues. The relatively small size of fungal chromosomes means that vectors as large as yeast artificial chromosomes (YACs) are often not needed. In fact, they may not provide a significant advantage over chromosomes in many cases, since the YACs may be of equivalent size. Cosmids (35–45 kb) are often very useful. Dr. S. I. Iwaguchi in our lab has found that cosmids containing repeated DNA tend to rearrange in *E. coli*. Bacteriophage λ is a useful vector; the size of the inserts (20 kb) means that more clones are required to cover a given stretch of DNA. Vectors with inserts smaller than those of λ are not useful for preparing contig maps in fungi.

We have chosen to use a vector that leads to clones called fosmids. This vector consists of the replication origin of the F plasmid from *E. coli*, a *cos* site, the selectable marker chloramphenical resistance, and two cloning sites, *Hind* III and *Bam* HI, flanked by *Not* I sites to excise the insert. In addition it has SP6 and T7 promoters on either side of the cloning sites. These sites allow the preparation of RNA copies for walking and also serve as primer sites for sequencing the insert. The advantage of this vector is that the F replication origin controls the copy number at 1 or 2 per bacterial cell. Repeated DNA is more stable in low-copy-number vectors.

B. The Library

The library we are using is composed of genomic DNA from strain 1161, a *Candida albicans* isolate we have chosen as a standard strain. We chose it because it has the most common *Candida* karyotype—eight pairs of chromosomes with the homologues, except for chromosome R, of nearly the same size (at least not separable under standard PFGE conditions). It contains five genetic markers, including *ura* 3, and derivatives have been prepared with markers on every chromosome [19]. Hence, this strain is the best available for eventual correlations of the physical map with the genetic map. In addition, a *Sfi* I restriction map genome of the parent strain, 1006, has been constructed.

The library contains 3840 clones of average size 40 kb, or 10 genomes' worth of DNA, stored in 40-microtiter dishes. We are using half the library to prepare our preliminary map, since any gene has a 99% chance of being represented, given randomness of the library. Statistically we would expect one or two gaps per chromosome using this half to build a map; we will use the second half to close the gaps and verify key structures.

3. PRACTICE

A. Constructing the Map

The principles of the mapping effort are to assign a sequence to the karyotype and the *Sfi* I restriction map as well as to the fosmid library. As the density of markers on the genome increases owing to the cloning of more genes, the gaps will close, because fosmids that contain two or more markers will be identified. The order of markers is simultaneously determined. The orientation of a contig can be determined with respect either to an *Sfi* I site or to the telomere repeats (for which we have a probe).

B. Assigning Genes

There are three ways to assign genes to the fosmids: direct probing of a filter containing a grid of DNA from the library in microtiter plates, use of the polymerase chain reaction (PCR) on pools of fosmids, and a recombination technique being developed by M. Strathman at Lawrence Berkeley Laboratory. The first technique is quite straightforward and will be used for most assignments. The second, PCR, is somewhat more complicated in that it requires knowledge of the sequence of the cloned gene so that the appropriate primers can be prepared. Furthermore, the technique requires the use of pools of fosmids; a positive pool is then subdivided and tested again until all positive fosmids are identified. Thus it is more time-consuming and more expensive than hybridization. The pools of DNA can of course be used to look for several different genes by PCR, thus increasing the efficiency of the process. The advantage of this approach is that it does not require the clone itself; the published sequence will suffice for preparation of the PCR primers. Furthermore, repeated DNA in the clone will not necessarily confuse the assignments to fosmids. The fosmids identified as containing the genes can then be used to identify the chromosome and *Sfi* I fragment on which the gene is located. An alternative, which is cheaper and quicker, is to use the PCR product to screen the library directly.

C. Identifying Genes

To be sure that sufficient markers exist to complete the contig map, we assign members of a library of random clones in the vector pUC to the library, first subjecting these clones to one-pass sequencing, a relatively inexpensive procedure which gives about 300 nucleotides of sequence. This means that we will build our physical map as much as possible with genes, not probes of unknown function. This sequence is then subjected to a blastx [20] search to see whether the clone contains an open reading frame that corresponds to a known gene (from any organism). The rapidly increasing database from the *S. cerevisiae* genome project will greatly aid this identification. About 1 in 5 of the sequences that have been determined appears to encode a homologue of a known gene. As more sequence data from the *Saccharomyces cerevisiae* genome product become available, we expect this frequency to rise to 1 in 4. As the map nears completion, the number of identified genes available as fosmids will be high enough that 10% of *Candida* genes will be available on known fosmids.

D. Completing the Map

We estimate that the map will be two-thirds complete by the time we have assigned 250 random clones to fosmids. Chromosome- or *Sfi* I fragment-specific libraries can be used to generate probes and saturate regions of interest. We calculate that 25 additional probes for any 1-Mb fragment (a small chromosome or an *Sfi* I fragment) will not only provide complete coverage but will also provide extensive ordering of the clones.

As the map approaches completion, use of random probes becomes more inefficient, since they are likely to come from regions already mapped. We will sequence the ends of fosmids to which no probes are assigned in order to make new probes, since by definition these fosmids are located in uncharted regions. Use of both ends means that a 45-kb region will be covered, thus ensuring efficient spacing as well. The sequence determination will also identify new genes as described above. We can also use the terminal sequences of assigned fosmids to order unordered regions and to walk across gaps by the same methods. By the time we have assigned 600 probes, we expect on statistical grounds to have a virtually complete map, except for one or two gaps per chromosome. We will close these gaps by using the second half of the library.

Regions of repeated DNA, such as the rDNA cluster and the tandem repeats of the RPS1 sequence, will not be part of our map. However, the rest of the genome should be covered at a density of about 1 marker every 25 kb, and a large fraction of the markers should be genes, many of them uncloned except in the fosmid library.

E. Data Sharing via the World Wide Web

We have put on line a server to provide up-to-date information on the progress of the map. The server is accessible via client programs such as MOSAIC. The server contains the chromosome numbering system of *Candida albicans*, the *Sfi* I restriction map, and a list of the known genes and markers. The list contains the name of the gene if it is known or the name of the probe if it does not appear to encode a gene; the fosmids to which it is assigned; the chromosomal and *Sfi* I fragment location if it is known; and the name of the person who isolated the gene. By clicking on the name of the gene or the probe, one can access the GenBank entry if it exists. There is a separate list of the fosmids and the genes assigned to them, so that one can determine all of the genes or sequences located on a fosmid of interest. The server also has a file of *Candida* literature references and a variety of other information, including information on methods and a blastx search program for new sequences.

The function of the server is to make the data obtained during the mapping process very rapidly accessible to the community of *Candida* investigators. New gene–fosmid assignments are added as soon as they are confirmed. In turn, the *Candida* community has been very helpful in sending us cloned sequences to assign to the map. We can assign any sequences which appear in GenBank via cloning by PCR, as described above.*

F. Distribution of the Library

The library will be distributed widely, as will filters with the library gridded out for gene mapping. Our plan is to enlist as much help as possible in building the map. Possession of the library will, of course, make quite simple the cloning of any mapped gene.

4. CONCLUSION

The rapid advances in genome analysis are making important changes in the way biology is being studied. Instead of having to start with the phenotype and move to the study of the gene, investigators can work in the opposite direction, from gene to physiology. *Candida albicans*, being diploid and lacking a sexual cycle, has proven a very difficult organism to study in the classical way. Our mapping of the genome will make possible the second approach in order to facilitate the understanding of the virulence of this important human pathogen.

*The address of the server is http://alces.med.umn.edu/candide.html

Within 5 years we expect to have completed the contig map of *Candida albicans* together with the assignment of more than 300 genes to it. The map will aid research on this organism in a variety of ways. The catalog of genes will provide a starting point for investigators of a great many aspects of *Candida* physiology, obviating in many cases the need for cumbersome approaches to cloning desired genes. The chromosome changes leading to variations in the electrophoretic karyotype can be determined. Specific chromosome translocations can be analyzed. Construction of strains designed to facilitate the analysis of pathogenesis will become much easier. In general, the map should provide a very large step forward toward understanding the properties of this important human pathogen.

ACKNOWLEDGMENTS

The work presented in this chapter was supported by USPHS grants AI16567 (to P.T.M.) and AI23850 (to S.S.).

REFERENCES

1. Macdonald, F., F. C. Odds (1983). Virulence for mice of a proteinase-secreting strain of *Candida albicans* and a proteinase-deficient mutant. *J. Gen. Microbiol. 129*:431–438.
2. Kwon-Chung, K.-J., D. Lehman, C. Good, P. T. Magee (1985). Genetic evidence for role of extracellular proteinase in virulence of *Candida albicans. Infect. Immun. 49*:571–575.
3. Shepherd, M. G. (1985). Pathogenicity of morphological and auxotrophic mutants of *Candida albicans* in experimental infections. *Infect. Immun. 50*:541–544.
4. Soll, D., B. Morrow, S. Srikantha (1993). High-frequency phenotypic switching in *Candida albicans. Trends Genet. 9*:61–65.
5. Kurtz, M. B., M. W. Cortelyou, D. R. Kirsch (1986). Integrative transformation using a cloned *Candida ADE2* gene. *Mol. Cell. Biol. 6*:142–149.
6. Kurtz, M. B., M. W. Cortelyou, S. M. Miller, M. Lai, D. R. Kirsch (1987). Development of autonomously replicating plasmids for *Candida albicans. Mol. Cell. Biol. 7*:209–217.
7. Kelly, R., S. M. Miller, M. B. Kurtz (1988). One-step gene disruption by cotransformation to isolate double auxotrophs in *Candida albicans. Mol. Gen. Genet. 214*: 24–31.
8. Magee, B. B., P. T. Magee (1987). Electrophoretic karyotypes and chromosome numbers in *Candida* species. *J. Gen. Microbiol. 133*:425–430.
9. Snell, R. G., R. J. Wilkins (1986). Separation of chromosomal DNA molecules from *C. albicans* by pulsed field gel electrophoresis. *Nucl. Acids Res. 14*:4401–4406.

10. Wickes, B., J. Staudinger, B. B. Magee, K. J. Kwon-Chung, P. T. Magee, S. Scherer (1991). Physical and genetic mapping of *Candida albicans*: several genes previously assigned to chromosome 1 map to chromosome R, the rDNA-containing linkage group. *Infect. Immun. 59*:2480–2484.
11. Chu, W. S., B. B. Magee, P. T. Magee (1993). Construction of an *Sfi*I macrorestriction map of the *Candida albicans* genome. *J. Bacteriol. 175*:6637–6651.
12. Barton, R. C., S. Scherer (1994). Induced chromosome rearrangements and morphologic variation in *Candida albicans. J. Bacteriol. 176*:756–763.
13. Kim, U.-J., H. Shizuya, P. J. D. Jong, B. Birren, M. Simon (1992). Stable propagation of cosmid sized human DNA inserts in an F factor based vector. *Nucl. Acids Res. 20*:1083–1085.
14. Green, E. D., M. V. Olson (1990). Chromosomal region of the cystic fibrosis gene in yeast artificial chromosomes: a model for human genome mapping. Science 250: 94–98.
15. Chumakov, I., P. Rigault, P. Ougen, et al. (1992). Continum of overlapping clones spanning the entire human chromosome 21q. *Nature 359*:380–387.
16. Magee, B. B., T. D. D'Souza, P. T. Magee (1987). Ribosomal DNA restriction fragment length polymorphism can be used to identify species and biotype of medically important yeasts. *J. Bacteriol. 169*:1639–1643.
17. Sadhu, C., M. J. McEachern, E. P. Rustchenko-Bulgac, J. Schmid, D. R. Soll, J. B. Hicks (1991). Telomeric and dispersed repeat sequences in *Candida* yeast and their use in strain identification. *J. Bacteriol. 173*:842–850.
18. Iwaguchi, S., M. Homma, H. Chibana, K. Tanaka (1992). Isolation and characterization of a repeated sequence (RPS1) of *Candida albicans. J. Gen. Microbiol. 138*:1893–1900.
19. Scherer, S., P. T. Magee (1990). Genetics of *Candida albicans. Microbiol. Rev. 54*: 226–241.
20. Altschul, S. F., W. Gish, W. Miller, E. W. Myers, D. J. Lipmann (1990). Basic local alignment search tool. *J. Mol. Biol. 215*:403–410.

12

Use of a Heterologous Gene as Marker for Genetic Analysis in *Aspergillus niger*

Fons Debets
Wageningen Agricultural University, Wageningen, The Netherlands

1. INTRODUCTION

Genetic analysis of the asexual fungus *Aspergillus niger* is based on genetic recombination in somatic cells (see Chapter 4) and requires proper genetic markers. Mostly such markers are induced by mutation. A great number of auxotrophic mutants have been isolated in this way, and genetic maps have been constructed [1]. Alternatively, genetic markers may be introduced by transformation with a heterologous gene. The heterologous *Aspergillus nidulans* gene coding for acetamidase (*amd*S) has proven to be an excellent marker for mitotic mapping of other markers in several linkage groups of *A. niger* [1,2]. The introduction of heterologous markers can be useful not only as extension to the amount of available markers, but also to trace other markers that are difficult to identify by their own phenotype. The *gla*A (glucoamylase) gene of *A. niger*, for example, does not have a simple phenotype for testing.

An insertional mutation of the *gla*A gene tagged with the heterologous *ble* gene (conferring resistance to bleomycin) in a transformational gene replacement, facilitated genetic analysis of the *glaA* gene [3]. In addition, transformation with a vector carrying both *gla*A and *amd*S enabled genetic analysis of the *gla*A/*amd*S inserts in various multicopy glucoamylase-overproducing transformants using the *amd*S gene as marker [4].

Key concepts for the use of a heterologous gene as marker for genetic analysis in *A. niger* are these:

1. Vectors with a heterologous marker may integrate at different sites in the genome, and multicopy transformants usually carry the plasmids as a single insert.
2. An insert with a heterologous marker can be used as marker in genetic mapping and for the localization of genes lacking a detectable phenotype.
3. Insertion of a plasmid with a heterologous dominant marker in a diploid can facilitate genetic analysis.

2. TRANSFORMATION WITH A PLASMID CARRYING A HETEROLOGOUS MARKER

Transformation in *Aspergillus* results from integration of the transforming DNA sequences in the genome of the recipient. Recombination of homologous transforming DNA may be at the homologous site either by single crossing over, by double crossing over, or by gene conversion. Ectopic integration apparently occurs by heterologous recombination with different integration events for one transformant to another [5,6]. If there is no significant homology between the transforming DNA and the recipient genome, transformation yields ectopic transformants.

This is the case for transformation of *A. niger* with the heterologous *A. nidulans* gene coding for acetamidase (*amd*S) and enabling growth on acetamide as sole carbon and nitrogen source. *A. niger* strains do not have an equivalent gene, but the *amd*S gene is subject to regulation if introduced by transformation [7]. *Amd*S$^+$ transformants of *A. niger* can be selected on the basis of acetamide utilization. On the other hand, loss of function (AmdS$^-$) can be monitored by selecting for resistance to fluoroacetamide (FA) [2,8]. Therefore, the heterologous *amd*S gene can be used as a two-way selectable marker.

More recently, autonomously replicating plasmids have been obtained that do not require integration into the genome upon transformation in *Aspergillus* [9].

3. MITOTIC MAPPING USING *AMD*S TRANSFORMANTS

In this case study we present the application of the two-way selection using *amd*S transformants in genetic analysis. We will see that the *amd*S insert of the different transformants can be localized rather easily and in fact provide an efficient mitotic mapping strategy for analysing other markers. As outlined in Chapter 4, a distal selective marker in cis position to the recessive markers on the chromosome facilitates the mitotic mapping analysis. The *amd*S gene provides such a marker. Growth of *amd*S transformants on acrylamide is due to constitutive expression of *amd*S gene(s) and indicative of multiple copies of the gene [2,7].

The main advantages of the mitotic mapping method using *amd*S transformants are:

1. The *amd*S gene may be introduced at different loci in the genome. Assuming random heterologous integration, each transformant would carry the dominant *amd*S$^+$ insert at a unique position.

2. The mitotic stability of the Amd$^+$ phenotype can be quantified, since revertants (*Amd*S$^-$ segregants) can be selected for on the basis of resistance to fluoroacetamide (FA).

3. Diploids constructed from such transformant strains and a strain carrying linked recessive markers are Amd$^+$ and thus sensitive to FA. From such diploids that are essentially hemizygous for the recessive FA resistance, homozygous crossing-over recombinants can be isolated quantitatively on the basis of resistance to FA.

4. The selective FA resistance "marker" is in the desired cis position to all recessive markers of the nontransformed strains. The method has been used to localize many recessive nonselective markers relative to the *amd*S insert of the different transformant strains [1,2].

The first step in the genetic analysis of *A. niger* transformants is to determine the linkage group of the insert on basis of haploidization. *A. niger* has eight linkage groups, and master strains with marked linkage groups are available [10]. In the first part (A) of the following experiment we do a transformation experiment with a plasmid containing the *A. nidulans amd*S gene. In the second part (B), some *amd*S$^+$ transformants are analyzed by haploidization. In the third part (C), we determine the frequency of reversion of the Amd$^+$ transformants to Amd$^-$. In the fourth part (D) we will determine the linear order of markers on a chromosome relative to the *amd*S insertion and the centromere.

In special cases it may be useful to start transformation with a diploid strain constructed from two haploid strains together carrying markers for each chromosome. In this way time-consuming isolation of heterozygous diploids from each individual transformant strain with a tester is avoided. An additional advantage may be that otherwise lethal transformants (insertional mutations in essential genes) can be obtained in a heterozygous diploid.

4. EXPERIMENTS

Strains and plasmid used. The *A. niger* strains used are listed in Table 1. The plasmid pGW325 carrying the *amd*S gene is a derivative of p3SR2 [11].

Table 1 Strains Used

	I	II	III	IV	V	VI	VII	VIII
N671	fwnA1	hisD4			metB1			
N716			bioA1	leuA1	nicA1	pabA1	oliA2	nirA2
N761	fwnA1					pyrB4 pabA1 cnxA1		
AT1					nicA1	amdS		

Media and growth conditions. The media (complete, minimal, and supplemented minimal medium; CM, MM, and SM, respectively) and the growth conditions are described in the Appendix. The Amd$^+$ character is isolated or tested on SM with 20 mM acetamide (SMA) as sole carbon and nitrogen source and the required auxotrophic supplements are added at 20% of the standard concentration. Growth on acrylamide is tested on SM containing 10 mM acrylamide as sole nitrogen source.

Chlorate resistance is tested on complete medium containing 150 mM KClO3 and 10 mM urea (CMC). Fluoroacetamide (FA)-resistant segregants are isolated on SM + FA (1 mg/mL) + 5 mM urea and 100 mM acetate as carbon source [8]. The techniques for genetic analysis are described in Chapter 4.

The osmotic buffer used for the preparation of protoplasts consists of 1.0 M sorbitol, 10 mM $CaCl_2$, 10 mM Tris (pH 7.5).

A. Transformation Experiment

1. Protoplasts are isolated from young mycelium of diploid strain N671/N716 grown for 15 h at 30°C on a rotary shaker. The mycelium is collected on a myracloth filter in a Büchnerfunnel, and 1 g is resuspended in osmotic buffer (0.7 M NaCl, 0.2 M $CaCl_2$) with 0.5–1 mg/mL Novozyme 234 (see also Chapter 8).
2. The protoplasts are washed and resuspended in osmotic buffer at a density of 10^7–10^8 protoplast per mL. Aliquots of 0.1 mL are mixed with an equal volume of the same buffer containing the vector DNA. This suspension is immediately mixed with 1 mL PEG buffer consisting of 30% PEG 6000, 10 mM $CaCl_2$, and 10 mM Tris (pH 7.50), and left at room temperature for 20 min.
3. The protoplasts are collected by centrifugation and resuspended in 0.4 mL osmotic buffer, and aliquots of 0.4 mL are plated in a 2-mL overlay on selective medium (MMA). The plates are incubated for 2 days at 30°C.
4. Amd$^+$ transformants are isolated and purified by transfer to a selective plate (MMA). About 20 independently derived single spore colonies are

preserved and tested for growth on acrylamide (growth indicates multiple copies of *amd*S).

B. Haploidization Experiment

1. Transfer small amounts of conidia from the different diploid transformants onto plates containing CM + benomyl (25 μg/mL) to induce haploidization. Incubate the plates at 30°C for 3–5 days.
2. Haploid segregants (either fawn-colored or black) are isolated with help of a dissecting microscope, purified by transfer to a plate of CM.
3. All segregant colonies are tested for each of the markers. Since *met*B1 and *nic*A1 are tightly linked on linkage group V, haploid segregants will be requiring either methionin or nicotinamide (the odd black prototrophic colony is therefore most likely diploid and should be discarded).

Results

1. Estimate the transformation frequency.
2. What is the fraction of multiple copy transformants?
3. Could viable haploid AmdS$^+$ segregants be recovered from all diploid transformants?
4. Calculate the recombination frequencies for the *amd*S gene to each of the other markers.
5. Assign the *amd*S inserts in the different transformants to linkage groups. Can all be assigned to a single chromosome, even the multiple copy transformants?

C. Estimating the Stability of the *amd*S Insert in Haploid and Diploid Culture

Since loss of the Amd$^+$ phenotype will result in resistance to FA, the mitotic stability of the Amd$^+$ phenotype can be quantified. Therefore, transformant AT1 and the hemizygous diploid AT1//N761 are grown on nonselective medium (CM). In principle FA-resistant segregants may result from loss of (part of) the *amd*S sequences in both haploids and hemizygous diploids ("reversion") and from crossing over between homologous chromosomes in diploids (homozygosity for the absence of the *amd*S sequences). Assuming that reversion events in diploids occur at the same rate as in haploids, we can estimate crossover frequencies between the *amd*S insert and the centromere. Haploid transformants that are relatively stable are good candidates for analyzing crossover recombinants in diploids. In order to obtain reproducible quantitative data the inocula used are first tested for the presence of revertants to avoid clonal bias.

Isolation of Amd⁻ Segregants

The quantitative selection of Amd^- segregants (revertants) from the haploid transformant is performed on SM + FA. Conidia of purified Amd^+ transformant AT1 are grown on CM for 3–4 days at 30°C; the inocula are also tested by plating on SM + FA. If no resistant colonies appear on the SM + FA plates, conidia of the CM plates are harvested and plated onto SM + FA and CMT. Plates are incubated for 3–4 days at 30°C, and the frequency of FA-resistant colonies is determined. FA-resistant colonies are tested for the inability to utilize acetamide as the sole nitrogen source. The quantitative selection of mitotic segregants from diploid strain AT1//N761 should be performed with similar precautions against clonal segregation. Suspensions of single conidial heads from primary diploid cultures are screened for mitotic recombinants, i.e., Amd^- segregants (on SM + FA) and chlorate-resistant $CnxA^-$ segregants (on CMC). Parallel cultures from the same suspensions are grown on CM to propagate the diploid. If no segregants resistant to FA or chlorate are found, the conidia are harvested from the corresponding CM plate and used to determine the frequency of FA or chlorate-resistant segregants.

Results

1. Determine the mitotic stability of the haploid transformant.
2. Determine the recombination frequencies in diploid culture for the *amd*S and the *cnx*A1 marker.

D. Mapping the *amd*S Insert by Mitotic Crossing Over

FA-resistant (FA^r) diploid recombinants can be isolated from diploids hemizygous for the dominant heterologous *amd*S sequence (i.e., the plasmid insert is present on only one of the two homologous chromosomes in the diploid). We choose transformants in which the *amd*S insert is in repulsion to the marker on that chromosome. In this case the diploid recombinants are most informative, since the genotypes can be inferred directly (otherwise each transformant would have to be haploidized to determine its genotype).

Alternatively, we can construct diploids from haploid *amd*S segregants and a tester strain carrying several markers linked to *amd*S insert. In this experiment we will further analyze the hemizygous diploid AT1//N761 (see exp. C). The selectable marker *cnx*A1 enables isolation of chlorate resistant recombinants in addition to FA-resistant recombinants. (*Note*: The frequency of mitotic recombination may be increased by UV irradiation. Heterozygous diploid conidia are germinated on CM plates for 16 h at 30°C, irradiated with UV to a final sublethal dose of 10 J/m^2, and further incubated.)

1. Construct a diploid AT//N761. Transfer small samples of conidiospores (preferably from a single conidial head) from primary diploid

colonies to plates containing MM + FA supplemented according to the requirements of that chromosome (20 inoculations per plate). Similarly chlorate-resistant colonies are selected for on CMC. Incubate for 3–5 days at 30°C.

2. Isolate (aconidial) FA-resistant recombinants by taking small mycelial fragments of the edge of the sectors and transfer them onto a CM plate (20 segregants per plate in a 4 × 5 pattern). Chlorate-resistant segregants can be transferred using small conidial inocula. Incubate the plates at 30°C for 3 days.

3. Replicate each plate on test plates to test for each marker on the chromosome. Score the test plates after 2 days at 30°C.

Results

1. Determine the relative frequency of the different genotypes.
2. Explain each recombinant genotype in terms of a recombinational event. Can all genotypes be explained by single crossovers?
3. Conclude the linear arrangement to the centromere and the marker.
4. Estimate the relative distance to the centromere.
5. Are the findings consistent with the results from experiment C?

REFERENCES

1. Debets, A. J. M., K. Swart, R. F. Hoekstra, C. J. Bos (1993). Genetic maps of eight linkage groups of *Aspergillus niger* based on mitotic mapping. *Curr. Genet. 23*:47.
2. Debets, A. J. M., K. Swart, E. F. Holub, T. Goosen, C. J. Bos (1990). Genetic analysis of *amd*S transformants of *Aspergillus niger* and their use in chromosome mapping. *Mol. Gen. Genet. 222*:284.
3. Verdoes, J. C., M. R. Calil, P. J. Punt, et al. (1994). The complete karyotype of *Aspergillus niger*: the use of introduced electrophoretic mobility variation of chromosomes for gene assignment. *Mol. Gen. Genet. 244*:75–80.
4. Verdoes, J. C., A. D. van Diepeningen, P. J. Punt, A. J. M. Debets, A. H. Stouthamer, C. A. M. J. J. van den Hondel (1994). Evaluation of molecular and genetic approaches to generate glucoamylase overproducing strains of *Aspergillus niger. J. Biotechnol. 36*:165–175.
5. Fincham, J. R. S. (1989). Transformation in fungi. *Microbiol. Rev. 53*:148–170.
6. Goosen, T., C. J. Bos, H. W. J. van den Broek (1990). Transformation and gene manipulation in filamentous fungi: an overview. In D. K. Arora, R. P. Elander (eds.). *Handbook of Applied Mycology*, Vol. IV: *Biotechnology*. Marcel Dekker, New York.
7. Kelly, J. M., M. J. Hynes (1985). Transformation of *Aspergillus niger* by the *amd*S gene of *Aspergillus nidulans. EMBO J.* 4:475.
8. Hynes, M. J., J. A. J. Pateman (1970). The genetic analysis of regulation of amidase synthesis in *Aspergillus nidulans.* II. Mutants resistant to fluoroacetamide. *Mol. Gen. Gent. 108*:107.

9. Gems, D., I. L. Johnstone, A. J. Clutterbuck (1991). An autonomously replicating plasmid transforms *Aspergillus nidulans* at high frequency. *Gene 98*:61–67.
10. Bos, C. J., S. M. Skakhorst, A. J. M. Debets, K. Swart (1993). Linkage group analysis in *Aspergillus niger. Appl. Microbiol. Biotechnol. 38*:742.
11. Hynes, M. J., C. M. Corrick, J. A. King (1983). Isolation of genomic clones containing the *amd*S gene of *Aspergillus nidulans* and their use in the analysis of structural and regulatory mutants. *Mol. Cell. Biol. 3*:1430–1439.

13

Mutation in *Neurospora crassa*

From X-Rays to RIP

Alan Radford, Jon J. P. Bruchez, Fawzi Taleb, and Paul J. Stone
The University of Leeds, Leeds, England

1. INTRODUCTION

Neurospora was the first organism in the development of molecular genetics, and has a 55-year history of continuous development. Its exploitation as a research tool has depended critically upon the ability to induce and select mutants at particular loci. For most of the first half-century, this depended on either brute force screening for particular mutant phenotypes in a population of randomly mutagenized propagules, or the exploitation of a variety of tricks for selection of particular phenotypes, again from a random mutagenized population. Because of the promiscuity of the organism in inserting any DNA taken up into predominantly ectopic locations, conventional in vitro insertional inactivation techniques are remarkably inefficient. Fortunately, the so-far *Neurospora*-specific RIP mechanism of in vivo site-directed mutagenesis permits potential inactivation of almost any gene for which a clone is available. Recent examples of our RIP inactivation of three new genes is described.

Some key concepts in the mutation of *N. crassa* are these:

1. In the course of over 50 years of genetic research on *Neurospora* more than 1000 genes have been described and mutants in them have been obtained. Of these, over 200 have now been cloned and sequenced.

2. In vivo mutagenesis, coupled with methods for general enrichment for mutants and subsequently specific enrichment/selection methods for mutants with particular phenotypes or in particular genes, provided powerful tools and led to the rapid development of the genetic analysis of the organism.

3. General in vitro methods for site-directed mutagenesis proved inefficient in *Neurospora*, due to efficiency of the organism in integrating transforming DNA in ectopic locations.

4. The process of RIP (rearrangement induced premeiotically), described first in *Neurospora* and unknown in any other major genetic species, provides a very powerful tool.

5. With RIP, a cloned sequence can be used to specificially inactivate the resident copy of the gene without the necessity for a homologous transformation integration event to cause gene disruption.

6. With RIP, one can not only do specific gene disruption, but can readily determine any function of a clone of unknown function.

7. Being specific to the process of meiosis, RIP is particularly simple to control experimentally, and without putting a transformed strain through a cross, it is totally protected from RIP inactivation.

2. *NEUROSPORA* AS A GENETIC RESEARCH ORGANISM

Neurospora crassa is the most extensively studied of all the filamentous fungi, with a literature dating back to the pioneering work of Shear and Dodge [1], detailed formal genetic analysis initiated by Lindegren [2], and the origin of microbial biochemical genetics including the induction of the first mutant, *pdx*(299), of *N. sitophila* by X-rays by Beadle and Tatum [3]. The Fungal Genetics Stock Center at University of Kansas Medical Center, Kansas City, publishes biennial lists of available stocks. The last major compilation of the loci of *Neurospora* was that of Perkins et al. [4], although many more loci have since been identified and characterized.

Approaching the molecular level, colinearity of gene and gene product was first demonstrated in eukaryotes in *Neurospora*, by Fincham and co-workers with the *am* gene for NADP-linked glutamate dehydrogenase [5]. This poineering work was based on a comparison of the in vivo gene fine-structure map and the determined amino acid sequence of the encoded enzyme.

The first Neurospora gene was cloned in 1977, when the *qa-2* gene for catabolic dehydroquinase was isolated by complementation of the anabolic activity of *Escherichia coli* [6]. Since that time, very many *Neurospora* genes, from nuclear and mitochondrial genomes and from plasmids, have been cloned. The latest compilation (*Fungal Genetics Newsletter* 41, 1994) shows that approximately 200 *Neurospora* genes have been cloned and sequenced and the sequences published, and that many other genes have been cloned but are not yet sequenced.

Neurospora is a very active research organism, with so much work in progress on in vitro genetics. This is aided by the efficient transformation systems available for the species, reviewed recently by Fincham [7]. Transformation frequencies in the range 10^{-3} to 10^{-4} per microgram of transforming DNA may be obtained using $CaCl_2$-PEG transformation of enzymatically prepared spheroplasts of germinating conidia. Transformation in *Neurospora* is essentially by integration of transforming sequences into the host genome, predominantly ectopically, as no effective autonomous vector exists for the organism. Experiments and calculations carried out some time ago suggest that the efficiency of integration of supplied DNA is such that persistence of autonomy would be difficult to achieve as selection could not be maintained once a functional integrated sequence was obtained [8]. Integrative transformation gives stability of transformants, but makes self-cloning very difficult. Self-cloning depends on either sib selection or inefficient marker recovery.

However, despite the advanced state of its in vitro genetics, the in vivo genetic systems of *Neurospora* still have major applications. The potential of meiotic analysis is greatly enhanced by the potential for analysis of ordered tetrads, as well as ready random meiotic products. The full exposition of formal genetic methods was published by Barratt et al. [9].

Formal, in vivo, genetic analysis including tetrad analysis lost some of its significance with the advent of in vitro genetics. However, with the elucidation of the RIP mechanism for in vivo genome protection by *Neurospora* [10], tetrad analysis was a necessary tool in the analysis of the inactivated products of duplicated, ectopic DNA sequences, just as important as the in vitro analysis of the effects of RIP on the resident and ectopic copies of the duplicated sequence.

The preceding is a very brief review of the development of *Neurospora* genetics. For most of the above developments, a critical aspect has been the availability of allelic differences. Initially these came from natural polymorphisms and spontaneous mutants. However, the work of Beadle and Tatum introduced the deliberate induction of mutation as a source of more variation. This case study is a summary of the development of methods of mutagenesis in *Neurospora*—many subsequently applied to other species, but others essentially restricted to the genus.

3. MUTATION DETECTION AND SELECTION IN *NEUROSPORA*

A. Nonselective Random Recovery and Screening

In the earliest work on mutation in *Neurospora*, both *N. crassa* and *N. sitophila*, random mutations were induced by physical and chemical mutagens. Growth on complete medium of total mutagenized ascospores or conidia revealed morphological mutants directly by visual inspection, and auxotrophic mutants by their inability to grow on minimal medium when subcultured. Identification of mutants at a particular locus then required determination of the biochemical requirement in the latter case, and in all cases follow-up complementation tests and crosses to determine allelism and map location.

B. Enrichment by Selective Removal of Nonmutants

Improvements were made to enrich for mutations, and for those with a particular phenotype, initially by filtration enrichment [11]. In this process, mutagenized conidia were incubated in liquid minimal medium, and all unmutated (and hence prototrophic) conidia germinated and were filtered out. After 3 or 4 days and numerous filtrations, the remaining conidia were plated on solid medium supplemented with the requirement for the desired auxotrophic phenotype. These then germinated and could be recovered.

C. Enrichment by Selective Death of Nonmutants

An alternative selective removal method of enrichment was inositolless death, based on the rapid death of inositol auxotrophs on minimal medium [12]. This required an inositol auxotrophy in the starting strain, which was mutagenized and plated on minimal medium. Unmutated *inl* conidia germinated and died, but those with a newly induced second auxotrophy remained ungerminated. After several days, the desired supplement was overlayered, and those with that auxotrophy germinated and formed colonies.

D. Locus-Specific Screening

The first visual selection method for a specific class of mutants was that exploiting the purple pigment accumulated by certain adenine auxotrophs, specifically at either the *ad-3A* or *ad-3B* locus [13]. In this method, conidia were mutagnized and then grown in large flasks of liquid sorbose medium. Individual conidia tended to form individual "snowball" colonies; those accumulating purple coloration were isolated.

A general method of screening for a specific auxotrophic phenotype is by replica plating, from colonies growing from mutagenized conidia on selectively supplemented medium onto minimal medium, and recovering from the former a colony that fails to grow on the latter [14].

E. Locus-Specific Selection

A number of special methods were developed to select from a random population of mutagenized conidia those that were mutant at a specific locus, using their ability to grow when all other conidia were unable to do so.

The most straightforward of these was selection for resistance to an inhibitor or antimetabolite. Among these are benomyl resistance [15], cycloheximide resistence [16], and 4-methyl-tryptophan resistance [17].

One of the most elegant methods was based on the common intermediate, carbamoyl phosphate, in the arginine and pyrimidine biosynthetic pathways. Although normally channeled, at least in eukaryotes, cross-feeding between the pathways is possible if the pool accumulates abnormally in one or other pathway. Exploiting this, forward mutants eliminating the ACT (aspartate carbamoyl transferase) function of the *pyr-3* locus but leaving its CPS (carbamoyl phosphate synthetase) function intact could be selected by their ability to suppress the arginine requirement brought about by loss of CPS activity in the arginine pathway due to an *arg-2* or *arg-3* mutation. Conversely, an *arg-12* (ornothine carbamoyltransferase) mutant strain could be selected by its suppression of a CPS-pgr3 allele [18]. In each direction, plating a large sample of mutagenized conidia on the appropriate medium selected for the rare mutants.

4. SITE-SPECIFIC MUTAGENESIS

A. Conventional Targeted Gene Disruption

Until recently *Neurospora crassa* has exhibited one major drawback for use in modern molecular biology when compared with other model organisms such as bacteria, yeast, and some other fungi. Unlike these other model organisms, homologous recombination and integration of transforming DNA are usually extremely rare and unstable, making isolation of mutants in specific genes of interest extremely difficult [19].

The standard method, by insertion of a piece of exogenous DNA into a gene to form a null mutant, has been demonstrated in the species, first by Paietta and Marzluf [20], who used linear DNA containing an *am* clone disrupted by *qa-2* (the selectable marker) to disrupt the resident copy of *am*. However, the method has shown limited success in *Neurospora* [21,22] due to its propensity for ectopic insertion of transforming DNA. Fortunately the timely discovery of the in-built in vivo RIP system in *Neurospora* [23] and targetable gene disruption machinery [24], and the recent novel developments in detecting homology-directed recombination [25,26] have overcome this problem, thus restoring *Neurospora* to its leading role in genetic research.

B. RIP—Rearrangement Induced Premeiotically/Repeat-Induced Point Mutations

Investigating the observation that transforming DNA, especially when present in multiple copies, was rarely stable through sexual crossing [27,28]. Selker et al. [10,30] used a number of transformant strains harboring single copies of the plasmid *p*ES174 as a parent in sexual crosses and examined the progeny. *p*ES174 consisted of the *Neurospora crassa am* (glutamate dehydrogenase) gene [29], *p*UC8 vector sequence, the *Neurospora crassa* ζ-η region—an imperfect tandem duplicate of a 0.8-kb segment, usually methylated and including a 5S RNA gene [30], and finally 6 kb of adjacent flank sequence. The transformant host strain (N24) contained a deletion of the entire *am* gene (am_{132}) [29] and a unique DNA sequence replacing and exhibiting partial homology to the native ζ-η region [30].

Restriction analysis and Southern hybridization of the progeny DNA indicated that expected restriction sites within the 6-kb flank region had been lost, suggesting local rearrangement of this DNA. Further analysis using CpG methylation-sensitive and insensitive isoschizomer restriction enzymes demonstrated that de novo hypermethylation of the flank sequences had occurred. The flank region was the only duplicated sequence in the transformant parental strain. The ζ-η region exhibited some methylation, but as this is also observed during normal vegetative growth, it was more likely a result of reversible inactivation [31] than sexual crossing, or intrinsic methylation signal as originally claimed [32]. Otherwise, none of the transforming DNA not previously present in the host genome (*am*, ζ-η and pUC8) appeared generally altered. Tetrad analysis revealed that rearrangement and de novo hypermethylation of the duplicated sequences occurred after fertilization but before karyogamy and meiosis. This phenomenon was termed RIP (rearrangement induced premeiotically).

To confirm that RIP was the result of DNA sequence duplication, Selker and Garrett [33] constructed strains housing single, duplicate, and multiple copies of the *am* gene at unlinked chromosomal locations and crossed them with other strains both harboring and not harboring the *am* gene. Only crosses in which one of the parents carried duplicate or multiple copies of the *am* gene, regardless of the mating partner, resulted in RIP. Furthermore, only even numbers of copies of the sequence were inactivated.

RIP affects both linked and unlinked copies of both *Neurospora crassa* and closely homologous foreign duplicate sequences [34], and sequences that have been altered by RIP remain susceptible to RIP after many further generations [35]. Direct neighboring duplications of the 6-kb flank region never survived a cross unaltered, but unlinked duplications of the same sequence escaped RIP in approximately 50% of cases [10].

To clarify the local duplicate sequence rearrangements resulting from RIP, Cambareri et al. [36] sequenced both linked and unlinked duplicate DNA segments before and after exposure to RIP. In both cases the rearrangements were identified as being point mutations. A number of the original sequence

G-C base pairs had been converted to A-T base pairs. A clear preference of RIP for certain G-C pairs could be seen; ~64% of all G-C base pairs mutated were those with an adenine 3' of the altered cytosine, and ~18%, ~13%, and ~5% of G-C base pairs mutated occurred in CpT, CpG, and CpC contexts. The acronym RIP was now redefined as "repeat-induced point mutations." In comparing the effects of RIP on the unlinked and linked duplications, it was seen that 10% of the G-C base pairs in the unlinked sequences had been altered whereas in 31% of the G-C base pairs the linked sequences had been changed, after being subjected to RIP through two meiotic cycles.

Using the bisulfite genomic sequencing technique developed by Frommer et al. [37], Selker et al. [38] were able to investigate the de novo methylation resulting from RIP. The bisulfite genomic sequencing technique allowed the identification of the precise bases that had become methylated as a result of the RIP in duplicated *am* gene sequences. Methylation occurred at cytosine bases only, on both DNA strands, and in apparently any sequence context. More than 80% of the cytosines present within the duplicated sequences were affected. The distribution of methylation showed no correlation to the base pair point mutation sites resulting from RIP.

In summary, RIP occurs associated with the process of meiosis and is triggered by the presence of a DNA duplication, with a higher efficiency for linked, tandem duplications (efficient with >1-kb duplication) than for unlinked, ectopic duplications (>2 kb required for efficient RIPing). Sequence homology needs to be at least 90%, but this does permit RIPing with closely similar but heterologous duplicate sequence [34]. RIP occurs until homology is <90%, perhaps taking several cycles of meiosis. At the end of RIP, a significant proportion of GC base pairs will have been changed to AT, giving a nonrevertible mutant.

RIP provides in *Neurospora*, and probably in some pyrenomycete relatives, a site-specific mechanism for mutagenesis, for almost any part of the genome for which a cloned fragment is available, and for any part of the genome for which a highly homologous sequence has been cloned from another species. As such, it provides a mechanism for inducing mutants at loci for which there is no selectable phenotype, and a way of potentially revealing the phenotype of a clone of unknown function.

A problem arises if the process of RIP disrupts an essential function that cannot be remedied by medium supplementation or other modification of the environment, as such disruption would be lethal, hence not recoverable. Identification of such a lethal phenotype would show up in tetrad analysis as 4 viable:4 nonviable spores per ascus, and could be mapped by segregation of the lethality. However, Metzenberg and Groteluschein [39] have devised a modified RIP procedure, *sheltered RIP*, which carries a RIP-lethal mutant nucleus in a heterokaryon.

The sheltered RIP method exploits the fact that in a cross homozygous for the *mei-2* mutant, meiotic recombination and normal chromosomal disjunction are inhibited [40,41]. Among the meiotic progeny are a range of aneuploid products, including disomics. A recessive lethal mutant may therefore be recovered from such a meiosis initially as a disomic partial heterozygote. As disomics in *Neurospora* break down within the first few mitotic divisions after meiosis, by random loss of the extra chromosome [42], the result is a heterokaryon with nuclei containing both a normal copy of the gene and the recessive lethal RIP-mutated allele. Sheltered RIP has been used to inactivate the *mom-19* gene specifying the mitochondrial protein import receptor MOM19 [43,44].

RIP has now been used to induce mutants in a number of loci, and the 1994 Catalogue of the Fungal Genetics Stock Center includes RIP-induced alleles at the *am, asd-1, eas, grg-1,* and *ro-11* loci.

5. RIP INACTIVATION OF NEW GENES

We have recently used RIP to induce mutation in previously uncharacterized genes, for a specific blue-light-inducible protein (*bli-4*), for the major extracellular glucoamylase (*gla-1*), and for cellobiohydrolase 1 (*cbh-1*), the major enzyme of the extracellular cellulase complex.

A. The RIP Inactivation of *bli-4*

Neurospora has proven to be an excellent system for the investigation of regulation by blue light induction. A number of blue-light-regulated genes have been identified, including *bli-3* [45], *bli-4* [46,47], and *bli-7* [48]. It has been shown that *bli-4* demonstrates increased mRNA transcription 2 minutes after blue light induction, reaching a maximum of 90-fold over the dark control after 30 minutes [46]. Transcriptional induction of *bli-3* and *bli-4* are the fastest blue-light-induced responses in nature. No *bli-4* mutant was known, the cloned wild-type allele being derived from differential cDNA libraries.

In order to RIP-inactivate the resident *bli-4* locus, a 6-kb *Hinc*II/*Kpn*I fragment containing the entire coding region and flanking sequences was isolated and inserted into the multicloning site of *p*UC19 to create *p*Bli4. The 2.4-kb *Sal*I fragment from pCSN44 containing the *hph* (hygromycin phosphotransferase) gene coding sequence flanked by the control regions of the *Aspergillus nidulans trpC* gene was isolated and inserted into an *Eco*RV site in the middle of the *bli-4* coding region in *p*Bli4, creating *p*BH1. Thus *p*BH1 contained the entire *bli-4* gene interrupted by a selectable *hph* tag conferring hygromycin resistance.

A forced heterokaryon, R179 (*al-1, arg-10*) + R186 (*al-2, arg-1*), was transformed with *p*BH1, using standard methods [49]. Note that the heterokaryon grows on minimal medium and has orange conidia, but if allowed to break down on medium supplemented with arginine, the two homokaryotic components would have either yellow conidia from *al-1* or white conidia from *al-2*. Stable transformants were selected on medium containing both hygromycin and arginine. These were then analyzed by Southern hybridization to identify those with only a single inserted copy of the transforming sequence.

One transformant, BA47-c10, was purified to homokaryosis, and crossed to wild-type strain R2. Subsequently, ascospores from this cross were plated onto medium containing arginine but not hygromycin, and germinated, and growing progeny colonies isolated. Genomic DNA was extracted from isolates RB47-1 to RB47-30, double-digested with *Cla*I and *Hin*dIII, run out, and probed with *p*BH1 DNA. Strains with altered patterns of restriction in comparison to parental patterns could be readily identified. Reprobing with the pCSN44 vector confirmed that these strains contained only a RIPed resident copy of the *bli-4* locus, so that any resulting phenotype would be attributable specifically to the *bli-4* mutational alteration.

Three isolates with altered restriction patterns, and hence sequence alterations in bli-4, RB47-3, RB47-13, and RB47-24, were selected for further study. Their growth, morphology, conidiation, protoperithecial production, and response to blue light were all examined, but on no criterion were they demonstrably different from wild-type.

B. The RIP Inactivation of *gla-1*

The major extracellular glucoamylase gene, *gla-1*, was cloned and sequenced by us recently [50]. No mutant in the structural gene had at that time been isolated.

A known, transformable *pyr-4*, a strain of *Neurospora crassa* was used in this experiment [8]. It was transformed with a chimeric plasmid containing the vector *p*CSN43, which has the *E. coli ampR*, and the *hph* gene for hygromycin resistance controlled by *Aspergillus nidulans* promoter and terminator sequences [51]. A 2-kb *Hin*dIII fragment of the *N. crassa gla-1* gene from *p*PS8, containing all except the 5' 10% of the coding region plus circa 500 bp of downstream flanking sequence [50] was cloned into the unique *Hin*dIII site of *p*CSN43.

This construct was then used to transform the recipient strain to hygromycin resistance, using a standard transformation method [49]. Hygromycin-resistant regenerants growing strongly after 4–5 days were isolated onto nonselective medium. Transformants were then crossed by using them to conidiate wild-type 74-OR32-1A as the protoperithecal parent. In due course, ascospores were harvested and germinated on plates containing 1% starch and 0.1% glycerol as carbon sources. Small pieces of slow-growing colonies (putative starch-nonutilizers) were subcultured onto slants of sucrose medium.

Extracellular amylase activity (combined glucoamylase and α-amylase) was visualized on these and other plates by flooding each plate with 3 mL of an aqueous solution of 0.15% iodine in 0.3% potassium iodide. This staining of the remaining starch revealed exported activity around colonies by the presence and diameter of the starch-free and therefore stain-free "halo."

Whereas glucoamylase releases glucose from starch, α-amylase liberates maltose. It is therefore simple to differentiate between the two activities. Quantitative assays of glucoamylase activity were carried out by inoculating $\sim 5 \times 10^6$ conidia into 50 mL Vogel's liquid medium containing 1% starch, and incubating with shaking at 30° for 3 days. Protein determination in culture filtrates was according to Bradford [52]. Glucose concentration was by the Sigma glucose determination kit, using the method of Bergmeyer and Bernt [53].

RIP derivatives were found with total glucoamylase-specific activity loss (compared to wild-type 74-OR23-1) of 80–90%. To confirm that these reductions correlated with actual RIP-induced changes in base sequence, DNA samples were extracted from three of the putative RIP derivatives. RFLP analysis of RIP derivatives and wild-type after restriction with the isoschizomers *Sau*3A and *Mbo*I indicated changes in both current methylation and primary sequence (total GATC targets and methylated GATC targets between different RIP progeny and between them and wild-type).

The *gla-1* mutant phenotype is characterized by reduced starch halo, reduced ability to grow on starch as sole carbon source, and reduced ability of culture filtrate to convert starch to glucose.

Representative RIP-induced mutant alleles of *gla-1* have been deposited in the Fungal Genetics Stock Center.

C. The RIP Inactivation of *cbh-1*

The major cellobiohydrolase gene, *cbh-1*, of *Neurospora crassa* has recently been cloned and sequenced [54]. A 2.7-kb fragment of the gene, containing 1.5 kb upstream and the first 1.2 kb of the coding sequence of *cbh-1* was subcloned into pBluescript to create *p*FT1. Into *p*FT1 was then inserted the 2.4-kb *Sal*I hygromycin resistance fragment of *p*CSN43 (see above) to create *p*CSN43OF.

*p*CSN43OF was then transformed into *Neurospora crassa* spheroplasts of strain 43a, using standard Vollmer and Yanofsky [49] methodology, selecting for hygromycin resistance. Several hygromycin-resistant transformants were then isolated for further study. These were used as microconidial male parents, fertilizing strain 74-OR23-1A as the protoperithecal parent. Ascospores were subsequently isolated and germinated on Vogel's sucrose medium. All showed normal growth on sucrose. Approximately half the progeny showed evidence of RIP inactivation of cellobiohydrolase activity, with severe or total inability to produce halos on medium containing carboxymethyl cellulose, revealed by staining with dilute Congo red.

Although the cellulase complex in *Neurospora* [55] resembles that in *Trichoderma* [56] in having multiple endoglucanases and cellobiohydrolases, synergism in the complex is such that the loss of the major cellobiohydrolase might be expected to have a drastic effect on the overall activity of the complex.

Representative RIP-induced alleles of *cbh-1* have been deposited in the Fungal Genetics Stock Center.

ACKNOWLEDGMENTS

J.J.P.B. was supported in part by the Max Planck Foundation and work was partly carried out in the laboratory of V.E.A. Russo, MPIMG, Berlin-Dahlem. F.T. was supported by a scholarship from the Government of Algeria. P.J.S. was supported by a CASE Studentship from BBSRC and the Wellcome Foundation.

REFERENCES

1. Shear, C. L., B. O. Dodge (1927). Life histories and heterothallism of the red bread-mold fungi of the *Monilia sitophila* group. *J. Agr. Res. 34*:1019–1042.
2. Lindegren, C. C. (1932). The genetics of *Neurospora*. I. The inheritance of response to heat-treatment. *Bull. Torrey Botan. Club 59*:85–102.
3. Beadle, G. W., E. L. Tatum (1941). Genetic control of biochemical reactions in *Neurospora. Proc. Natl. Acad. Sci. USA 27*:499–506.
4. Perkins, D. D., A. Radford, D. Newmeyer, M. Björkmann (1982). Chromosomal loci of *Neurospora crassa. Micro. Rev. 46*:426–570.
5. Holder, A. A., J. C. Wootton, A. J. Baron, G. K. Chambers, J. R. S. Fincham (1975). The amino-acid sequence of *Neurospora* NADP-specific glutamate dehydrogenase. Peptic and chymotryptic peptides and the complete sequence. *Biochem. J. 149*: 757–773.
6. Vapnek, D., J. A. Hautala, J. W. Jacobson, N. H. Giles, S. R. Kushner (1977). Expression in *Escherichia coli* K-12 of the structural gene for catabolic dehydroquinase of *Neurospora crassa. Proc. Natl. Acad. Sci. USA 74*:3508–3512.
7. Fincham, J. R. S. (1989). Transformation in fungi. *Micro. Rev. 53*:148–170.
8. Buxton, F. P., A. Radford (1984). The transformation of mycelial spheroplasts of *Neurospora crassa* and the attempted isolation of an autonomous replicator. *Mol. Gen. Genet. 196*:339–344.

9. Barratt, R. W., D. Newmeyer, D. D. Perkins, L. Garnjobst (1952). Map construction in Neurospora crassa. *Adv. Genet. 6*:1–93.
10. Selker, E. U., E. B. Cambareri, B. C. Jensen, K. R. Haack (1987). Rearrangement of duplicated DNA in specialised cells of *Neurospora. Cell 51*:741–752.
11. Woodward, V. W., J. R. de Zeeuw, A. M. Srb (1954). The separation and isolation of particular biochemical mutants by differential germination of conidia, followed by filtration and plating. *Proc. Natl. Acad. Sci. USA 40*:192–200.
12. Lester, H. E., S. R. Gross (1959). Efficient method for selection of auxotrophic mutants. *Science 129*:572.
13. de Serres, F. J., H. G. Kølmark (1958). A direct method for the determination of forward mutation rates in *N. crassa. Nature 182*:1249–1250.
14. Maling, B. (1960). Replica plating and rapid ascus collection of *Neurospora. J. Gen. Microbiol. 23*:257–260.
15. Borck, K., H. D. Braymer (1974). The genetic analysis of resistance to Benomyl in *Neurospora crassa. J. Gen. Microbiol. 85*:51–56.
16. Hsu, K. S. (1963). The genetic basis of actidione resistance in *Neurospora crassa. J. Gen. Microbiol. 32*:341–347.
17. Stadler, D. R. (1967). Suppressors of amino-acid uptake mutants of *Neurospora. Genetics 57*:935–942.
18. Reissig, J. L. (1960). Forward and back mutation in the pyr-3 locus of *Neurospora crassa*. I. Mutations from arginine-dependence to prototrophy. *Genet. Res. 1*:356–374.
19. Asch, D. K., J. A. Kinsey (1990). Relationship of vector insert size to homologous integration during transformation of *Neurospora crassa* with the cloned *am* (GDH) gene. *Mol. Gen. Genet. 221*:37–43.
20. Paietta, J., G. A. Marzluf (1985). Development of shuttle vectors and gene manipulation techniques for *Neurospora crassa*. In: *Molecular Genetics of Filamentous Fungi* (ed. W. W. Timberlake). UCLA Symposia on Molecular and Cellular Biology, new series, Vol. 34. Alan R Liss, New York.
21. Frederick, G. D., D. K. Asch, J. A. Kinsey (1989). Use of transformation to make targeted sequence alterations at the *am* (GDH) locus of *Neurospora. Mol. Gen. Genet. 217*:294–300.
22. Nehls, U., T. Friederich, A. Schmiede, T. Ohnishi, H. Weiss (1992). Characterization of assembly intermediates of NADH: ubiquinone oxidoreductase (complex I) accumulated in *Neurospora* mitochondria by gene disruption. *J. Mol. Biol. 227*:1032–1042.
23. Selker, E. U., J. N. Stevens (1985). DNA methylation at asymmetric sites is associated with numerous transition mutations. *Proc. Natl. Acad. Sci. USA 82*: 8114–8118.
24. Aaronson, B. D., J. C. Lindgren, J. C. Dunlap, J. J. Loros (1994). An efficient method for gene disruption in *Neurospora crassa. Mol. Gen. Genet. 242*:490–494.
25. Case, M. E., R. F. Geever, D. K. Asch (1992). Use of gene replacement transformation to elucidate function of the *qa-2* gene cluster of *Neurospora crassa. Genetics 130*:729–736.
26. Chang, S., C. Staben (1994). Direct replacement of *mtA* by *mta-1* affects a mating type switch in *Neurospora crassa. Genetics 138*:75–81.

27. Grant, D. M., A. M. Lambowitz, J. A. Rambosek, J. A. Kinsey (1984). Transformation of *Neurospora crassa* with recombinant plasmids containing the cloned glutamate dehydrogenase (*am*) gene: evidence for autonomous replication of the transforming plasmid. *Mol. Cell. Biol. 4*:2041–2051.
28. Case, M. E. (1986). Genetical and molecular analyses of the *qa-2* transformants in *Neurospora crassa. Genetics 113*:569–587.
29. Kinnaird, J. H., M. A. Keighren, J. A. Kinsey, M. Eaton, J. R. S. Fincham (1982). Cloning of the *am* (glutamate dehydrogenase) gene of *Neurospora crassa* through the use of a synthetic DNA probe. *Gene 20*:387–396.
30. Selker, E. U., J. N. Stevens (1987). Signal for DNA methylation associated with tandem duplication in *Neurospora crassa. Mol. Cell. Biol. 7*:1032–1038.
31. Pandit, N. N., V. E. A. Russo (1992). Reversible inactivation of a foreign gene, *hph*, during the asexual cycle in *Neurospora crassa* transformants. *Mol. Gen. Genet. 234*:412–422.
32. Selker, E. U., B. C. Jensen, G. A. Richardson (1987). A portable signal causing faithful DNA methylation *de novo* in *Neurospora crassa. Science 238*:48–53.
33. Selker, E. U., P. W. Garrett (1988). DNA sequence duplications trigger gene inactivation in *Neurospora crassa. Proc. Natl. Acad. Sci. USA 85*:6870–6874.
34. Foss, E. J., P. W. Garrett, J. A. Kinsey, E. U. Selker (1991). Specificity of repeat-induced point mutation (RIP) in *Neurospora*: sensitivity of non-*Neurospora* sequences, a natural diverged tandem duplication, and unique DNA adjacent to a duplicated region. *Genetics 127*:711–717.
35. Cambareri, E. G., M. J. Singer, E. U. Selker (1991). Recurrence of repeat-induced point mutation (RIP) in *Neurospora crassa. Genetics 127*:699–710.
36. Cambareri, E. B., B. C. Jensen, E. Schabtach, E. U. Selker(1989). Repeat-induced G-C to A-T mutations in *Neurospora. Science 244*:1571–1575.
37. Frommer, M., L. E. McDonald, D. S. Millar, et al. (1992). A genomic sequencing tool that yields positive display of 5-methylcytosine residues in individual DNA strands. *Proc. Natl. Acad. Sci. USA 89*:1827–1831.
38. Selker, E. U., D. Y. Fritz, M. J. Singer (1993). Dense nonsymmetrical DNA methylation resulting from repeat-induced point mutation in *Neurospora. Science 262*:1724–1728.
39. Metzenberg, R. L., J. S. Groteluschein (1992). Disruption of essential genes in *Neurospora* by RIP. *Fungal Genet. Newsl. 39*:50–58.
40. Smith, D. A. (1974). Unstable diploids of *Neurospora* and a model for their somatic behaviour. *Genetics 76*:1–17.
41. Smith, D. A. (1975). A mutant affecting meiosis in *Neurospora. Genetics 80*:125–133.
42. Pittenger, T. H. (1958). Mitotic instability of pseudo-wild types in *Neurospora. Proc. 10th Int. Cong. Genet. 2*:218–219.
43. Harkness, T. A. A., R. L. Metzenberg, H. Schneider, R. Lill, W. Neupert, F. E. Nargang (1994). Inactivation of the *Neurospora crassa* gene encoding the mitochondrial protein import receptor MOM19 by the technique of sheltered RIP. *Genetics 136*:107–188.

44. Sollner, T., G. Groiffiths, R. Pfaller, N. Pfanner, W. Neupert (1989). MOM19, an import receptor for mitochondrial precursor proteins. *Cell 59*:1061–1070.
45. Eberle, J., V. E. A. Russo (1994). *Neurospora crassa* blue-light-inducible gene bli-3. *Biochem. Mol. Biol. Int. 34*:737–744.
46. Sommer, T., J. A. A. Chambers, J. Eberle, F. R. Lauter, V. E. A. Russo (1989). Fast light-regulated genes of *Neurospora crassa. Nucl. Acids Res. 17*:5713–5723.
47. Bruchez, J. J. P., J. Eberle, W. Kohler, V. Kraft, V. E. A. Russo (1995). *bli-4*, A fast blue-light-regulated gene of *Neurospora crassa*. (In preparation.)
48. Eberle, J., V. E. A. Russo (1992). *Neurospora crassa* blue-light-inducible gene *bli-7* encodes a short hydrophobic protein. *DNA Sequence 3*:131–141.
49. Vollmer, S. J., C. Yanofsky (1986). Efficient cloning of genes of *Neurospora crassa. Proc. Natl. Acad. Sci. USA 83*:4867–4873.
50. Stone, P. J., A. J. Makoff, J. H. Parish, A. Radford (1993). Cloning and sequence analysis of the glucoamylase gene of *Neurospora crassa. Curr. Genet. 24*:205–211.
51. Staben, C., B. Jensen, M. Singer, et al. (1989). Use of bacterial *hygromycinB*-resistance gene as a dominant selectable marker in *Neurospora crassa* transformation. *Fungal Genet. Newslett. 36*:79–81.
52. Bradford, M. M. (1976). A simple colorimetric assay for protein. *Anal. Biochem. 72*:248–254.
53. Bergmayer, H. U., E. Bernt (1974). In Bergmeyer, H. U. (ed.): *Methods of Enzymatic Analysis*, 2nd ed. Verlag Chemie, Weinheim, Germany, pp. 1205–1225.
54. Taleb, F. (1993). Studies on the cellulase genes of *Neurospora crassa*. Ph.D. Thesis, University of Leeds, Leeds, U.K., xvii + 196 pp.
55. Yazdi, M. T., J. R. Woodward, A. Radford (1990). Cellulase production by *Neurospora crassa*: purification and characterization of cellulolytic enzymes. *Enzyme Microb. Technol. 12*:120–123.
56. Knowles, J. K. C., P. Lehtovaara, T. Teeri (1987). Cellulase families and their genes. *Trends Biotechnol. 5*:255–260.

14

Genetic Analysis and Mapping of Avirulence Genes in *Magnaporthe grisea*

Mark L. Farman
University of Wisconsin, Madison, Wisconsin

Sally Ann Leong
U.S. Department of Agriculture, University of Wisconsin, Madison, Wisconsin

1. INTRODUCTION

Magnaporthe grisea, the causal agent of rice blast disease, has a wide host range among graminaceous species [1], among which rice (*Oryza sativa*) is the most economically important. *M. grisea* was originally classified as two species, *Pyricularia oryzae* and *P. grisea*, based on their apparently mutually exclusive hosts, rice and grasses, respectively. However, the two species were morphologically indistinguishable and, more importantly, were interfertile. Thus both species are now classified under the name *Pyricularia grisea* Sacc., teleomorph, *Magnaporthe grisea* Barr.

M. grisea is a heterothallic ascomycete (class Pyrenomycetes) with a bipolar mating-type locus [2]. The sexual stage was first described by Hebert [3], who crossed isolates from crabgrass (*Digitaria sanguinalis*). The club-shaped ascus contains eight unordered ascospores which are crescent-shaped, consisting of four cells each of which contains a single haploid nucleus. Hermaphroditism is common in grass-infecting strains of *M. grisea*, but isolates from rice that are mating-competent are usually female-sterile. Furthermore, the majority of isolates from rice are apparently incapable of crossing at all [4,5]. Consequently it is not usually possible to cross rice pathogens among themselves. However, Kato et al. [6] found that isolates from rice were capable of crossing with isolates from grasses, although the fertility of these crosses was poor. Workers from several laboratories have since invested significant efforts into developing fertile rice-pathogenic strains to simplify genetic analysis of pathogenicity and host/cultivar specificity [5,7–10]. The rewards of such efforts are manifest in the current availability of many fertile strains of *M. grisea* that will enable genetic analysis of various aspects of *M. grisea* biology.

Key concepts regarding the genetic analysis and mapping of avirulence genes in *Magnaporthe grisea* are the following:

1. *M. grisea* is an ascomycete fungus with a haploid genome. Fertile strains have been developed to facilitate genetic analysis based on meiotic crosses.
2. *M. grisea* is readily cultured on artificial media, is easily mutated, and is amenable to genetic transformation using a variety of positively selectable markers.
3. *M. grisea* exemplifies many other plant-pathogenic fungi in terms of pathogenic strategies. It forms appressoria; produces toxins and degradative enzymes; and sporulates on host tissues.
4. As a species, *M. grisea* infects many graminaceous hosts. Individual isolates display specificity toward particular hosts; therefore, the *M. grisea*/rice pathosystem provides a good model for studying the molecular genetic basis of pathogen/host specificity.
5. Cloning avirulence genes by transformation requires screening transformants individually for an acquired inability to grow on particular hosts. Instability of transforming DNA, or incompleteness of gene libraries may cause such approaches to fail. Map-based cloning procedures provide a proven alternative.

2. *MAGNAPORTHE GRISEA* AS RESEARCH OBJECT

M. grisea is an ideal system in which to study fungal phytopathology. It is easily cultured on defined artificial media and grows rapidly. Being haploid, it is readily amenable to mutagenesis. Ultraviolet irradiation [11,12] and chemical treatment [13] have both been used successfully to generate mutants. Genetic transformation of *M. grisea* was established using the *arg*B gene of *Aspergillus nidulans* [14], and more recently Leung et al. reported transformation using the positively selectable hygromycin B resistance gene [15]. Therefore, potentially any *M. grisea* isolate may be transformed. We have demonstrated that the phleomycin resistance gene from *Streptomyces hindustanus* in vector pAN8-1 [16] and the *GUSA* gene from *Escherichia coli* in pNOM102 [17] are also efficiently expressed in *M. grisea* [18]. In this study, we also demonstrated histochemical staining of fungal tissues expressing GUS in infected rice leaves.

The ready availability of genetic methods and molecular tools for *M. grisea* opens many avenues of research, and several aspects of *M. grisea* pathogenicity provide attractive research possibilities. Development of infection structures, growth within host tissues, conidiation, and the genetic basis of host and cultivar specificity are all amenable to rigorous genetic and molecular analysis [for review, see 19]. Moreover, *M. grisea* exemplifies many phytopathogenic fungi in terms of its pathogenic strategies. Like many other phytopathogenic fungi, *M. grisea* develops appressoria [20] which are involved in penetration of the host epidermis. It also produces toxins [21,22], degradative enzymes [23–26], and conidiates within lesions on host tissue.

Another prominent feature of *M. grisea* pathogenicity is its host specificity. Despite being a pathogen of many gramineous hosts [1,27], the host range of any given *M. grisea* isolate is often very limited. Genetic analysis has suggested that host specificity toward weeping lovegrass (*Eragrostis curvula*), finger millet (*Eleusine coracana*) and goosegrass (*Eleusine indica*) is determined by single genetic loci [5,28]. *M. grisea* strains infecting rice (*Oryza sativa*) show cultivar specificity which is also often determined by single genetic factors [8,29–31]. Genes affecting host and cultivar specificity in *M. grisea* are termed *a*virulence genes, since strains possessing them are unable to infect specific hosts/cultivars which presumably possess a corresponding resistance gene.

A practical limitation to genetic and molecular studies of traits associated with phytopathogenicity is that mutant production and subsequent complementation by transformation may require screening transformed strains on many thousands of plants. Fortunately, *M. grisea* can be induced to develop infection structures on artificial surfaces [32,33] and it sporulates on agar medium. This enables mutants deficient in these processes to be identified and screened for complementation in vitro. Testing the pathogenicity of these mutants and complemented strains should require inoculation of only a small number of plants. However, phenotypes associated with host and cultivar specificity genes are manifest only by inoculation of potential hosts. Therefore, cloning these genes by screening transformants for the avirulence phenotype would require inoculation of tens of thousands of plants. This practical burden is likely to preclude the success of "shotgun cloning" approaches. However, information gained from genetic analysis and mapping of avirulence genes provides a strong theoretical foundation which supports a systematic approach for cloning these genes by map-based strategies. Two *M. grisea* avirulence genes have been cloned thus far [34]. Both of them were isolated through knowledge of their chromosomal locations. *PWL2* maps close to a cosmid RFLP marker and was isolated by chromosome walking from this marker. *AVR2-YAMO* maps close to a telomeric RFLP. Spontaneous mutants that had gained virulence on this cultivar were found to have suffered deletions within a telomeric restriction fragment, suggesting that *AVR2-YAMO* resided immediately adjacent to the telomere. Cloning a telomeric restriction fragment resulted in the isolation of *AVR2-YAMO* [34].

In this chapter, we describe the genetic analysis and mapping of an avirulence gene to rice cultivar CO39 to illustrate the principles and practices involved in such studies.

3. PRINCIPLES

M. grisea displays a wealth of natural variability with respect to host and cultivar specificity. Consequently, field isolates of *M. grisea* are expected to possess many different complements of avirulence genes. For this reason they are attractive candidates for genetic analysis of specificity. Grass-infecting isolates have proven to be readily amenable to crossing, enabling investigation of host specificity [5,28]. Furthermore, fertility of grass-infecting strains can be improved by selecting for hermaphroditic progeny [5]. However, it has proven difficult to establish fertile rice-pathogenic strains by crossing with fertile, hermaphroditic strains because these strains are usually nonpathogens of rice, and using them in crossing schemes compromises virulence toward rice in the progeny. The complex genetic bases of fertility and pathogenicity necessitate an empirical breeding approach, involving a certain amount of trial and error. Attempts to coselect for fertility and pathogenicity on rice have only met with partial success [8,35]. Leung and co-workers reported that fertility was established in early generations while attempting to coselect for fertility and pathogenicity [9]. Although the fertile progeny were poorly pathogenic on rice and produced small lesions, pathogenicity was improved by intercrossing these isolates. By contrast, Ellingboe and co-workers established pathogenicity to rice in early generations of a breeding line [36]. Subsequent selection for fertility resulted in a cross in which every progeny was a pathogen of rice and was capable of mating with each sib of the opposite mating type. Furthermore, every progeny except one was capable of acting as a female in crosses. The cross that produced progeny that were all pathogenic on rice, and two successive crosses, which ultimately produced the fertile progeny, involved Guy11 as one of the parental strains. Guy11 is a field isolate of *M. grisea* collected in French Guyana by J. L. Notteghem [9]. It is relatively fertile, hermaphroditic, and virulent on many rice cultivars. Therefore, Guy11 is extremely useful in strain development because it will rarely contribute additional avirulence genes. Moreover, crosses involving Guy11 are likely to produce progeny with reduced numbers of avirulence genes, simplifying subsequent genetic analyses.

It may not be possible to achieve 100% viability in progeny of some crosses. Presumably, crosses having significantly reduced viability do not yield entirely random combinations of parental genotypes within surviving progeny. For this reason, the ratio of avirulent:virulent progeny may not be a good indicator of how many avirulence genes are segregating. This is most reliably determined by intercrossing avirulent F1 progeny. If a single avirulence gene segregated in the initial cross, crosses among avirulent sibs should yield only avirulent F2 progeny. If any of these crosses segregate virulent progeny, the original avirulent parent likely contained more than one avirulence gene. By crossing avirulent progeny in all possible combinations, it should be possible to identify each avirulence gene that segregated in the initial cross. It is also informative to make test crosses between virulent isolates. The appearance of avirulent progeny may indicate the presence of additional loci that interact to induce or suppress avirulence. For example, in a study of avirulence toward rice cultivars K1 and Aichi asahi, Silué and Notteghem [30] crossed two isolates that were virulent on these cultivars. One-quarter of the progeny from this cross were avirulent on each cultivar. The authors hypothesized that two genes interacted to produce avirulence. Ellingboe recovered avirulent progeny from crosses of isolates that were virulent on rice cultivar Katy [37]. He postulated that one of the virulent parents contained an avirulence gene whose expression was suppressed. In the event that avirulent progenies appear in crosses between two virulent isolates, it is necessary to perform judicious crosses to test different "interacting gene" models. While it is not possible to distinguish between a suppressor and a positively acting effector of avirulence by crossing, the presence of either would interfere with a mapping effort. Therefore, for an avirulence gene to be reliably mapped, test crosses should be reiterated until it is reasonably certain that the mapping cross segregates for a single locus determining avirulence.

It is important to note that procedures for inoculation and disease assessment vary among laboratories. Differences in the interpretation of the results of such inoculation experiments will affect the conclusions of a genetic analysis. By mapping a genetic locus determining avirulence/virulence, we have gained qualitative evidence that supports the conclusions of the inoculation experiments described below.

We illustrate below the genetic analysis and mapping of an avirulence gene in a cross between two *M. grisea* strains [31]: Strain 2539 was constructed by intercrossing various wild and laboratory isolates [9] (Fig. 1) and was avirulent when inoculated onto rice cultivar CO39. This strain presumably contains an avirulence gene which prevents infection of CO39 and we have designated this gene *AVR*CO39. It is postulated that CO39 contains a corresponding resistance gene [8,31]. Guy11, described above, was the virulent parent used to cross with 2539. Rice cultivar CO39 is susceptible to most rice-infecting isolates of *M. grisea* and was used as a recurrent parent for the

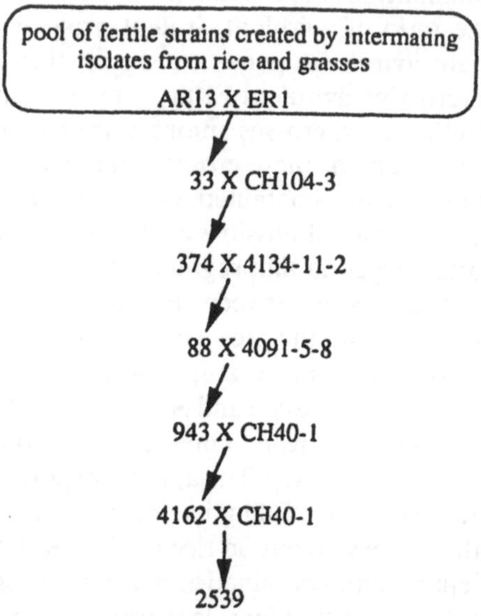

Figure 1 Crossing scheme used to construct *M. grisea* strain 2539 (adapted from Leung et al. [9]). Isolates on the left are progeny from the previous cross. The isolates on the right are wild isolates and laboratory strains as follows: CH104-3, rice pathogen (China [35]); 4134-11-2, grass pathogen (laboratory strain [5]); 4091-5-8, grass pathogen (laboratory strain [5]); CH40-1, rice pathogen (China [35]).

introgression of resistance genes from four donor cultivars [38]. This resulted in five near-isogenic lines, each of which possessed a single additional resistance gene. The resistance genes introgressed into the CO39 background potentially could identify additional avirulence genes in *M. grisea* strain 2539. It would not be possible to detect these genes by inoculation of the isolines with 2539 owing to the epistatic effect from *AVR*CO39. However, inoculation of the isolines with progeny from the cross between 2539 and Guy11, which is virulent on all the isolines, would reveal the presence of additional avirulence genes segregating in the cross.

Mapping avirulence genes segregating in the cross used to construct the genetic map [39] is straightforward, requiring only the addition of the new data to an existing RFLP data set before analysis with the mapping program, MAPMAKER v1.0 [40].

4. PRACTICE

A. Experimental Approach

Construction of a Fertile Strain for Analysis of Cultivar Specificity

Strain 2539 was constructed as shown in Figure 1: Five isolates (two from rice and one each from goosegrass, weeping lovegrass, and finger millet) were intercrossed randomly. Fertile ascospore cultures that were pathogenic on rice were selected from this mating population and used to establish a breeding line. At each generation, cultures derived from whole asci or single ascospores were inoculated onto rice. Conidial isolates were recovered from lesions and used for the next cycle of mating. This scheme resulted in strains that were highly fertile but poorly pathogenic on rice. Only 3% of ascus cultures formed lesions when inoculated on rice. Nevertheless, strain 2539, which was recovered after the two final crosses to the rice pathogen strain, CH40-1, has been shown to be moderately virulent to 29 rice lines [29]. The inability of 2539 to infect many other rice lines and cultivars suggested that it may be a rich source of avirulence genes.

Genetic Analysis of Specificity Toward Rice Cultivar CO39 and Five Isoline Derivatives

Strain 2539 is avirulent on CO39 and five isolines derived from this cultivar (Table 1). This strain was crossed with the field isolate Guy11, which is virulent on CO39 (Table 1). Sixty-one progeny were selected for analysis of their virulence. Seventeen progeny exhibited a meiosis-induced *buf*⁻ mutation, causing a pigmentation defect. *buf*⁻ mutants are nonpathogenic [41] due to

Table 1 Interaction phenotypes of Guy11 and 2539 on cultivar CO39 and five derived isolines.

Cultivar/isoline	Resistance gene*	Interaction phenotype of *M. grisea* parental strains†		Segregation in progeny A:V
		Guy11	2539	
CO39		V	A	25:18
C101lac	Pi-1	V	A	25:18
C101A51	Pi-2	V	A	25:18
C104PKT	Pi-3	V	A	25:18
C101PKT	Pi-4a	V	A	25:18
C105TTP-4	Pi-4b	V	A	25:18

*Reference 38.

†Strains were considered virulent if they could sporulate on the leaf surface (see Experimental Procedures).

their inability to penetrate the host [42]. These progeny isolates were therefore not expected to be informative with respect to their virulence, and were eliminated from the analysis. The remaining 44 progeny were inoculated on 51583, a cultivar that is susceptible to both parents, to ensure that they were capable of infecting rice. Forty-three formed lesions on 51583, and one failed to infect. The latter progeny also failed to infect CO39 as well as 29 other rice varieties known to be susceptible to 2539, indicating that it is likely defective in general pathogenicity. This progeny was also eliminated from the analysis.

Inoculation of the 43 progeny onto CO39 yielded a segregation ratio of 25 avirulent:18 virulent progeny. A χ^2 value of 1.14 indicated that segregation was consistent with a 1:1 ratio at the $p < .05$ level of significance. However, the data also fitted a 5:3 ratio, expected if two avirulence systems were involved, one requiring two interacting loci for expression [43]. For this reason we expanded the progeny population by another 58 isolates to more accurately assess the segregation ratio. Combining the ratios from both populations gave 53 avirulent:48 virulent, which caused us to reject a 5:3 ratio at the $p < .05$ level. The combined ratio remained consistent with a 1:1 segregation (χ^2 = 0.25). Subsequent analysis by crossing among sibs and by backcrossing was used to support this interpretation. In most crosses, 30–40 progeny were tested to ensure statistical confidence in the segregation ratios. In all test crosses between avirulent isolates, only avirulent progeny were recovered; in test crosses between virulent isolates, only virulent progeny were produced. Crosses between avirulent and virulent strains produced avirulent and virulent progeny in approximately 1:1 ratios. These data are consistent with avirulence/virulence to cultivar CO39 being determined by a single gene difference between Guy11 and 2539.

Five isolines of CO39, each containing an additional resistance gene were also inoculated with the original 43 progeny isolates used in the CO39 analysis. Segregation of avirulence/virulence toward these isolines coincided exactly with that on CO39; therefore, we concluded that 2539 did not possess additional avirulence genes corresponding to the resistance genes introgressed into these rice lines.

Mapping an Avirulence Gene to Cultivar CO39

Mapping *AVR*CO39 was simplified by the fact that it segregated as a single gene in the progeny population used for RFLP mapping. Segregation data for *AVR*CO39 was treated in the same manner as for the other RFLP markers. Markers were sorted into linkage groups with the MAPMAKER GROUP command using high-stringency parameters (L.O.D. = 4.0; theta = 0.2). Once *AVR*CO39 was assigned to a linkage group, the stringency of the maximum recombination fraction was relaxed to join this linkage group to others that

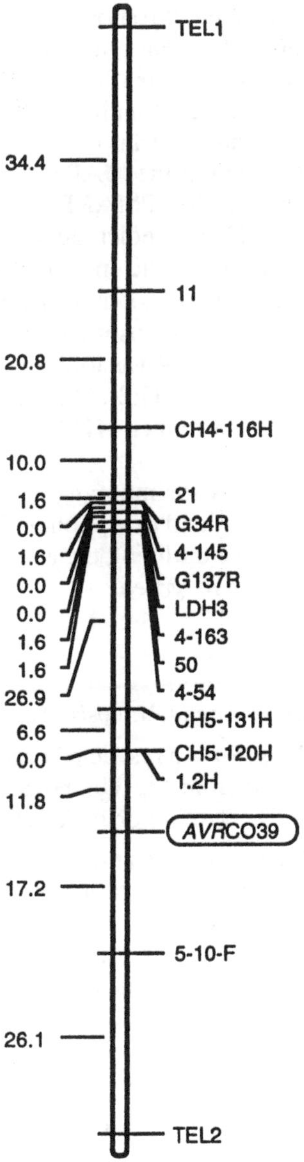

Figure 2 Map location of the *AVR*CO39 gene. The map was constructed using MAPMAKER v1.0, using a L.O.D. score of 4.0 and a theta value of 0.3. Recombination fractions and centimorgan distances between *avr*CO39 and its flanking markers were calculated by hand to account for nonpathogenic progeny. The validity of loose linkage associations was tested by chi-square analysis ($p < .05$).

were known to be on the same chromosome from Southern hybridization studies of CHEF gels [36] (M. L. Farman and S. A. Leong, in preparation). The validity of joining loosely linked markers was tested by chi-square analysis at the $p < .05$ level of significance. Finally, a map was constructed using the MAPMAKER MAP command. Centrimorgan distances were calculated using the Kosambi mapping function in the MAPMAKER program. When calculating recombination fractions, MAPMAKER used the segregation data of the markers flanking *AVR*CO39 to predict the genotypes of the nonpathogenic progeny that were excluded from the inoculation experiments. This could introduce errors in the determination of map distance if any double crossovers had occurred in this region. For this reason, the map distances were calculated by hand. By this analysis, *AVR*CO39 was located 11.8 cM from cosegregating markers CH5-120H and 1.2H, and 17.2 cM from marker 5-10-F (Fig. 2). By comparison, MAPMAKER positioned *AVR*CO39 8.5 cM from CH5-120H and 1.2H, and 11.6 cM from 5-10-F.

Experimental approaches for mapping genes that have been characterized in other crosses are outlined in section 6 (Extensions).

The following is a scheme outlining the genetic analysis of avirulence in a cross between two isolates of *M. grisea*.

B. Materials Needed

Construction of Strains and Genetic Analysis of Avirulence

Fungal strains. *M. grisea* is a restricted organism and requires permits for shipment within, to, and from the United States. Regulations for each country should be checked before importing or exporting strains.

Containment facilities (standards may vary in different countries). *M. grisea* has a high-level containment rating in the United States, which necessitates the use of a transfer hood (flow cabinet) with a BL-1 physical containment level rating. All manipulations of the fungus must be performed in this hood. Cultures must always be disposed of by autoclaving.

Media. *M. grisea* is usually grown on oatmeal agar for isolation of conidia and for crosses [5]. Rolled oats (50 g) are suspended in 400–500 mL water and heated to 65°C for 1 hour. The suspension is filtered through two layers of cheesecloth, and the volume is adjusted to 1 L. Agar (15 g) is added and the medium is autoclaved for 50 min.

Complete medium is used for growing large quantities of mycelium for DNA isolation, and contains 6 g casamino acids, 6 g yeast extract, and 10 g sucrose dissolved in 900 mL water. The pH is adjusted to 6.5 with NaOH, and the volume is then adjusted to 1 L. This medium is autoclaved for 20 min. For solid medium, 15 g of agar is added prior to autoclaving.

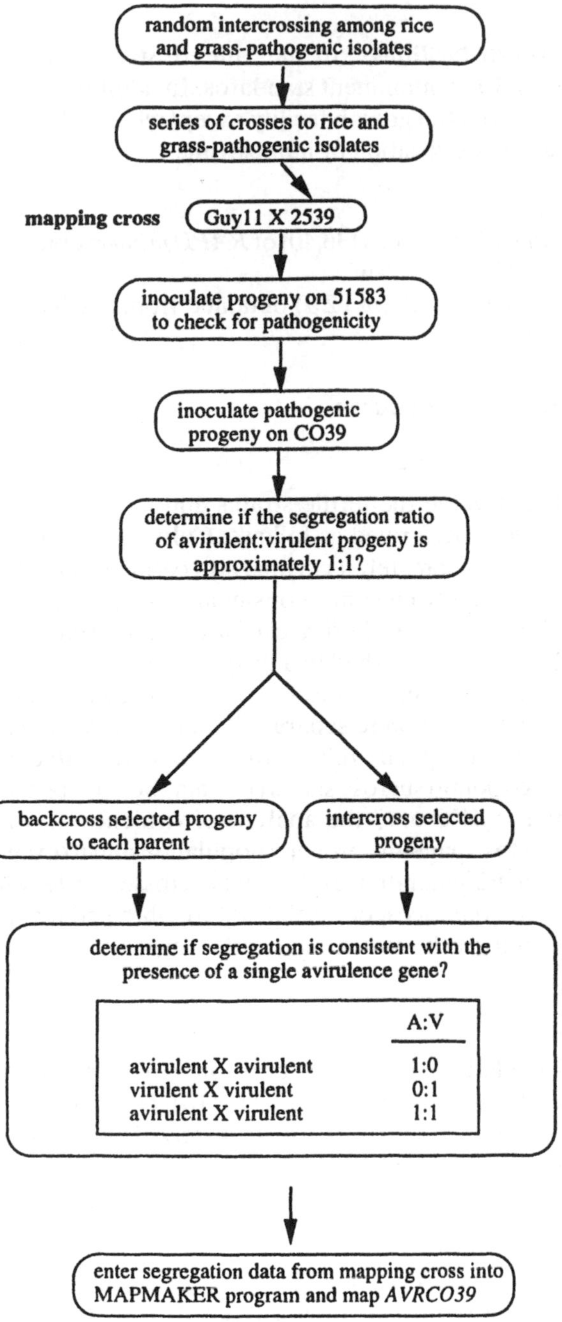
random intercrossing among rice and grass-pathogenic isolates
series of crosses to rice and grass-pathogenic isolates
mapping cross
Guy11 X 2539
inoculate progeny on 51583 to check for pathogenicity
inoculate pathogenic progeny on CO39
determine if the segregation ratio of avirulent:virulent progeny is approximately 1:1?
backcross selected progeny to each parent
intercross selected progeny
determine if segregation is consistent with the presence of a single avirulence gene?
A:V
avirulent X avirulent 1:0
virulent X virulent 0:1
avirulent X virulent 1:1
enter segregation data from mapping cross into MAPMAKER program and map *AVRCO39*

Scheme 1

Plant growth facilities. In the United States plant growth facilities need to meet BL3-P containment standards. In addition, infected plants may not be removed from the growth facility except when tightly sealed in biohazard safety bags for sterilization by autoclaving.

Mapping

1. *A set of RFLP markers* [36,40] or *RAPD primers* obtained from Operon Technologies,* Alameda, Calif.
2. *MAPMAKER v1.0 (or v2.0)* obtained from S V Tingey, DuPont Co., Wilmington, Del.

C. Experimental Procedures

Crossing Strains of M. grisea

Crosses used to develop the fertile strains and to generate progeny for the mapping cross were performed as follows: Cultures of opposite mating-type were inoculated approximately 4 cm apart on oatmeal agar. They were grown at a distance of 30 cm from continuous near UV light (360 nm, 40 W black light, Westinghouse) at 25°C. Perithecia formed 16–20 days after inoculation. Mature perithecia were squashed in a drop of sterile water on 4% water agar. The asci were separated and allowed to dry on the agar surface to loosen the ascus wall. Individual asci were separated under a stereo microscope (250×) by using a drawn-out capillary tube. After 12 h, germinated ascospores were transferred to rice potato sucrose slants (this may be replaced by oatmeal agar) and maintained at 24°C until the analysis of virulence. In the later crosses, which were used to expand the progeny population and to examine *AVR*CO39 segregation in the F2 generation, isolates were crossed under white fluorescent light at 18°C. Random ascospore isolates were obtained by picking an entire germinated ascus to fresh oatmeal agar and recovering a single conidial isolate from this culture [36].

Rice Inoculation

Growth and inoculation of rice were performed in a chamber equipped with white fluorescent light (230 μE/m/sec) set for 16 h photoperiods. Day and night temperatures were 28°C and 21°C, respectively. The relative humidity was 33%. Rice seeds were planted in Bacto potting soil (Michigan Peat Co., Houston, Tex.) at a density of 5–6 seeds per pot (3"). The pots were placed in

*Names are necessary to report factually on available data; however, the USDA neither guarantees nor warrants the standard of the product, and the use of the name by the USDA implies no approval of the product to the exclusion of others that may also be suitable.

a tray and were watered daily by adding water until it drained out of the pots. When the seedlings had emerged, about 7 days after planting, oatmeal agar plates were inoculated with four evenly spaced agar plugs of each progeny culture. These plates were incubated at room temperature (approximately 22°C) for 14 days under constant illumination from white fluorescent light (55 μE/m/sec). On day 14 the plants were watered and sealed in plastic bags (12 plants per bag). A sterile glass pipette was placed in one of the pots to support the bag and prevent it from damaging the leaves. The bags were then placed in the dark for approximately 1 h to increase the humidity. Agar plates containing cultures of each progeny were flooded with a sterile 0.2% gelatin solution and gently massaged with a sterile, bent glass rod to release the conidia. A sample of conidia was checked for germination. The conidia were counted using a hemacytometer and the density was adjusted to 10^4 conidia mL^{-1}. After approximately 2 h in the dark, the bags were opened and the plants were inoculated by spraying with 12 mL of conidial suspension (1 mL/ plant). The bags were resealed tightly and placed in the dark. After 24 h, the plants were removed from the bags and placed in trays in the light as before. After 7 days, representative leaves were detached and fixed onto cards using clear adhesive tape. If kept in the dark, these cards serve as permanent records of the infection phenotype.

The following rating scheme was used to classify each infection type: 0—no visible symptoms; 1—small brown pinpoint-size, nonsporulating lesions; 2—dark brown, nonsporulating lesions 2–3 mm in length; 3—circular, sporulating lesions with tan centers and dark brown margins; and 4—large, diamond-shaped, sporulating lesions with tan centers and dark brown margins. The different rating schemes used by various investigators and their interpretation may affect the genetic analysis. In our experiment, isolates that were able to sporulate on the leaf surface after the detached leaves were incubated overnight on water agar (types 3 and 4) were considered virulent; nonsporulating lesions (Types 0, 1, and 2) were considered avirulent.

Mapping Using MAPMAKER

Full instructions for using MAPMAKER are included in a manual supplied with the program. From our experience, we suggest using the GROUP command of MAPMAKER with fairly stringent mapping parameters of L.O.D. = 4.0 and a maximum theta value (recombination fraction) of 0.2 to define linkage groups. We found that a larger theta value tended to link markers from separate chromosomes. Once linkage groups have been identified using these parameters and their constituent markers have been ordered, attempts can be made to join linkage groups that are on the same chromosome. This can be achieved using the NEAR command of MAPMAKER to identify markers that are "near" to markers on the ends of linkage groups. The validity of joining linkage groups should be checked using chi-square analysis. As a guideline, in a mapping population of 61 progeny, we found that markers having greater than approximately 30% recombination between them should be classified as unlinked at the $p < .05$ level of significance.

5. QUESTIONS

Why perform genetic analysis and use map-based strategies to clone avirulence genes? An alternative approach would be to clone these genes by "shotgun" methods, involving transforming a virulent strain with a gene library of an avirulent strain and screening transformants for avirulence. After all, these methods were used to clone avirulence genes from bacterial phytopathogens without preliminary genetic analysis [43]. Genetic analysis of *M. grisea* avirulence has revealed some systems that may not always be amenable to shotgun cloning approaches. For example, Silué and Nottheghem [30] described systems where two genetic loci are required for the expression of avirulence. Genes that suppress the expression of avirulence have also been postulated [37,44]. If two genes are required for avirulence, it will not be possible to confer avirulence by transformation of a virulent strain with DNA from avirulent strains. Likewise, an avirulence gene borne on a cosmid clone will not be expressed if the recipient contains a dominant suppressor of that gene [44]. The success of shotgun cloning also depends on the avirulence gene being dominant. In the absence of classical methods to test dominance relationships in *M. grisea* [11], this can only be assumed.

Several inherent problems in the transformation may also compromise the shotgun approach. Expression of transforming DNA may be sensitive to position effects and copy number. To improve the chances of identifying an avirulence gene in a library, more transformants than expected may have to be screened. This could require inoculation of tens of thousands of rice plants. Stability of transforming DNA is a critical concern in shotgun cloning because antibiotic selection cannot be maintained on the plant. Orbach and co-workers initiated an attempt to clone several avirulence genes by screening transformants. However, it was discovered that the cosmid vector possessed a sequence that caused autonomous replication of many library clones in *M. grisea* transformants. Fifty percent of the transformants were found to be mitotically unstable, significantly compromising the feasibility of the experiment [45]. Even integrated DNA in *M. grisea* transformants may be unstable as transforming DNA can be lost or rearranged after *M. grisea* is passed through a plant host [46].

Cloning by "complementation" can lead to the cloning of unexpected genes. For example, Mann et al. repeatedly cloned *pho-4*$^+$ in an experiment to complement a *Neurospora crassa nuc-1* mutation by transformation [47]. By analogy, it is possible that avirulence may be induced by overexpression of a second locus. Finally, it is desirable that a gene library be fully representative of the fungal genome. This is often difficult to assure as some sequences may be "poisonous" to *Escherichia coli*. Such a sequence was linked to the *nuc-1*$^+$ gene, causing it to be absent from a cosmid-based library. The *nuc-1*$^+$ gene was eventually cloned by chromosome walking, using a bacteriophage lambda library [48].

It is important to note that even if such problems are encountered and lead to the failure of attempts to clone avirulence genes, the effort may not be wasted if the results tell us something about plant–pathogen interactions.

How do genetic analysis and map-based cloning obviate the potential problems outlined above? Genetic analysis can establish how many genes are required to produce avirulence and whether additional genes modify the expression of avirulence [13]. Mapping of the gene can provide qualitative evidence that a trait is truly controlled by a single locus and provides a route to cloning the gene. When a candidate avirulence gene clone is identified, knowledge of its genetic behavior will enable the design of rational transformation experiments. In particular, appropriate donor and recipient strains can be chosen, enabling confirmation of gene identity and function. Once a genetic locus associated with avirulence is identified, it is not necessary to know the dominance relationships between avirulence and virulence. These can be tested by reciprocal gene transfers.

6. EXTENSIONS

A. Mapping Genes Identified in Other Crosses

Five genetic maps have been constructed for *M. grisea* [39,49–51] (J. L. Notteghem, pers. comm.). Integration of two of these maps [39,49] is in progress in our laboratory. From the integrated map, it should be possible to pick a subset of markers spanning the entire genome. For example, if each marker in the subset were separated from its adjacent marker(s) by approximately 30 cM, then any newly identified gene would be linked to another marker. The genetic distance between the new gene and its nearest associated marker would be no greater than 15 cM. This degree of linkage would be easy to detect even by rudimentary linkage analysis. Once linked markers are identified in the subset, analyzing the segregation of additional markers in this region will lead to a finer resolution map in the region of interest.

RFLP mapping of avirulence genes is useful for establishing a position in the genetic map. However, the ability to rapidly clone the gene based on its map position will depend on how close the nearest associated markers lie. Chromosome walking and physical mapping studies performed in our laboratory indicate that 1 cM may average 30–60 kb over extended regions of the genome but that specific recombination points may be more than 100 kb apart (M. Farman and S. Leong, in preparation; and unpublished results). Consequently, if the target gene is flanked by markers that are both at least 10 cM away, it would be more expeditious to identify nearer markers.

Mapping techniques based on randomly amplified polymorphic DNA (RAPD) fragments [52] can be used to identify PCR amplicons linked to a gene of interest. Specific chromosomal regions can be targeted for marker identification by using bulk segregant methods [53]. DNA of progeny strains is pooled according to a chosen phenotype (in this case, avirulence or virulence), creating DNA pools that are homozygous for the chosen locus and linked markers but are apparently heterozygous for unlinked loci. Therefore RAPD markers that are unlinked to the gene of interest will appear in both pools, while markers linked to either allele of the chosen gene will appear only in one pool. Fine-structure mapping of the targeted region is then accomplished by performing the amplification reaction individually on each member of the complete progeny set.

Shi and Leung used RAPD analysis of bulked DNA pools from segregants to identify RAPD markers linked to *con1* and *con2*, loci involved in *M. grisea* conidiation [54]. Two hundred random 10-mer primers were tested. Approximately 65% of them actually primed amplification reactions on *M. grisea* DNA, and 30% of 125 primers identified polymorphisms between parental DNAs. Bulked segregant analysis was performed on pooled DNAs (15 DNA samples per pool) according to *con1* phenotype. Out of 125 primers, four genetic markers were identified that were linked to the *con1* locus.

Prior knowledge of RFLP marker segregation could be employed to increase the power of bulked segregant methods. If the pools of DNA are made to include progeny that exhibit recombination between the target gene and the closest flanking markers, any RAPD markers amplified specifically in one pool will be more closely linked than the existing markers.

Representational difference analysis (RDA) is a PCR-based method that enables regions of DNA that are polymorphic between two genomes to be specifically amplified [55]. This method works on the same principle as genomic subtraction [56], except that the subtraction is not physical but effected by rendering "common" DNA segments unamplifiable. This method can accommodate pooled DNA samples consisting of segregating individuals, resulting in specific amplification of polymorphic DNA linked to the gene of interest. Subtraction techniques have already been used successfully in *M. grisea* to clone the mating-type locus [57], suggesting that such approaches may be extremely fruitful.

RFLPs can be converted to amplified fragment length polymorphisms (AFLPs) [58] by ligating adaptors to restriction fragments and using primers complementary to the adaptors for PCR amplification. The amplification products are electrophoresed in a sequencing gel, enabling many markers to be analyzed in a single gel. Although this method does not target specific areas of the genome for marker identification, the number of loci that can be analyzed in a single experiment improves the chance of identifying a marker linked to the chosen genetic locus.

B. Mutational Analysis of Avirulence

Lau and Ellingboe [13] successfully isolated *M. grisea* strains that showed cultivar-specific mutations to increased virulence. Identification of such mutant strains is potentially simple since an avirulent strain can be mutagenized and mass-inoculated onto resistant cultivars. Mutants affected in genes involved in avirulence may produce lesions and can be recovered from the infected tissues. By mass inoculation, Ellingboe and co-workers have found that it is possible to detect a single virulent strain among approximately 1000 avirulent individuals (A. Ellingboe, pers. comm.). This experimental approach is attractive because it may also identify additional genes that are intimately involved in the expression of avirulence but that do not usually segregate in crosses. For example, the two cloned *M. grisea* avirulence genes appear to be expressed only *in planta* (B. Valent, pers. comm.). Consequently, there must be several genes involved in sensing the plant environment and activating avirulence gene expression. Mutation in any one of these genes could lead to increased virulence as long as they are not also necessary for the expression of other pathogenicity functions.

Transformation-mediated insertional mutagenesis has been used for gene tagging in *Neurospora crassa* [59]. More recently, Sweigard et al. [60] and Leung (pers. comm.) have used this approach to tag *M. grisea* genes involved in pathogenicity and conidiation, respectively. The advantage of transformation-mediated mutagenesis is that once a mutant has been identified and a genetic analysis confirms that the mutation was caused by plasmid integration, the mutated gene may be easily recovered. This is achieved by isolating DNA sequences surrounding the plasmid integration site and using it to identify the intact gene in a genomic library. In this manner, Sweigard et al. cloned four genes required for pathogenicity [60].

Genetic analysis has played an important role in the study of *M. grisea* phytopathology. It is essential for full characterization of populations in terms of their avirulence gene complements. This is the principal means by which different avirulence genes may be identified in *M. grisea* populations. Along with studies on the genetics of other aspects of *M. grisea* pathogenicity, such as appressorium development, growth, and conidiation, such studies will lay the foundations for and complement analysis of the molecular biology of pathogenicity and avirulence. By combining these studies with corresponding experiments on the genetics of host resistance, we hope to establish an integrated concept of how these genes interact and evolve.

REFERENCES

1. Ou, S. H. (1985). *Rice Diseases*, 2nd Ed. Commonwealth Myc. Inst. Kew, Surrey, U.K.
2. Barr, M. E. (1977). *Magnaporthe*, *Telimenella*, and *Hyponectria* (Phycosporellaceae). *Mycologia 69*:952.
3. Hebert, T. T. (1971). The perfect stage of *Pyricularia grisea. Phytopathology 61*: 83.
4. Itoi, S., T. Mishima, S. Arase, M. Nozu (1983). Mating behaviour of Japanese isolates of *Pyricularia oryzae. Phytopathology 73*:155.
5. Valent, B., M. S. Crawford, C. G. Weaver, F. G. Chumley (1986). Genetic studies of fertility and pathogenicity in *Magnaporthe grisea* (*Pyricularia oryzae*). *Iowa State J. Res. 60*:569.
6. Kato, H., T. Yamaguchi, N. Nishihara (1976). The perfect state of *Pyricularia oryzae* Cav. from rice plants in culture. *Ann. Phytopathol. Soc. Jpn. 42*:507.
7. Valent, B., F. G. Chumley (1987). Genetic analysis of host species specificity in *Magnaportha grisea*. In *Molecular Strategies for Crop Protection* (C. J. Arntzen, C. Ryan, eds.). Liss, New York, p. 83.
8. Valent, B., L. Farrall, F. G. Chumley (1991). *Magnaporthe grisea* genes for pathogenicity and virulence identified through a series of backcrosses. *Genetics 127*:87.
9. Leung, H., E. S. Borromeo, M. A. Bernado, J. L. Notteghem (1988). Genetic analysis of virulence in the rice blast fungus *Magnaporthe grisea. Phytopathology 78*:1227.
10. Chao, C.-C. T., A. H. Ellingboe (1991). Selection for mating competence in *Magnaporthe grisea* pathogenic to rice. *Can. J. Bot. 69*:2130.
11. Crawford, M. S., F. G. Chumley, C. G. Weaver, B. Valent (1986). Characterization of the heterokaryotic and vegetative diploid phases of *Magnaporthe grisea. Genetics 114*:1111.
12. Hamer, J. E., B. Valent, F. G. Chumley (1989). Mutations at the *SMO* genetic locus affect the shape of diverse cell types in the rice blast fungus. *Genetics 122*:351.
13. Lau, G. W., A. H. Ellingboe (1993). Genetic analysis of mutations to increased virulence in *Magnaporthe grisea. Phytopathology 83*:1093.
14. Parsons, K. A., F. G. Chumley, B. Valent (1987). Genetic transformation of the fungal pathogen responsible for rice blast disease. *Proc. Natl. Acad. Sci. USA 84*:4161.
15. Leung, H., U. Lehtinen, R. Karjalainen, et al. (1992). Transformation of the rice blast fungus *Magnaporthe grisea* to hygromycin B resistance. *Curr. Genet. 17*:409.

16. Mattern, I. E., P. J. Punt, C. A. M. J. J. van den Hondel (1989). A vector of *Aspergillus* conferring phleomycin resistance. *Fungal Genet. Newsl. 35*:25.
17. Roberts, I. N., R. P. Oliver, P. J. Punt, C. A. M. J. J. van den Hondel (1989). Expression of the *Escherichia coli* β-glucuronidase gene in industrial and phytopathogenic filamentous fungi. *Curr. Genet. 15*:177.
18. Farman, M. L., S. A. Leong (1991). Molecular techniques for analysis of pathogenesis in *Magnaporthe grisea*. Abstracts from 16th Fungal Genetics Conference, Asilomar, Calif.
19. Valent, B., F. G. Chumley (1991). Molecular genetic analysis of the rice blast fungus, *Magnaporthe grisea. Annu. Rev. Phytopathol. 29*:443.
20. Emmett, R. W., D. C. Parberry (1975). Appressoria. *Annu. Rev. Phytopathol. 13*: 147.
21. Lebrun, M.-H., J. Orcival, C. Duchartre (1990). Resistance of rice to tenuazonic acid, a toxin from *Pyricularia oryzae. Rev. Cytol. Biol. Veget. Bot. 7*:249.
22. Lebrun, M.-H., F. Dutfoy, F. Guademer, G. Kunesch, A. Gaudemer (1990). Detection and quantification of the fungal phytotoxin tenuazonic acid produced by *Pyricularia oryzae. Phytochemistry 29*:3777.
23. Singh, N., I. S. Kunene (1980). Cellulose decomposition by four isolates of *Pyricularia oryzae. Mycologia 72*:182.
24. Hirayama, T., H. Nagayama, K. Matsuda (1980). Studies on cellulases of a phytopathogenic fungus, *Pyricularia oryzae* Cavara. IV. Kinetic studies on beta-glucosidases. *J. Biochem. Tokyo*:1177.
25. Wu, S.-C., B. Valent, A. G. Darvill, P. Albersheim (1993). Disruption of an endo-β-1,4-D-xylanase gene reduces the virulence of *Magnaporthe grisea*, the rice blast pathogen. Abstracts from Symposium on Molecular Genetics of Plant-Microbe Interactions, Rutgers, N.J.
26. Sweigard, J. A., F. G. Chumley, B. Valent (1992). Cloning and analysis of *CUT1*, a cutinase gene from *Magnaporthe grisea. Mol. Gen. Genet. 232*:174.
27. Borromeo, E. S., R. J. Nelson, J. M. Bonman, H. Leung (1993). Genetic differentiation among isolates of *Pyricularia* infecting rice and weed hosts. *Phytopathology 83*:393.
28. Yaegashi, H. (1978). Inheritance of pathogenicity in crosses of *Pyricularia* isolates from weeping lovegrass and finger millet. *Ann. Phytopathol. Soc. Jpn. 44*:626.
29. Silué, D., J. L. Notteghem, D. Tharreau (1992). Evidence of a gene-for-gene relationship in the *Orzya sativa–Magnaporthe grisea* pathosystem. *Phytopathology 82*:577.
30. Silué, D., J. L. Notteghem (1992). Identification of a cross between two compatible isolates of *Magnaporthe grisea* (Hebert) Barr and genetic analysis of avirulence/virulence of rice. *J. Phytopathol. 135*:77.
31. Smith, J. R., S. A. Leong (in press). Mapping of a *Magnaporthe grisea* locus affecting rice (*Oryza sativa*) cultivar specificity. *Theor. Appl. Genet.*
32. Bourett, T. M., R. J. Howard (1990). In vitro development of penetration structures in the rice blast fungus *Magnaporthe grisea. Can. J. Bot 68*:329.
33. Lee, Y.-H., R. A. Dean (1993). cAMP regulates infection structure formation in the plant pathogenic fungus *Magnaporthe grisea. Plant Cell 5*:693.

34. Valent, B., F. G. Chumley (1993). Avirulence genes and mechanisms of genetic instability in the rice blast fungus. In *Rice Blast Disease*, Proceedings of the International Symposium on Rice Blast Disease, Madison, Wisc.
35. Kolmer, J. A., A. H. Ellingboe (1988). Genetic relationships between fertility and pathogenicity and virulence to rice in *Magnaporthe grisea. Can. J. Bot 66*:891.
36. Ellingboe, A. H., B.-C. Wu, W. Robertson (1988). Fertility and pathogenicity to rice in *Magnaporthe grisea*. In *Molecular Genetics of Plant-Microbe Interactions* (R. Placios, D. P. S. Verma, eds.). American Phytopathological Society, St. Paul, Minn., p. 235.
37. Ellingboe, A. H. (1992). Segregation of avirulence/virulence on three rice cultivars in 16 crosses of *Magnaporthe grisea. Phytopathology 82*:597.
38. MacKill, D. J., J. M. Bonman (1992). Inheritance of blast resistance in near-isogenic lines of rice. *Phytopathology 82*:746.
39. Skinner, D. Z., A. D. Budde, M. L. Farman, J. R. Smith, H. Leung, S. A. Leong (1993). Genome organization of *Magnaporthe grisea*: genetic map, electrophoretic karyotype, and occurrence of repeated DNAs. *Theor. Appl. Genet. 87*: 545.
40. Lander, E. S., P. Green, J. Abrahamson, et al. (1987). Mapmaker: an interactive computer package for constructing primary genetic linkage maps of experimental and natural populations. *Genomica 1*:174.
41. Chumley, F. G., B. Valent (1990). Genetic analysis of melanin-deficient, nonpathogenic mutants of *Magnaporthe grisea. Mol. Plant-Micr. Interact. 3*:135.
42. Howard, R. J., M. A. Ferrari (1989). Role of melanin in appressorium function. *Exp.Mycol. 13*:403.
43. Lau, G. W., C. T. Chao, A. H. Ellingboe (1993). Interaction of genes controlling avirulence/virulence of *Magnaporthe grisea* on rice cultivar Katy. *Phytopathology 83*:375.
44. Keen, N. T. (1990). Gene-for-gene complementarity in plant-pathogen interactions. *Annu. Rev. Genet. 24*:447.
45. Orbach, M., J. Sweigard, A. Walter, L. Farrall, F. Chumley, B. Valent (1991). Strategies for isolation of avirulence genes from the rice blast fungus *Magnaporthe grisea.* Abstracts from 16th Fungal Genetics Conference, Asilomar, Calif.
46. Tooley, P. W., H. Leung, S. A. Leong (1992). Meiotic and mitotic stability of transforming DNA in the phytopathogenic fungus *Magnaporthe grisea. Curr. Genet. 21*:55.
47. Mann, B. J., R. A. Akins, A. M. Lambowitz, R. L. Metzenberg (1988). The structural component in a phosphorus-repressible phosphate permease of *Neurospora crassa* can complement a mutation in a positive regulatory gene. *Mol. Cell. Biol. 8*:1376.
48. Kang, S., R. L. Metzenberg (1990). Molecular analysis of *nuc-1*$^{+}$, a gene controlling phosphorus acquisition in *Neurospora crassa. Mol. Cell. Biol. 10*:5839.
49. Sweigard, J. A., B. Valent, M. J. Orbach, A. M. Walter, A. Rafalski, F. G. Chumley (1993). Genetic map of the rice blast fungus *Magnaporthe grisea* (n=7). In *Genetic Maps* (O'Brien, ed.). Cold Spring Harbor Laboratory Press, Cold Spring Harbor, NY, p. 3.112.

50. Romao, J., J. E. Hamer (1992). Genetic organization of a repeated DNA sequence family in the rice blast fungus. *Proc. Natl. Acad. Sci. USA 89*:5316.
51. Hayashi, N., H. Naito (1993). Genome mapping of *Magnaporthe grisea.* Abstracts from the International Symposium on Rice Blast Disease, Madison, Wisc.
52. Williams, J. G. K., A. R. Kubelik, J. A. Rafalski, S. V. Tingey (1991). Genetic analysis with RAPD markers. In *More Gene Manipulations in Fungi* (J. W. Bennett, L. L. Lasure, eds.). Academic Press, San Diego, p. 431.
53. Michelmore, R. W., I. Paran, R. V. Kesseli (1991) Identification of markers linked to disease-resistance genes by bulked segregant analysis: a rapid method to detect markers in specific genomic regions by using segregating populations. *Proc. Natl. Acad. Sci. USA 88*:9828.
54. Shi, Z., H. Leung (in press). Genetic analysis and rapid mapping of a sporulation mutation in *Magnaporthe grisea. Mol. Plant-Microbe Interact.*
55. Lisitsyn, N., N. Lisitsyn, M. Wigler (1993). Cloning the differences between two complex genomes. *Science 259*:946.
56. Strauss, D., F. M. Ausubel (1990). Genomic subtraction for cloning DNA corresponding to deletion mutations. *Proc. Natl. Acad. Sci. USA 87*:1889.
57. Kang, S., F. G. Chumley, B. Valent (1993). Cloning of the mating type genes of *M. grisea* by genomic subtraction. Abstracts from 17th Fungal Genetics Conference, Asilomar, Calif.
58. Zabeau, M. (1994). Applications of AFLP. Presentation at Plant Genome II Conference, San Diego, Calif.
59. Kang, S., R. L. Metzenberg (1993). Insertional mutagenesis in *Neurospora crassa*: cloning and molecular analysis of the *preg+* gene controlling the activity of the transcriptional activator NUC1. *Genetics*:193.
60. Sweigard, J., A. Walter, B. Valent (1993). *Magnaporthe grisea* pathogenicity genes cloned by insertional inactivation mutagenesis. Abstracts from International Symposium on Rice Blast Disease, Madison, Wisc.

15

Genetic Analysis of Senescence in *Podospora anserina*

Heinz D. Osiewacz
Johann Wolfgang Goethe-Universität, Frankfurt am Main, Germany

1. INTRODUCTION

Podospora anserina is an ascomycete belonging to the family of the Sordariaceae. Its natural habitat is the dung of herbivores. The sexual cycle (Fig. 1) is controlled by two mating-type alleles, termed *mat*– and *mat*+. After fertilization four binucleate ascospores are formed which are arranged in a linear order. Usually, these spores contain both mating types and give rise to self-fertile mycelia (*pseudocompatibility*). It therefore appears as if *P. anserina* would be homothallic. However, in about 1–2% of all asci, two smaller mononucleate ascospores are produced instead of one binucleate spore. These spores give rise to mycelia which are not fertile. Perithecia are only formed when mycelia of one mating type are crossed with mycelia of the opposite mating type (see Fig. 1) demonstrating that *P. anserina* is in fact *heterothallic*. The behavior of the binucleate, heterokaryotic ascospores is described by the term *secondary homothallism* [1,2]. This breeding system has also been termed monoecism superimposed by incompatibility [3]. Irregular asci with mononucleate spores (e.g., five-spored, six-spored asci) are well suited for tetrad analysis (see Chapter 3). For genetic experiments the smaller mononucleate spores, which can easily be isolated using a dissection microscope, are of particular interest since these cells possess only one mating type and give rise

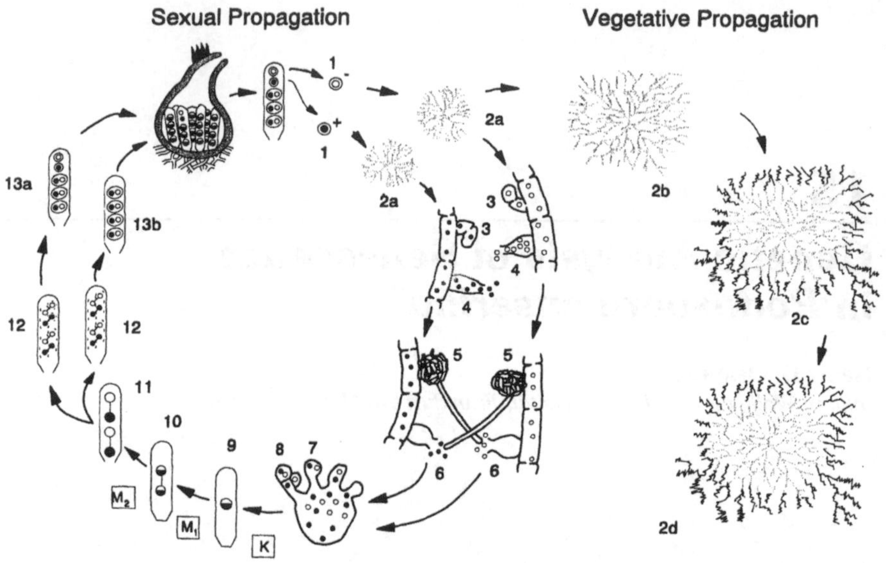

Figure 1 Sexual and vegetative reproduction of *Podospora anserina*. After germination of ascospores (1) a highly branched mycelium arises. If mycelia arise from mononucleate spores, cultures (2) of two genetical specificities (*mat*+ and *mat*–, respectively) are formed. On both types of cultures, ascogonia (3) and spermogonia (4) develop. The latter release the male gametes, the spermatia (6). Spermatia of one mating type fuse with the trichogyne of the opposite mating type. After a number of conjugative divisions of the female and the male nucleus, the tip of an ascogenous hypha recurves and forms a crozier (7). An ascus initial cell (8) is formed which contains one female and one male nucleus. After karyogamy, the two meiotic and one postmitotic nuclear divisions (9–12) lead to the formation of an ascus in which, in the majority of all cases, four binucleate ascospores (13b) are located in a linear order. However, a few asci contain five (13a) or even more spores. Two of the five-spored asci are smaller than the other ascospores and contain only one nucleus. Vegetative propagation of *P. anserina* is restricted to mycelial growth. The hypha develop a highly branched mycelium (2a–2d). After a strain-specific growth time, the growth rate of a mycelium decreases, the morphology of a senescent culture changes (e.g., the pigmentation increases), and the colonial growth arrests completely.

to self-incompatibile mycelia which produce two types of sexual organs: ascogonia serving as female gametangia, and spermogonia in which microconidia are produced. These latter cells are also termed spermatia and function as male gametes. In contrast to the closely related ascomycete *Neurospora crassa*, no macroconidia are produced. Fertilization of ascogonia occurs when spermatia of one mating type fuse with the trichogyne of an ascogonium of the opposite mating type. Subsequently, multinucleate ascogenous hyphae develop. The tip of these hyphae recurves and forms a crozier. After nuclear movement and the formation of cell walls, an ascus initial cell is formed that contains the two nuclei of the opposite mating type. Subsequently, after karyogamy, the two meiotic nuclear divisions and one postmeiotic mitosis occur. Usually, during spore formation, two nuclei become surrounded by one cell wall. Depending on whether the alleles of a particular gene exhibit pre- or postreduction (first- or second-division segregation), different types of asci are formed (see Chapter 3). The frequency of postreduction depends on the distance of the corresponding genetic marker from the centromere, thus providing a basis for the construction of genetic maps. In the case of the mating-type locus, the two alleles undergo postreduction in about 98% of all cases since the mating-type locus is located rather on the end of linkage group 1.

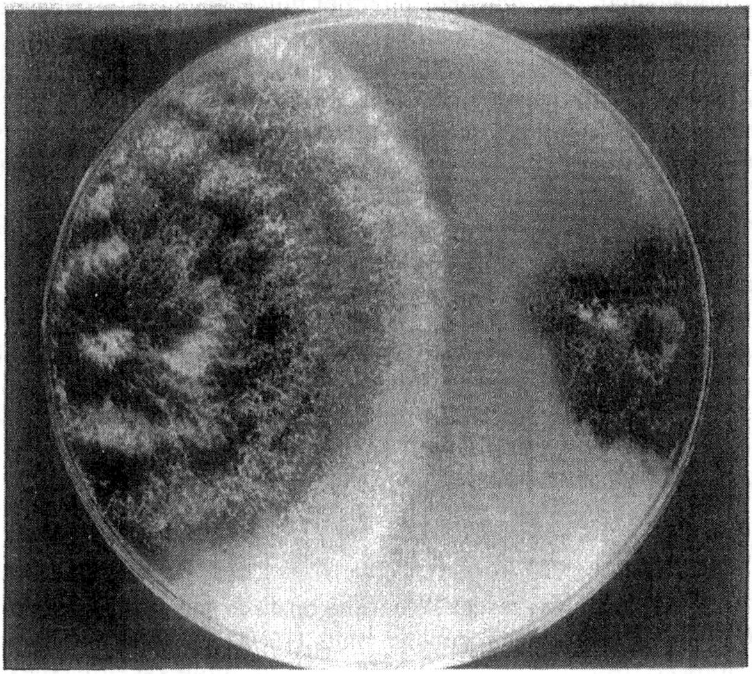

Figure 2 Comparison of a juvenile (left) and a senescent (right) culture of *P. anserina* grown on solid medium. The two colonies were cultured for the same time. The senescent culture, which appears darker, due to a reduced formation of aerial hyphae, stopped growing.

Vegetative propagation of *P. anserina* is restricted to mycelial growth. No specialized cells (e.g., vegetative spores) are involved. In contrast to most other fungi, mycelial growth is restricted to a characteristical period of time. At the end of this time, the morphology of the corresponding culture changes at the growth front (e.g., increased pigmentation), and mycelial growth stops (Fig. 2). This "senescence syndrome" was first described by Rizet in the early 1950s [4] and has subsequently been thoroughly analyzed by different laboratories.

2. KEYNOTES

Podospora anserina and later on a number of related filamentous fungi (e.g., *Neurospora crassa, N. intermedia*) have proved to be ideal model systems to unravel the complex mechanisms controlling the onset of senescence. *P. anserina* is now one of the best-analyzed models in experimental gerontology [for reviews see 5–9]. This is due to a number of characteristics: 1.) Filamentous fungi have a simple organization; specialized cells or "organs" are only formed during propagation. 2.) *P. anserina* and other fungi can easily be cultivated in simple liquid and on solid medium. 3.) The life span of all wild-type strains of *P. anserina* is rather short (e.g., wild-type strain s, 25 days). 4.) The generation time is short due to the completion of the sexual cycle in about 10–12 days. 5.) *P. anserina* is accessible to a formal genetic analysis; defined mutants are available and ascospores can easily be isolated allowing the analysis of a large number of progeny of well-defined genetic crosses in a rather short time. 6.) *P. anserina* is accessible to molecular genetics; nucleic acids and proteins can be isolated and can be analyzed by standard molecular techniques. In addition, foreign DNA can be introduced into protoplasts via transformation allowing a defined genetic manipulation.

In this chapter, approaches to unravel the genetic mechanisms involved in the control of the life span of this simple eukaryotic microorganism are described. In particular, experiments are outlined aimed to isolate mutants in which the life span is affected and to characterize these mutants on the genetical and the molecular level.

3. PRINCIPLES

A formal genetic analysis revealed that the onset of senescence is under the control of extrachromosomal genetic traits [10,11]. This became apparent from reciprocal crosses using juvenile and senescent cultures as either the male or the female parent. As outlined in Figure 3, fertilization of ascogonia of a culture by a suspension of spermatia leads to the formation of perithecia. Ascospores that are formed after spermatization of juvenile cultures with spermatia of a senescent culture ("senescent spermatia") give rise to juvenile

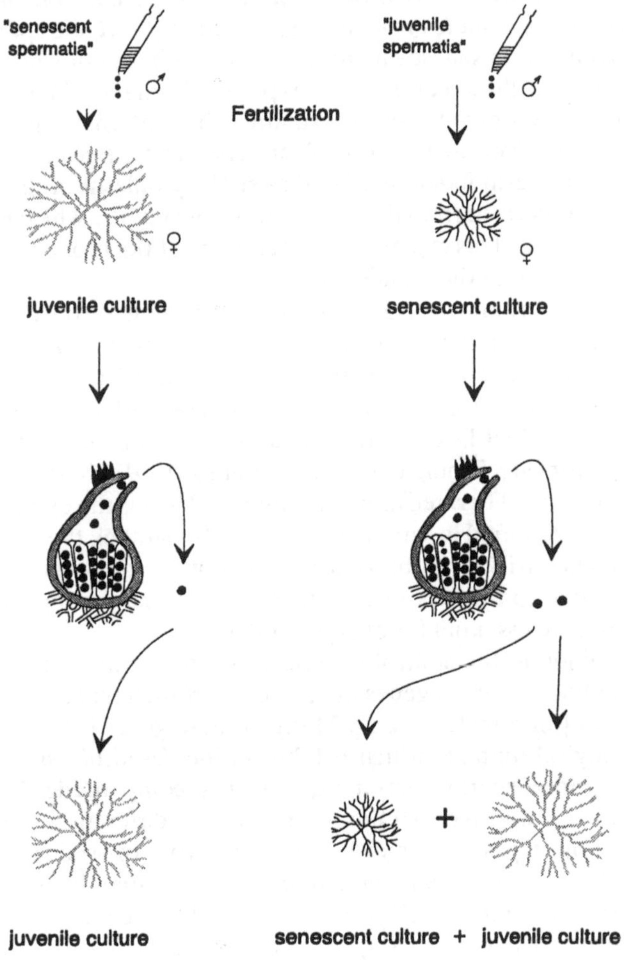

Figure 3 Reciprocal spermatization experiments demonstrating the genetic control of the onset of senescence by extrachromosomal genetic traits. Spermatia isolated from a senescent ("senescent spermatia") and a juvenile *P. anserina* culture ("juvenile spermatia") are used to fertilize ascogonia from juvenile and senescent cultures, respectively. Single ascospores are isolated and germinated, and the life span of the derived culture is determined. In this type of experiment reciprocal differences (juvenile cultures exclusively vs. juvenile and senescent cultures) are observed.

cultures. In contrast, ascospores isolated after fertilization of a senescent culture with male gametes from a juvenile strain ("juvenile spermatia") lead to both juvenile and senescent cultures, the frequency of which depends on the age of the fertilized culture. This type of inheritance, in which reciprocal differences are observed, can be explained by maternal inheritance of the genetic trait involved in the control of senescence. Because of a very low amount of cytoplasm in spermatia, almost all cytoplasm in the product of a genetic cross is derived from the ascogonia of the culture which is fertilized by isolated spermatia. This explains the phenotype of the progeny, which resembles the phenotype of the female parent.

The extrachromosomal genetic trait controlling the onset of senescence has been identified in the mitochondria of senescent *Podospora* cultures. It is a covalently closed circular DNA species, termed plDNA or αsenDNA [12,13]. In juvenile cultures this DNA species is an integral part of the high-molecular-weight mitochondrial DNA (mtDNA) and represents the first intron of the gene coding for cytochrome c oxidase subunit I *(COI)* [14–17]. During aging the intron becomes liberated and amplified, and forms the extrachromosomal DNA species identified in senescent cultures. In parallel, the high-molecular-weight mitochondrial DNA becomes disintegrated. Consequently, senescent cultures die due to the lack of certain mitochondrial enzyme activities [18], activities that are essential for energy production.

In addition to the control of senescence by extrachromosomal genetic traits, a number of nuclear genes have been identified that affect the life span of *P. anserina* [for a review see 5]. Many of these genes have been identified in morphological mutants which exhibit an increased life span. In the two possible reciprocal spermatization experiments (compare Fig. 3) between the mutant and the wild-type strains there are no differences observed. The progeny of both types of crosses leads to two types of strains: strains with a wild-type life span, and strains with an increased, mutant life span, indicating that the corresponding genetic traits are located in the nucleus.

Some examples of nuclear genes affecting the life span of *P. anserina* are indicated in Table 1. It may be seen that in mutants such as *incoloris, grisea*, or *vivax* the life span is increased for a few days to about a few weeks. Interestingly, double mutations of the corresponding genes appear to have a synergistic effect. Some double mutants grow for several years and seem to be immortal [19,20].

4. PRACTICE

A. Experimental Approach

In the following section, experiments are outlined to isolate novel life-span-affected mutants of *P. anserina*, to characterize these mutants on the genetic

Table 1 Characteristics of the wild-type strain *s* of *P. anserina* and a selected number of pleiotropic, long-lived mutant.

Strain	Site of mutation (linkage group)	Life span	Characteristics	References
wild type (*s*)	—	25 d	mycelium: green to black	Esser [22]
incoloris (*i*)	III	42 d	mycelium: colorless	Esser [22] Prillinger and Esser [27]
vivax (*viv*)	III	66 d	mycelium: green to black, rhythmic growth	Esser and Keller [19]
grisea (*gr*)	IV	39 d	mycelium: green to black, ascospores: gray, aerial hyphae reduced	Prillinger and Esser [27]
lannosa (*la*)	V	33 d	mycelium: velvety	Esser [22]
tarda (*ta*)	VI	66 d	slow growing	Esser [22]
crispa (*cr*)	VII	44 d	aerial hyphae curly	Prillinger and Esser [27]
grisea vivax (*gr viv*)	IV+III	> 9 a	mycelium: green to black, ascospores gray	Tudzynski et al. [44]
incoloris vivax (*i viv*)	III	>12 a	mycelium: yellow to reddish; aerial hyphae reduced	Esser and Keller [19] Tudzynski et al. [44]

level, and to clone and characterize the genetic factors that are mutated in the corresponding strains. This approach combines experiments of classical and molecular genetics. Experiments of this type can be expected to provide novel clues about the genetic control of degenerative processes in *P. anserina* and will provide data to unravel the complex molecular mechanisms that modulate the aging process and the life span of this rather simple biological system.

Figure 4 represents a scheme outlining experiments to select and characterize novel life span affected mutants of *P. anserina*.

B. Material Needed

Strains

Wild-type strains of *Podospora anserina* can be easily isolated from dung of various herbivores—e.g., horses, sheep, or rabbits. Perithecia of about 0.5 mm with a characteric morphology, in particular with a bundle of dark hairs at the

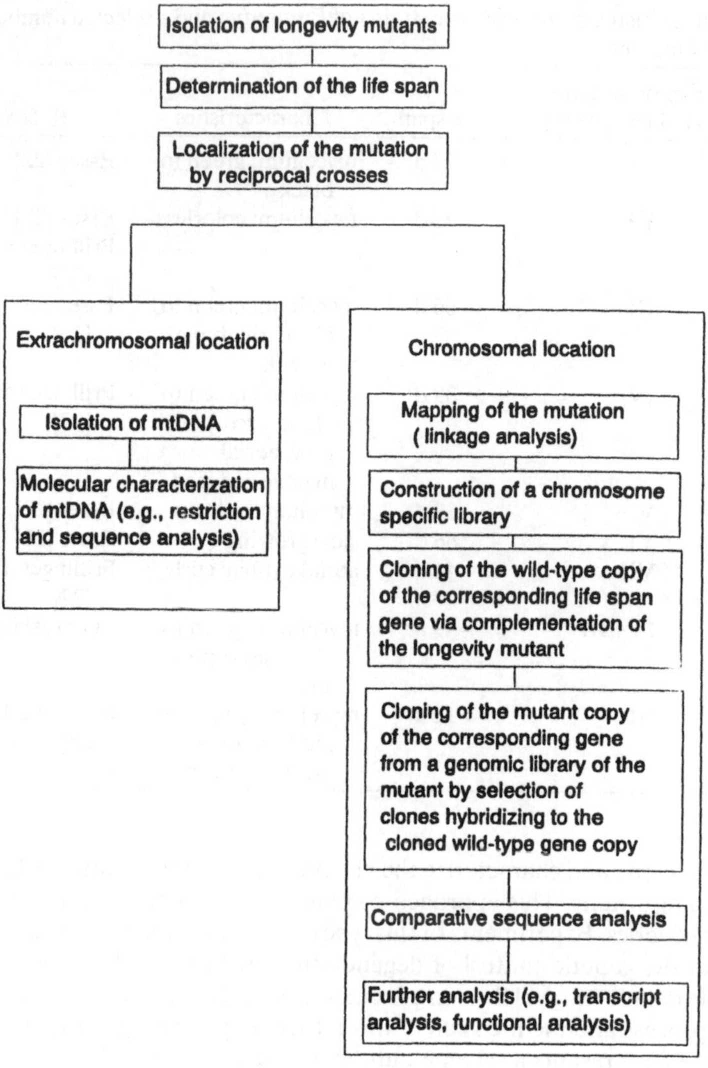

Figure 4 Outline of experiments to select and to characterize novel life-span-affected mutants of *P. anserina*.

ostiolum, appear on dung usually after about 2 weeks' incubation of fresh dung in a large Petri dish. Since the ascospores are actively ejaculated from the perithecium, they can be isolated by placing the top of an agar plate containing medium which allows the germination of ascospores upside-down above the ripe perithecia. The ejaculation of spores is enhanced by light, and cultures should therefore be incubated in the light. After a few hours' incubation of ascospores on germination medium, small colonies arise. Pieces of these colonies are transferred to complete medium and incubated at 27°C in the light. After about 10–14 days perithecia of the typical morphology are formed. From these fruiting bodies mononucleate spores of either mating type can be isolated, and colonies arising from these spores can be used in further experiments.

Wild-type strain *s* and a large collection of mutant strains are available from different laboratories [for references see 21]. The site of mutation has been mapped (Fig. 5), and the corresponding strains can be used as tester strains to localize the site of mutation of a novel mutant by linkage analysis.

Media

For a detailed description of the different minimal, complete, and special media see Esser [22]. Here only the three media are mentioned that are of particular importance for the experiments described below.

Complete medium. One liter of solid medium contains 1.5 g malt extract and 20 g agar in 1 L cornmeal extract. Cornmeal extract is obtained from 250 g cornmeal incubated in 10 L of water at 60°C overnight. After this time the supernatant is filtered through several layers of cheesecloth, and the cornmeal is discarded.

Spore-germinating medium. Complete medium is supplemented with 0.44% ammonium acetate.

Transformation medium. One liter contains 3.7 g NH_4Cl, 2 g tryptone, 1 g casamino acids, 1 g yeast extract, 1.5 g KH_2PO_4, 0.5 g KCl, 0.5 g $MgSO_4 \times 7\ H_2O$, 1 mg $CuSO_4 \times 5\ H_2O$, 1 mg $FeCl_3 \times 6\ H_2O$, 1 mg $MnSO_4 \times 7\ H_2O$, 10 g glucose, 1 M sucrose, 1.2 g agar.

Liquid medium. For the cultivation of cultures for different purposes liquid medium, generally complete medium without agar, is used.

C. Experimental Procedures

Selection of Long-Lived Mutants

Many of the different long-lived mutants of *P. anserina* have been isolated from spontaneous outgrowths from the edge of the growth front of a senescent culture [23–25]. A scheme for the systematic isolation of spontaneous long-lived mutants has been reported by Schulte et al. 1988 [26] (Fig. 6).

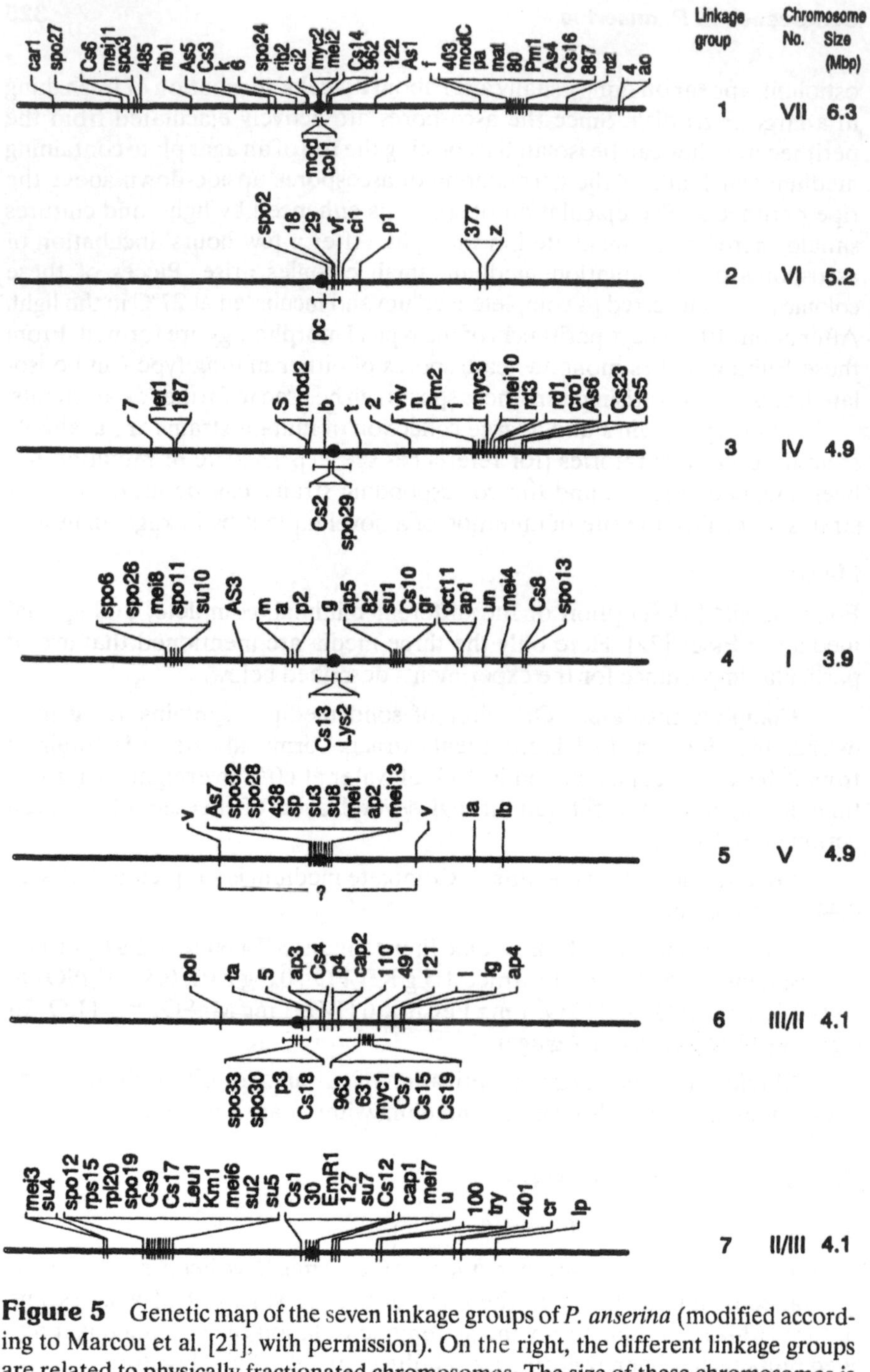

Figure 5 Genetic map of the seven linkage groups of *P. anserina* (modified according to Marcou et al. [21], with permission). On the right, the different linkage groups are related to physically fractionated chromosomes. The size of these chromosomes is indicated according to Osiewacz et al. [30].

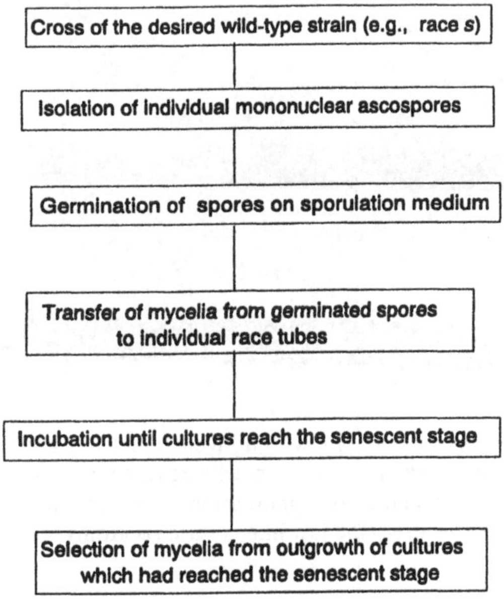

Figure 6 Scheme to experimentally select for life-span-affected mutants of *P. anserina*.

In addition, the inducation of mutations may be performed by chemical mutagenesis of microconidia. Using ethyl-methane-sulfonate (EMS) and other chemical mutagens, a number of mutants were selected that are affected in the formation of laccases, a group of enzymes belonging to the phenoloxidases [27]. Some of these mutants turned out to represent pleiotropic mutants in which mycelium morphology, differentiation of sex organs, and life span are affected.

Measuring the Life Span

The determination of the life span is performed in race tubes which contain the desired solid medium, usually complete medium. These glass tubes have a length of 30–50 cm and a diameter of about 2 cm (Fig. 7).

From 50 to about 200 independent, freshly germinated ascospores are transferred to separate race tubes. The cultures are incubated at the desired temperature (e.g., 27°C) and examined every day. The growth rate is recorded at the distance of the mycelial growth front extended per day. After a specific time of growth, the growth rate declines, the morphology of the colony becomes altered (e.g., increased pigmentation), and finally the growth arrests. This time is recorded, and the mean life span of a particular strain is calculated

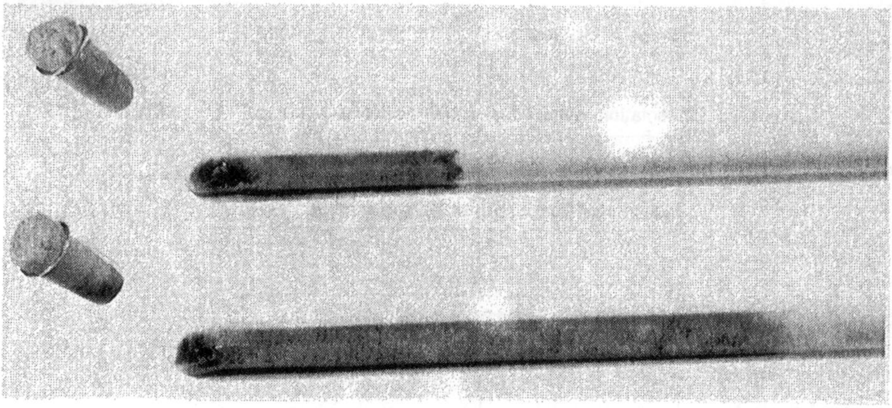

Figure 7 Race tubes of *P. anserina* with a juvenile, actively growing culture and a senescent culture, respectively. The senescent culture (race tube on the top), which has stopped growing, shows an increased pigmentation at the growth front. The growth of this colony has arrested, whereas the juvenile culture (bottom) is still growing.

as the time at which growth arrest is observed in 50% of all parallel cultures. In addition, the maximal life span can be determined as the time at which all parallel cultures stopped growing.

Genetic Characterization of Life-Span Mutants

After a mutant with an affected (usually increased) life span has been identified, the site of mutation needs to be determined. For this reason reciprocal crosses are performed as outlined in Figure 3. As the female parent a culture derived from a mononucleate ascospore of either mating type + or – is used. Onto the surface of such a culture grown for about 7–10 days on complete medium, spermatia of the complementary parent are poured. These spermatia were isolated from agar cultures of the strain of interest by pouring about 5 mL of sterile water onto the surface of the culture and moving the drop of water thoroughly until the spermatia were resuspended.

After the formation of perithecia, individual mononucleate ascospores are isolated from the different crosses and germinated, and the mean life span is determined. As introduced above, different outcomes can be expected:

1. The life span may be maternally inherited. In this case the two reciprocal crosses between a long-lived mutant and the wild-strain lead to different results (reciprocal differences), indicating that the genetic trait involved is located in the cytoplasm.

2. There may be no differences in the outcome of reciprocal crosses. About 50% of the progeny are characterized by a long-lived phenotype, and about 50% by the wild-type phenotype. In this case the genetic factors are located in the nucleus of the two parents.

Molecular Characterization of Extrachromosomal Mutants

Mitochondrial DNA is prepared from mycelia of both the wild-type strain and the long-lived mutant by centrifugation of cleared cell extracts in a CsCl gradient. After extraction of the isolated DNA and dialysis overnight, the undigested mtDNA and the mtDNA digested with different restriction endonucleases are fractionated in agarose gels. Thereafter the gel is stained with ethidiumbromide and analyzed under UV light. Mitochondrial DNA of the wild-type strain may differ from that of the mutant. This may be quite obvious in cases where an additional DNA species, such as a linear plasmid, is present in one strain but missing in the other [25,28,29]. Alternatively, rearranged mtDNA with large deletions, insertions, or inversions may become obvious by a different restriction pattern of the two different DNA preparations [25,26]. A detailed molecular characterization, including a restriction analysis, a Southern blot analysis, or finally a sequence analysis may be performed to identify the exact nature of the mutation in the novel long-lived mutant.

Analysis of Chromosomal Long-Lived Mutants

By conventional linkage analysis of crosses between the novel long-lived mutant and different tester strains containing a mutation mapped to different chromosomes, the novel life-span-affecting gene can be located on one of the seven individual linkage groups. The experimental steps of such an analysis are:

1. Crosses between a culture of the long-lived mutant isolated from a mononucleate ascospore of one mating type and a culture of a characterized nuclear mutant (tester strain) of the opposite mating type. Different crosses with different tester strains representing marker strains for the different linkage groups (Fig. 5) need to be performed.

2. After the formation of perithecia a tetrad analysis is performed. Spores from individual irregular, five-spored asci are isolated and analyzed. In an initial analyses it is sufficient to isolate about 10–20 asci and to analyze the progeny for linkage of the long-lived phenotype and the phenotype of the tester strain. If linkage becomes apparent, that is, when the number of the parental phenotype does not differ significantly from that of the recombinant phenotype, a more detailed analyses has to be performed. In this analysis more asci (e.g., 100–500) are isolated and the outcome of this cross can be statistically evaluated as outlined in Chapter 3. An analysis of this type leads to the

localization and to the relative mapping of the novel long-lived gene on one of the seven linkage groups of *P. anserina.*

Cloning of Life-Span-Affecting Genes by Complementation

After mapping the gene of interest on a particular linkage group, cloning of the corresponding wild-type copy is possible by complementation of the long-lived mutant to wild-type characteristics using a genomic cosmid library of the wild-type strain. Since transformation efficiencies of most filamentous fungi, including *P. anserina*, are rather low, ranging from a few transformants to about 80 transformants per μg of transforming DNA, the use of specific gene libraries of reduced complexity is desirable. The construction of this type of libraries is possible because individual chromosomes of *P. anserina* can be resolved on pulsed-field gels [30,31] and the DNA from these gels can be reextracted and used to construct chromosome-specific libraries. In addition, from previous investigations it is clear which of the size-fractionated chromosomes corresponds to which of the seven linkage groups (see Fig. 5). Using the reextracted DNA derived from the chromosome corresponding to the linkage group on which the gene of interest is located, it is possible to construct a genomic library which is enriched for cloned sequences derived for this particular chromosome. The use of this type of library has the advantage that the number of transformants that have to be obtained and analyzed to isolate a clone containing the gene of interest is rather low. In fact, using this strategy, we were recently able to clone a nuclear gene by complementation of a specific *P. anserina* mutant from a genomic library of about 400 cosmid clones (unpublished).

After a transformant with wild-type characteristics has been identified which originated from the transformation with a pool of about 100 individual hybrid cosmids, the corresponding hybrid cosmid can be identified by a subsequent use of subsets of the cosmid library pool used for transformation. This procedure, called "sib selection" (Fig. 8), which was successfully used in *Neurospora* [32,33], is very time-consuming and cost-intensive, at least in *P. anserina*. However, it may be possible to recover the complementing plasmid directly from the wild-type transformant via *in vitro* packaging of high-molecular-weight chromosomal DNA of the corresponding mutant into phage lambda and a subsequent *E. coli* transfection. In a few cases, this strategy led to a successful selection of antibiotic resistant *E. coli* colonies and to the reisolation of a hybrid plasmid or cosmid, which contained the gene of interest [34,36] (Osiewacz, unpublished).

Once the wild-type gene has been recovered, the mutant copy can be isolated from a genomic library of the mutant via hybridization using the cloned wild-type copy as a specific probe. Subsequently, a wide avenue of experiments opens in which the cloned gene and its mutated copy can be

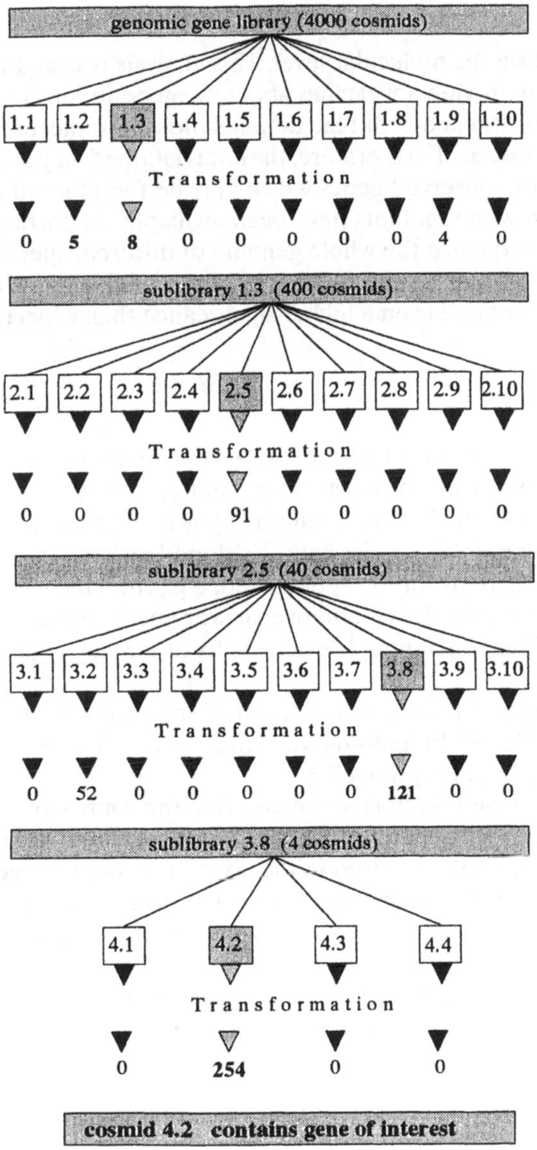

Figure 8 Scheme outlining cloning of genes by complementation of fungal mutants ("sib selection"). A complete cosmid library (e.g., 4000 individual cosmids) is divided into several pools (1.1–1.10). DNA of these pools is used to transform protoplasts of a fungal mutant. The number of wild-type transformants is determined as indicated below the arrowheads (e.g., eight colonies occur after transformation with cosmid pool 1.3). The cosmid pool (1.3) giving rise to the largest number of wild-type transformants is subdivided into sublibraries of lower complexity (e.g., 40 cosmids each). After several rounds of transformation the complementing capacity can be attributed to a single hybrid cosmid (4.2), which contains the wild-type gene of interest.

characterized on the molecular level. This analysis is straightforward and can be expected to provide novel clues about the genetic control of senescence in *Podospora anserina* and about the detailed molecular mechanisms involved in this complex process. Furthermore, the data obtained may also provide insight into the role of conserved genes which may be found in other organisms for which, until now, no mutants have been available. In particular, considering the efforts to sequence the whole genome of different species including *Saccharomyces cerevisiae*, *Caenorhabditis elegans*, and humans, the approach outlined above may be of even a higher significance than expected.

5. QUESTIONS

1. Aging is a biological process which occurs in almost all biological systems. However, can it be expected that aging of the various, diverse systems is due to one basic, unifying mechanism? Or is it not more likely that all groups of organisms or even all species have developed different mechanisms? These questions can theoretically be approached, e.g., from the view of evolutionary biology. However, the final, definitive answer needs a lot more of firm experimental data. These data should be derived from as many organisms of different taxa as possible.
2. Dealing with fungi, it is of particular interest to ask why most fungi are propagatable indefinitely and why other species (e.g., *Podospora anserina*) senesce regularly and reproducibly?
3. What genetic factors are involved in the control of aging processes?
4. Are mobile genetic elements, e.g., linear and circular plasmids or transposable elements, involved in the control of aging processes in a larger number of biological systems, or is their significance restricted to only a few species? If there is a more general importance of this type of elements, how do they act?
5. Are there many or only a few particular genes that control aging processes? What do these genes code for and how did they evolve?

6. EXTENSIONS

The data obtained from investigations of the aging process in *Podospora anserina*, in particular in the last two decades, when the powerful tools of molecular biology have become available, have stimulated investigations to identify similar processes and their genetic basis in other related fungi. In fact, at least in certain strains of *Neurospora crassa*, *N. intermedia*, *Podospora curvicolla*, *Aspergillus amstelodami*, and *Cochliobolus heterostrophus* degenerative processes have been also described [for review see 37]. In *Neurospora*, as in *P.*

anserina, these processes are caused by instabilities of the mitochondrial genome. Also in higher organisms like humans, mtDNA instabilities appear to play a major role in different degenerative processes, including certain neuromuscular diseases or aging [38–43]. However, the mechanisms leading to these instabilities appear to vary in detail, suggesting that in fact the mechanisms leading to aging may be rather more variable than conserved.

Finally, it should be noted that the experimental approaches described in this chapter may be used to approach a number of other questions that have been investigated in this fungal system in the past. Among these, questions of basic genetics, cytology, or developmental biology may be addressed in detail.

REFERENCES

1. Whitehouse, H. L. K. (1949). Heterothallism and sex in fungi. *Biol. Rev. 24*:411–447.
2. Whitehouse, H. L. K. (1949). Multiple-allelomorph heterothallism in the fungi. *New Phytol. 48*:212–224.
3. Esser, K., B. Böckelman (1985). Fungi. In *Non-Mammalian Models for Research on Aging* (F. A. Lints, ed.). Karger, Basel, pp. 231–246.
4. Rizet, G. (1953). Sur l'impossibilite d'obtenir la multiplication vegetative ininterromue et illimite de l'ascomycete *Podospora anserina. C. R. Acad. Sci. Paris 237*: 838–855.
5. Esser, K., P. Tudzynski (1980). Senescence in fungi. In *Senescence in Plants* (K. V. Thimann, ed.). CRC Press, Boca Raton, pp. 67–63.
6. Esser, K. (1985). Genetic control of ageing: the mobile intron model. In *The 1984 Sandoz Lectures in Gerontology* (M. Bergener, M. Ermini, H. B. Stähelin, eds.). Academic Press, London, pp. 3–20.
7. Kück, U. (1989). Mitochondrial DNA rearrangements in *Podospora anserina. Exp. Mycol. 13*:111–120.
8. Osiewacz, H. D. (1992). The genetic control of aging in the ascomycete *Podospora anserina.* In *Biology of Aging* (R. Zwilling, C. Balduini, eds.). Springer-Verlag, Heidelberg, pp. 153–164.
9. Osiewacz, H. D., J. Hermanns (1992). The role of mitochondrial DNA rearrangements in human diseases and aging. *Aging Clin. Exp. Res. 4*:273–286.
10. Rizet, G., D. Marcou (1954). Longevite et senescence chez l'ascomycete *Podospora anserina. C. R. 8. Int. Congr. Bot. Paris Sec. 10*:121–128.
11. Marcou, D. (1961). Notion de longevite et nature cytoplasmique du determinant de la senescence chez quelques champignons. Ann. Sci. Nat. Bot. Ser. 12(2):653–764.
12. Stahl, U., P. Lemke, P. Tudzynski, U. Kück, K. Esser (1978). Evidence for plasmid like DNA in a filemantous fungus, the ascomycete *Podospora anserina. Mol. Gen. Genet. 162*:341–343.
13. Cummings, D. J., L. Belcour, C. Grandchamps (1979). Mitochondrial DNA from *Podospora anserina.* II. Properties of mutant DNA and multimeric circular DNA from senescent cultures. *Mol. Gen. Genet. 171*:239–250.

14. Kück, U., U. Stahl, K. Esser (1981). Plasmid-like DNA is part of mitochondrial DNA in *Podospora anserina. Curr. Genet. 3*:151–156.
15. Belcour, L., O. Begel, M. O. Mosse, C. Vierny (1981). Mitochondrial DNA amplification in senescent cultures of *Podospora anserina. Curr. Genet. 3*:13–21.
16. Osiewacz, H. D., K. Esser (1984). The mitochondrial plasmid of *Podospora anserina*: a mobile intron of a mitochondrial gene. *Curr. Genet. 8*:299–305.
17. Cummings, J. M., I. A. MacNeil, J. M. Domingo, E. T. Matsuura (1985). Excision-amplification of mitochondrial DNA during senescence in *Podospora anserina*. DNA sequence analysis of three unique "plasmids." *J. Mol. Biol. 185*:659–690.
18. Belcour, L., O. Begel, A. M. Keller, C. Vierny (1982). Does senescence in *Podospora anserina* result from instability of the mitochondrial genome? In *Mitochondrial Genes* (P. P. Slonimski, P. Borst, G. Attardi, eds.). Cold Spring Harbor Laboratory Press, Cold Spring Harbor, NY, pp. 415–422.
19. Esser, K., W. Keller (1976). Genes inhibiting senescence in the ascomycete *Podospora anserina. Mol. Gen. Genet. 144*:107–110.
20. Tudzynski, P., K. Esser (1979). Chromosomal and extrachromosomal control of senescence in the ascomycete *Podospora anserina. Mol. Gen. Genet. 173*:71–84.
21. Marcou, D., M. Picard-Bennoun, J. M. Simonet (1991). Genetic map of *Podospora anserina*. In *Genetic Maps* (S. J. O'Brien, ed.). Cold Spring Harbor Laboratory Press, Cold Spring Harbor, NY, pp. 3.58–3.67.
22. Esser K. (1974). *Podospora anserina*. In *Handbook of Genetics I* (R. C. King, ed.). Plenum Press, New York, pp. 531–551.
23. Kück, U., H. D. Osiewacz, U. Schmidt, et al. (1985). The onset of senescence is affected by DNA rearrangements of a discontinuous mitochondrial gene in *Podospora anserina. Curr. Genet. 9*:373–382.
24. Koll, F., L. Belcour, C. Vierny (1985). A 1100-bp sequence of mitochondrial DNA is involved in senescence process in *Podospora*: study of senescent and mutant cultures. *Plasmid 14*:106–117.
25. Osiewacz, H. D., J. Hermanns, D. Marcou, M. Triffi, K. Esser (1989). Mitochondrial DNA rearrangements are correlated with a delayed amplification of the mobile intron (plDNA) in a long-lived mutant of *Podospora anserina. Mut. Res. 219*:1–7.
26. Schulte, E., U. Kück, K. Esser (1988). Extrachromosomal mutants from *Podospora anserina*: permanent vegetative growth in spite of multiple recombination events in the mitochondrial genome. *Mol. Gen. Genet. 211*:342–349.
27. Prillinger, H.-J., K. Esser (1977). The phenoloxidases of the ascomycete *Podospora anserina*. XIII. Action and interaction of genes controlling the formation of laccase. *Mol. Gen. Genet. 156*:333–345.
28. Hermanns, J., H. D. Osiewacz (1992). The mitochondrial plasmid pAL2-1 of a long-lived *Podospora anserina* mutant is an invertron encoding a DNA and RNA polymerase. *Curr. Genet. 22*:491–500.
29. Hermanns, J., A. Asseburg, H. D. Osiewacz (1994). Evidence for a life span prolonging effect of a linear plasmid in longevity mutant *Podospora anserina. Mol. Gen. Genet. 243*:297–307.
30. Osiewacz, H. D., A. Clairmont, M. Huth (1991). Electrophoretic karyotype of the ascomycete *Podospora anserina. Curr. Genet. 18*:481–483.

31. Javerzat, J.-P., C. Jacquier, C. Barreau (1993). Assignment of linkage groups of electrophoretically separated chromosomes of the fungus *Podospora anserina*. *Curr. Genet.* *24*:219–222.
32. Akins, R. A., A. M. Lambowitz (1985). General method for cloning *Neurospora crassa* nuclear genes by complementation of mutants. *Mol. Cell. Biol.* *5*:2272–2267,
33. Vollmer, S. J., C. Yanofsky (1986). Efficient cloning of genes of *Neurospora crassa*. *Proc. Natl. Acad. Sci. USA* *83*:4869–4873.
34. Yelton, M. M., W. E. Timberlake, C. A. M. J. J. van den Hondel (1985). A cosmid for selecting genes by complementation in *Aspergillus nidulans*: selection of the developmentally related *yA* locus. *Proc. Natl. Acad. Sci. USA* *82*:834–838.
35. Weltring, K.-M., B. G. Turgeon, O. C. Yoder, H. D. van Etten (1988). Isolation of a phytoalexin-detoxification gene from the plant pathogenic fungus *Nectria haematococca* by detection its expression in *Aspergillus nidulans*. *Gene* *68*:335–344.
36. Turgeon, K.-M., H. Bohlmann, L. M. Ciufetti, et al. (1993). Cloning and analysis of the mating type genes from *Cochliobolus heterostrophus*. *Mol. Gen. Genet.* *238*: 270–284.
37. Esser, K., U. Kück, C. Lang-Hinrichs, et al. (1986). *Plasmids of Eukaryotes*. Springer-Verlag, Heidelberg.
38. Linnane, A. W., S. Marzuki, T. Ozawa, M. Tanaka (1989). Mitochondrial DNA mutations as an important contributor to ageing and degenerative diseases. *Lancet* *I*:642–645.
39. Linnane, A. W., A. Baumer, R. J. Maxwell, H. Preston, C. Zhang, S. Marzuki (1990). Mitochondrial gene mutation: the ageing process and degenerative diseases. *Biochem. Int.* *22*:1067–1076.
40. Corral-Debrinski, M., T. Horton, M. T. Lott, J. M. Shoffner, M. F. Beal, D. C. Wallace (1991). Mitochondrial DNA deletions in human brain: regional variability and increase with advanced age. *Nature Genetics* *2*:324–329.
41. Katayama, M., M. Tanaka, H. Yamamoto, T. Ohbayashi, Y. Nimura, T. Ozawa (1991). Deleted mitochondrial DNA in the skeleton muscle of aged individuals. *Biochem. Int.* *25*:47–56.
42. Yen, T. C., C. H. Su, K. L. King, Y. H. Wei (1991). Ageing-associated 5 kb deletion in human liver mitochondrial DNA. *Biochem. Biophys. Res. Commun.* *178*:124–131.
43. Wallace, D. C. (1992). Diseases of the mitochondrial DNA. *Annu. Rev. Biochem.* *6*:1175–1212.
44. Tudzynski, P., U. Stahl, K. Esser (1982). Development of a eukaryotic cloning system in *Podospora anserina*. I. Long-lived mutants as potential recipients. *Curr. Genet.* *6*:219–222.

16

Horizontal Transmission in Fungal Populations

Rolf F. Hoekstra
Wageningen Agricultural University, Wageningen, The Netherlands

1. INTRODUCTION

A thorough understanding of many aspects of the biology of fungi should include knowledge of their consequences at the population level. This applies to the mode of reproduction (sexual or asexual; predominantly selfing or outcrossing), to the phenomenon of anastomosis between genetically different individuals, and also to many details of the genetic transmission system itself, to mention just a few aspects of fungal biology that clearly have relevance at the population level.

Population genetics is the branch of genetics that studies the population level consequences of the nature and transmission of genetic information. At the descriptive level this involves studying the amount of genetic variability present in populations, and of the spatial and temporal distribution of these variants. At the causal level the underlying processes responsible for the generation, maintenance, and distribution of genetic variation are studied, as well as those responsible for changes in the genetic composition of populations. Population genetics is therefore important for understanding evolutionary processes, since evolution is basically progressive change in the genetic composition of populations. Understanding the selective conditions that favor the spread of a certain genotype in populations will provide an understanding of the adaptive significance of the trait affected by this genotype.

This chapter does not aim to give a broad review of results from fungal population genetics. One reason is that the field is relatively young, and pioneering studies in the 1970s are only recently being followed up, mainly because molecular methodologies offer new possibilities [1]. Instead, a particular feature which is quite special for fungi as opposed to plants and animals will be singled out and considered from a population genetic point of view. This is the inherent potential capacity of fungi for hyphal anastomosis, which if occurring between genetically different individuals may lead to so-called horizontal transmission of genetic information (i.e., between individuals other than from parent to offspring).

2. ANASTOMOSIS AND VEGETATIVE (IN)COMPATIBILITY

A. Population Genetic Significance

In filamentous fungi between hyphae of the same individual anastomoses may be formed with consequent possibility of movement of cytoplasmic entities and nuclei between these hyphae. Anastomosis within an individual remains without genetic consequences, except when by mutation a nuclear or cytoplasmic type would occur which is somehow more effective in replication than the resident genotype and therefore increases in frequency within the mycelium. Sometimes, however, anastomoses may be formed between hyphae from different individuals, leading to the formation of heteroplasmons and/or heterokaryons. Such cases of vegetative compatibility are believed to be rare in most species under natural conditions [2], due to the occurrence of genetically determined vegetative incompatibility. The significance in natural populations of the so-called parasexual cycle in generating recombinant nuclear genotypes (in addition to, or—in imperfect species—as a substitute of, sexually produced recombinants) is therefore probably limited, contrary to what has been suggested earlier [3,4]. However, even a low rate of interindividual anastomosis resulting in horizontal transmission of (nuclear or cytoplasmic) DNA may have important population genetic and evolutionary consequences.

To understand the possible consequences at the population level of horizontal gene transfer, it is important to realize the crucial importance of the regulation of genetic segregation. Following sexual or somatic fusion, genetic information from different lines of descent is brought together in a diploid, respectively heterokaryotic or heteroplasmic structure. At some following stage genetic segregation occurs, either to restore the original ploidy level or as a consequence of random assortment during successive mitotic divisions. In the vertical route of genetic transmission (from parent(s) to offspring), segregation appears to be strictly regulated, as is apparent from the well-known Mendelian segregation laws for nuclear genes and from the uniparental transmission mechanism for cytoplasmic genes. The essence of both segregation mechanisms is that they are fair, in the sense that on average each allele present has an equal chance of ending up in a reproductive cell. This fairness of the genetic segregation in vertical transmission prevents the spread of mutants that combine a segregation advantage with a negative effect on the fitness of their "host" individual. There is, however, no regulated segregation mechanism to separate genetic information that has been combined horizontally. (In cases of horizontal transmission of viruses or plasmids following anastomosis between a strain with and a strain without virus or plasmids, fair segregation in this context would mean that half the progeny should be produced virus- or plasmid-free.) As a consequence, the horizontal transmission route allows the spread of sequences that are harmful but superior in segregation.

B. Theoretical and Empirical Population Genetic Studies on Vegetative Incompatibility

Various authors have suggested defense against the detrimental genetic elements referred to above as a functional explanation of vegetative incompatibility [5–8]. This explanation (defense against "parasitic" DNA) has been explored theoretically in a population genetic model by Hartl et al. [7]. They considered a nuclear genetic factor, comparable to a gene in *Neurospora crassa* described by Pittenger and Brawner [9], with a segregation advantage in a heterokaryon but a selective disadvantage in a homokaryon. Hartl et al. [7] concluded that such parasitic genes could well be responsible for the selective pressure causing the evolution of vegetative incompatibility. However, they made some rather restrictive assumptions and considered only populations consisting of two vegetative incompatibility groups (VCGs). Recently the problem has been analyzed in a more general model, allowing many VCGs [10]. This is a realistic feature since numbers of VCGs found in samples of strains from various ascomycete species appear to be very high, varying between 10% and 80% of the number of strains tested in samples from natural populations (overview in Nauta and Hoekstra [10]). The model assumes a homogeneous large population where encounters between individual mycelia belonging to different VCGs occur proportional to the product of the relative frequencies of these VCG types in the population. This is equivalent to random dispersal of conidia, which germinate and grow out to new mycelia. These mycelia meet in pairs, forming a heterokaryon (or heteroplasmon) in case they belong to the same VCG and growing along side by side as homokaryons if they are in different VCGs.

The main conclusion from this theoretical analysis is that populations should contain large numbers of VCGs only under very restricted conditions. The mean number of encounters between different mycelia must be high, the selective disadvantage of the parasitic genes must be small, and the frequencies of the existing VCGs must be distributed in a special way. Once a limited number of VCGs has been established, selection pressure for additional VCGs becomes very weak and eventually vanishes due to the very small probability for a mycelium containing a parasitic gene to meet (and fuse with) a compatible individual. This study therefore casts some doubts on the idea that vegetative incompatibility has evolved primarily as a defense against "parasitic" genes. However, more work is needed, both theoretical and empirical, to decide this issue. In particular, a population structure with strong spatial differentiation (which violates the model assumption of random encounters between mycelia) might be conducive to the maintenance of higher numbers of VCGs.

There is another reason to doubt whether "genetic defense" is the sole selective force driving the evolution of vegetative incompatibility. Evidence is accumulating that it is not very effective in prohibiting the exchange of cytoplasmic DNA, such as mitochondrial genes [11,12], plasmids [13,14], and viruses [15,16]. An interesting question in this connection is how much effective horizontal transfer of deleterious genetic elements could be tolerated in a fungal population.

3. POPULATION GENETICS OF FUNGAL SENESCENCE

A good example of the potential relevance of the rate of horizontal transfer is the phenomenon of fungal senescence. As has been described in Chapter 7, senescence is known to occur in a number of fungal species and is particularly well investigated in *Podospora* and *Neurospora*. Typically the senescence is associated with an increasing concentration of particular mitochondrial plasmids, and the latter are thought to be the causative agents of the phenomenon. Usually senescence starts after sporulation and results in progressive slowing down of mycelial growth accompanied with typical changes in mycelial morphology, eventually leading to death.

In this section we will take a closer look at the population genetical aspects of senescence. It may serve as a good example of the population genetic approach, and it nicely illustrates what has been said above about horizontal transmission.

The first question to ask is what is known about the frequency of occurrence of senescence in natural populations of fungi. For the two species referred to above, this answer turns out to be different. In *Podospora anserina*, so far no strains have been isolated from nature that do *not* senesce, which seems to suggest that the frequency of mitochondria containing the particular plasmid insert is very high in natural populations. In *Neurospora intermedia* about 30% of the strains in a sample from Hawaii appeared to contain the *kalilo* plasmid which is associated with senescence in this species. In a sample from the same area taken 4 years later about the same frequency was found (Debets, pers. comm.). How can we explain that the senescence factor in *Podospora* has spread to such an extent through the whole species? If anything, one would imagine senescence to be disadvantageous, since nonsenescent strains could go on growing (provided sufficient substrate is available) and still produce spores when senescent strains would already be dead. But how could a disadvantageous trait increase in frequency in natural populations? In the case of Neurospora senescence may have originated recently in this population and may still be spreading, eventually approaching a similar situation as in Podospora, or the population may perhaps have reached a stable state with selective forces favoring the spread of the *kalilo* plasmid balancing selection acting against the plasmid.

The next thing to do is to try and get some more precise and quantitative insights in what may be the relevant processes in determining the frequency of occurrence of senescence in fungal populations. This we do by constructing and analyzing a simple mathematical model.

In order to formulate a simple population genetic model, let us make the following simplifying assumptions (see Fig. 1). Different genotypes are classified into groups depending on whether or not they carry the senescence plasmid with respective relative frequencies in the population of x and $y = 1 - x$. We suppose there is a fixed order of events in the life cycle of the fungal colonies. Upon germinating they will grow, then possibly encounter conspecifics, and finally sporulate. When growing close to each other two individuals may come into contact, and perhaps some anastomosis between them takes place. We assume that the population is well mixed so that the probability that two senescent strains meet is equal to x^2, for a meeting between a senescent and a nonsenescent strain this is $2xy$, and for two nonsenescent strains y^2. As a consequence of these meetings we assume that horizontal transfer from a senescent strain to a nonsenescent strain may occur with probability h. Furthermore, suppose that senescent strains produce on average a fraction s fewer spores than nonsenescent strains, and that a fraction p of the spores produced by a senescent strain may have lost the plasmid.

Actually we have no information on the occurrence and extent of the latter two postulated phenomena, but they seem at least conceivable and if true could have important population genetic consequences. In this simple model we disregard the occurrence of sexual reproduction since sex would not affect the population frequency of the senescence plasmids, if 1.) transmission

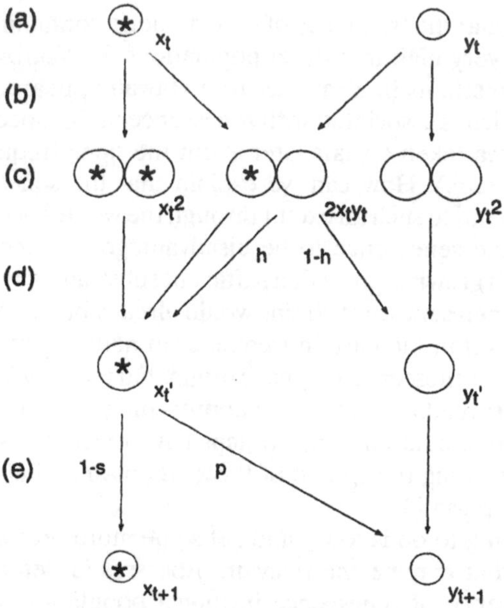

Figure 1 Schematic representation of assumptions and notation of population genetic model of senescence caused by mitochondrial plasmids. Asterisk indicates presence of plasmids. (a) Spores forming generation t; (b) germination; (c) encounters between different mycelia; (d) horizontal transfer of plamids; (e) production of spores forming generation $t + 1$. For further explanation, see text.

of mitochondria is strictly uniparental (maternal), and 2.) senescent individuals have the same probabilities as nonsenescent individuals of acting as a male or female parent. Of course, if either or both of these two assumptions are not true, the model should be adapted to accommodate this.

We now deduce from these assumptions a formal model as follows. As a consequence of horizontal transfer the relative frequencies of colonies with and without senescence plasmids will have changed from resp. x and y to x' *and* y' (see Fig. 1), such that in generation t

$$x'_t = x_t^2 + (1 + h)\, x_t\, y_t$$

and

$$y'_t = y_t^2 + (1 - h)\, x_t\, y_t$$

Then the relative frequencies of the two types among the newly produced spores that will form generation t + 1 will be

$$x_{t+1} = \frac{(1-s)(1-p)x'_t}{(1-s)(1-p)x'_t + px'_t + y'_t}$$

and

$$y_{t+1} + 1 - x_{t+1}$$

These equations describe how the relative frequencies of senescent and nonsenescent strains in a population will change over successive generations under the assumptions specified by the model. From an analysis of these equations we may thus predict under what conditions (in terms of the values of the parameters h, s, and p) senescence plasmids will spread through a population or will tend to disappear. Such an analysis uses standard mathematical techniques which are explained for example in population genetic textbooks. The complete analysis will not be presented here, but the results are summarized below.

1. If $p > 0$ ("spontaneous" loss of plasmids or occasional production of plasmid-free spores by a senescent strain is possible) and $s > 0$ (senescent strains have a lower growth rate and/or spore production than nonsenescent strains), then two different outcomes are possible:

(a) if $(1-p)(1-s)(1+h) > 1$ (that is, if the rate of horizontal transfer h is sufficiently high), then a stable coexistence of senescent and nonsenescent strains in the population is expected. In any case, however, a stable coexistence of uninfected and infected strains requires the horizontal transfer rate h to be greater than the fitness reduction s.

(b) If $(1-p)(1-s)(1+h) < 1$ (the rate of horizontal transfer is too low), then senescence plasmids are not expected to spread through a population, and—if initially present—will disappear.

2. If $p = 0$ (senescence plasmids have a 100% transmission in spores), then a stable coexistence of senescent and nonsenescent strains does not appear to be possible. Eventually a population will consist exclusively of senescent strains or exclusively of nonsenescent strains.

To what extent may this simple model be helpful in better understanding the occurrence of senescence? One important point is that it may serve as a quantitatively precise hypothesis about the selective forces acting upon senescence in natural populations. Thus the model suggests the importance of measuring the rate of horizontal transfer of senescence plasmids between strains, since this appears to be a crucial parameter in the model. In this respect the finding of Debets et al. [14] that in *Neurospora* even between strains that were vegetatively incompatible some horizontal transfer of the senescence plasmid could be detected, is relevant. Other very important parameters to measure are the loss rate p, on which so far very little experimental information is available and the selective disadvantage of senescent strains relative to nonselescent strains.

Summing up, a population genetic model may be useful in hinting at the most important factors responsible for the selection for or against (or for the stable maintenance) of traits and may suggest useful experiments for gaining a better understanding of the adaptive and selective aspects of traits.

4. EXPERIMENT ON HORIZONTAL TRANSFER

A. General Discussion

In *Neurospora*, like most other ascomycetes, vegetative incompatibility results from allelic differences at one or more of several vegetative incompatibility loci (so-called *het*-loci). Strains are compatible only if they carry identical alleles for all these *het*-loci. In *N. crassa* vegetative incompatibility has been studied using heterokaryon tests and by using chromosome duplications. In this experiment we will study the effect of *het*-loci on the lateral spread of senescence from one colony to the other (see also Debets et al. [14]). Senescence in the donor strain is due to the mitochondrial Kalilo plasmid (kalDNA) that integrates into the mitochondrial DNA thus causing cell death.

B. Demonstrating the Effect of Vegetative Incompatibility in *Neurospora* on Horizontal Transfer of Senescence

Strains

The *N. crassa* het-tester strains can be obtained from the Fungal Genetics Stock Center (F.G.S.C., Department of Microbiology, University of Kansas Medical School, Kansas City, Kansas). These strains require either inositol or pantothenic acid for growth. The KalDNA containing strain 3.7 is prototrophic and al-2 (albino conidiospores).

Media and Growth Conditions

Standard *Neurospora* growth protocols can be used as described by Davis and DeSerres [17]. For general hints and precautions for working with *Neurospora* see Perkins [18]. Cultures are grown in small test tubes containing 1 mL Vogel's minimal medium at 25°C. The subculture of the kalDNA containing strain is given as the extension of the strain number: 3.7-3 is the third subculture. The lifespan of strain a strain or the mixture of strains is expressed as the number of subcultures till death occurs.

Experiment

At first the lifespan of the kalDNA carrying strain 3.7 is determined. Conidia from the available culture of 3.7 are transferred to fresh media, grown for 2–3 days at 25°C and further subcultured till the culture dies, and no longer produces conidia. Conidia of a late subculture of 3.7 (a few subcultures before death, still well sporulating) of genotype *hetCd3, a* (i.e., mating type *a*) are mixed with conidia of each 16 het-testers (see Table), including all combinations of the *het*-loci *het-c*, *het-d*, *het-e* and the two mating-type idiomorphs (the mating type (alleles *a* and *A*) also acts as a *het*-gene). The mixtures are subcultured to test for senescence on nonselective media (i.e. Vogel's supplemented with inositol and pantothenic acid). As a control 3.7 is subcultured too. Continue subculturing until the culture dies, or until 15 subcultures after the death of the control culture 3.7. Test the culture for the presence of 3.7 genotype by inoculating conidia from the 10th subculture on Vogel's minimal medium (3.7 is prototrophic, whereas the *het*-testers are auxotrophic).

Three possible outcomes of the experiment can be distinguished:

1. Transfer of the senescent state: the mixture of the senescent strain 3.7 and the non-senescent *het*-tester dies at the same time as does the senescent strain by itself.

2. Transfer of the senescence causing plasmid without the senescent stage: the mixture has a longer lifespan than 3.7 alone, after several subcultures, 3.7 is no longer detectable but eventually the infected *het*-tester strain dies.*

3. No transfer of the senescent state and/or the senescence plasmid: after several subcultures strain 3.7 is no longer detectable. Further subculturing does not result in senescence.*

het-tester Growth on MM	Genotype	Dead at subculture
1427	*pan-1,al-2, het-CDE, a*	
1439	*inl, het-CDe, a*	
1428	*pan-1,al-2, het-CdE, a*	
1438	*inl, het-Cde, a*	
1429	*pan-1,al-2, het-cDE, a*	
1437	*inl, het-cDe, a*	
1430	*pan-1,al-2, het-cdE, a*	
1436	*inl, het-cde, a*	
1423	*pan-1,al-2, het-CDE, A*	
1454	*inl, het-CDe, A*	
1424	*pan-1,al-2, het-CdE, A*	
1453	*inl, het-Cde, A,*	

*Detection of the plasmid in the het-tester can also be done by Southern hybridization.

het-tester Growth on MM	Genotype	Dead at subculture
1425	*pan-1,al-2, het-cDE, A*	
1455	*inl, het-cDe, A*	
1426	*pan1,al-2, het-cdE, A*	
1422	*inl, het-cde, A*	

C. Questions

1. To what *het*-tester is the senescent state of the kalDNA strain transferred?

2. What is the effect of allelic differences at *het*-loci on horizontal transfer of the kalilo plasmid?

D. Appendix of Media

Vogel's vegetative medium (minimal medium) is made as follows:

- 20 mL of 50× Vogel's solution (see below)
- 20 g glucose
- 20 g agar
- 940 mL H_2O (bidest)
- add required supplements (250 mg/L for amino acids and nucleotides;
- 50 mg/L for vitamins), adjust to pH 5.5–6.0 and sterilize by autoclaving.

Vogel's solution 50× is made as follows:

- 650 mL H_2O
- 125 g Na_3citrate.$2H_2O$
- 250 g KH_2PO_4
- 100 g NH_4NO_3
- 5 g CaCl.$2H_2O$ (previously dissolved in 50 mL H_2O)
- 10 g mgSO_4.$7H_2O$ (previously dissolved in 50 mL H_2O)
- 5 mL trace element solution (see below)
- 0.5 mg biotin
- Sterilize filter and store in the refrigerator.

Vogel's trace element solution:

- 95 mL H_2O
- 5 g citric acid.H_2O
- 5 g $ZnSO_4$.$7H_2O$
- 1 g Fe(NH_4)2(SO_4).$6H_2O$

- 0.25 g $CuSO_4.5H_2O$
- 0.05 g $MnSO_4.H_2O$
- 0.05 g H_3BO_3
- 0.05 g $Na_2MoO_4.2H_2O$
- Sterilize filter and store in the refrigerator.

REFERENCES

1. Hoekstra, R. F. (1994). Population genetics of filamentous fungi. *Antoni van Leeuwenhoek 65*:199–204.
2. Caten, C. E., J. L. Jinks (1966). Heterokaryosis: its significance in wild homothallic ascomycetes and fungi imperfecti. *Trans. Br. Mycol. Soc. 49*:81–93.
3. Pontecorvo, G. (1956). The parasexual cycle in fungi. *Annu. Rev. Microbiol. 10*: 393–400.
4. Raper, J. R. (1966). Life cycles, basic patterns of sexuality and sexual mechanisms. In *The Fungi, An Advanced Treatise*, Vol. II (G. C. Ainsworth, A. S. Sussman, eds.). Academic Press, New York, pp. 473–511.
5. Day, P. R. (1968). The significance of genetic mechanisms in soil fungi. In: *Root Diseases and Soil-borne Pathogens* (T. A. Tousson, R. V. Bega, P. E. Nelson, eds.). University of California Press, Berkeley, pp. 69–74.
6. Caten, C. E. (1972). Vegetative incompatibility and cytoplasmic infection in fungi. *J. Gen. Microbiol. 72*:221–229.
7. Hartl, D. L., E. R. Dempster, S. W. Brown (1975). Adaptive significance of vegetative incompability in *Neurospora crassa. Genetics 81*:553–569.
8. Todd, N. K., A. D. M. Rayner (1980). Fungal individualism. *Science Progress 66*: 331–334.
9. Pittenger, T. H., T. G. Brawner (1961). Genetic control of nuclear selection in *Neurospora* heterokaryons. *Genetics 46*:1645–1663.
10. Nauta, M. H., R. F. Hoekstra (1994). The evolution of vegetative incompatibility. *Evolution 48*:979–995.
11. Collins, R. A., B. J. Saville (1990). Independent transfer of mitochondrial chromosomes and plasmids during unstable vegetative fusion in *Neurospora. Nature 345*: 177–179.
12. Coenen, A., A. J. M. Debets, R. F. Hoekstra (1994). Additive action of partial heterokaryon incompatibility genes in *Aspergillus nidulans. Curr. Genet. 26*:233–237.
13. Griffiths, A. J. F., S. R. Kraus, R. Barton, D. A. Court, C. J. Myers, H. Bertrand (1990). Heterokaryotic transmission of senescence plasmid DNA in *Neurospora. Curr. Genet. 17*:139–145.
14. Debets, A. J. M., X. Yang, A. J. F. Griffiths (1994). Vegetative incompatibility in *Neurospora*: its effect on horizontal transfer of mitochondrial plasmids and senescence in natural populations. *Curr. Genet. 26*:113–119.
15. Brasier, C. M. (1983). A cytoplasmically transmitted disease of *Ceratocystis ulmi. Nature 305*:220–223.

16. Anagnostakis, S. L. (1983). Conversion to curative morphology in *Endothia prasitica* and its restriction by vegetative compatibility. *Mycologia 75*:777–780.
17. Davis, R. H., F. J. DeSerres (1970). Genetic and microbiological research techniques for *Neurospora crassa. Methods in Enzymology 17A*:79–143.
18. Perkins, D. D. (1986). Hints and precautions for the care feeding and breeding of *Neurospora. Fungal Genetics Newsletter 33*:35–41.

17

Genetic Analysis in the Imperfect Fungus *Cladosporium fulvum*

The Development of a Map-Based Strategy for the Cloning of Genes Involved in Pathogenicity

José Arnau* and Richard P. Oliver
University of East Anglia, Norwich, England

1. INTRODUCTION

Cladosporium fulvum is a biotrophic imperfect fungal pathogen of tomato that colonizes the intercellular spaces between mesophyll cells and is confined to the apoplast for most of its life cycle. The fungus does not differentiate any specialized infection structure and causes no damage to the mesophyll cells [1]. This plant-pathogen interaction has been assumed to be a gene-for-gene relationship, whereby a dominant fungal avirulence gene is recognized by a resistance plant gene. The absence of a functional avirulence (avr) gene leads to successful infection, resulting in mold leaf. Alternatively, the presence of a functional avr gene is recognized by the plant carrying the corresponding resistance gene and fungal growth is arrested as a result of a hypersensitive response (HR) which involves localized cell death and accumulation of plant defence proteins.

Genes for resistance to *C. fulvum*, termed Cf, have been identified in tomato and a number of them are available in near isogenic lines. The

**Current affiliation*: Biotechnological Institute, Lyngby, Denmark.

genotypes of different fungal isolates can be unequivocally assigned by infection of the different tomato cultivars. As an example, a *C. fulvum* isolate able to successfully infect tomato plants carrying the resistance gene Cf4 is considered race 4, and its genotype is assumed to be avr4 (recessive allele). This is also called a compatible interaction. An isolate failing to infect such a tomato line (in an incompatible interaction) is considered to carry the dominant allele (Avr4). In many gene-for-gene systems which include obligate pathogens, the inheritance of resistance genes in plants and avr genes in the fungus is well documented. The *C. fulvum*-tomato interaction presents several advantages to be chosen as a model system. The communication between plant and pathogen is limited to the apoplast, since *C. fulvum* does not penetrate host cells. The study of the protein composition of the apoplastic fluids (also called intercellular fluids, IF) has permitted the identification of genes thought to be responsible for the establishment of basic compatibility (i.e., pathogenicity genes) and race-specific elicitors (i.e., avirulence genes) directly involved in the induction of the HR.

One of these genes, avr9, has been recently shown to be directly responsible for the HR on plants carrying the corresponding resistance gene Cf9. Molecular analysis of this gene, by gene disruption of a resident functional allele, resulted in a race 9 isolate [2]. Additionally, a race 9 isolate transformed with the cloned gene is recognized by Cf9 plants [3]. The study of the avr9-Cf9 system has shown fully agreement with the gene-for-gene hypothesis.

It is clear that the analysis of the proteins present in IF from different plant–pathogen combinations will allow the isolation of genes involved in this interaction. In fact, another of these genes, avr4, has recently been isolated and characterized using this strategy (de Wit, personal communication).

However, the use of this approach for the isolation of basic compatibility genes has proven more difficult. A fungal extracellular protein (Ecp2) present only in IF obtained from compatible interactions, therefore a good candidate for a role in pathogenicity, have been shown to be non-essential upon gene disruption and subsequent infection (de Wit, personal communication). It is possible that pathogenicity genes do not code directly for the proteins present in the IF, and this may represent a limitation to the above described approach for the cloning of those genes.

We are following a complementary approach for the study of this plant-pathogen interaction. It is based on the establishment of a genetic map as a tool for the cloning of genes involved in the interaction. An important step towards this goal is the isolation of mutations that affect the infection process, resulting on non-pathogenic isolates. Such mutants, with no altered phenotype on axenic culture, have been obtained and characterized in *C. fulvum*. They have been assigned into two groups, depending on the degree of completion of the vegetative cycle on plants. "Pathogenicity" mutants which stop growth at different stages after stomatal penetration and failed to sporulate; "Aggression" mutants which exhibited reduced sporulation [4].

C. fulvum is an imperfect fungus, and therefore classical genetic analysis based on meiotic recombination, involving sexual crosses of different isolates, is not possible. Additionally, no other natural means of genetic exchange, otherwise widespread in other fungi, like fungal anastomosis or a parasexual cycle has been observed in *C. fulvum*, despite early reports of heterokaryon formation [5,6]. This problem has been overcome by utilizing somatic hybridization in what is called an induced parasexual cycle, where only chromosome reassortment and, at a lower frequency, mitotic recombination is feasible. Protoplasts of genetically marked strains of *C. fulvum* are fused and selection against the parental strains is imposed. The ploidy of the fusion products obtained has been studied using DAPI-stained nuclei [7]. Vegetative, non-selective growth typically leads to haploidy, although heterokaryosis cannot be ruled out using this method.

Mutants and transformed strains, suitable for use in the induced parasexual cycle, can be readily generated in *C. fulvum*, using positive selection systems [8,9]. More recently, molecular evidence has been obtained for a high level of recombination associated with this induced parasexual cycle in an interracial (race 4 × race 5) fusion, by studying the inheritance in the progeny of vector sequences, present in one of the parental strains [10]. Phenotypic, RFLP and RAPD markers are being used to generate a genetic map, using one hundred haploid progeny from this interracial fusion. Additionally, the molecular karyotype of *C. fulvum* has been studied, and therefore the assignment of established linkage groups to individual chromosomes is now possible [11,12]. When genes involved in the pathogenic process are shown linked to a molecular marker, cloning of the gene of interest can be achieved by chromosome walking.

2. KEY NOTES

C. fulvum is poorly characterized genetically, a feature common to most plant pathogenic fungi [1]. *C. fulvum* can easily be grown in axenic culture and a number of molecular tools have been developed in recent years [8,9,11,13, 14]. A small-scale pathogenicity test has also been developed, allowing the screening of large numbers of isolates [4]. The infection process can be studied using strains transformed with the GUS gene and staining infected tissues to develop a color reaction directly related to fungal biomass [15].

Somatic hybridization has been used in a number of imperfect fungi and permitted the establishment of a genetic map of eight linkage groups in *Aspergillus niger* [16]. Genetic analysis based on mitotic recombination has also been used as a complementary approach in organisms which exhibit meiosis, either by means of a natural or induced parasexual cycle [17,18]. In these examples, consistency between the genetic organization derived from meiotic and mitotic recombination has been demonstrated.

The use of an induced parasexual cycle in *C. fulvum* can be regarded as an alternative approach to establish a genetic map if marker reassortment and haploidization of the fusion products is demonstrated.

2. PRINCIPLES

Several lines of research have been followed to characterize the parasexual cycle in *C. fulvum* and to establish a genetic map. Firstly, it is necessary to demonstrate that the progeny to be studied are haploid. Although no information can be obtained about the number of protoplasts involved in each fusion event or the number of nuclei present in each intervening protoplast, the study of marker segregation using a large number of haploid progeny may ensure the detection of mitotic recombination between linked markers as distinct from the random reassortment of unlinked markers, allowing the establishment of linkage groups. In *C. fulvum*, the haploidy of fusion products can be established using phenotypic and RAPD markers. A recessive color mutation, which results in red mycelium, has been shown to yield red sectoring colonies from originally dark green (wild type) fusion products (Fig. 1A). This observation indicated that haploidization occurred, leading to mycelial sectors containing only the mutant allele. An alternative strategy to study ploidy at a molecular level is provided by the use of RAPD markers [19]. RAPDs are dominant markers obtained through the use of single short oligonucleotides in a PCR reaction. Differences in amplified products from the parental strains can be used as markers, and their segregation can be studied in the progeny. Diploid or polyploid progeny will always exhibit all RAPD markers identified in the parental strain, since they may contain both parental genetic complements. In *C. fulvum*, the study of the segregation of 49 RAPD markers in the progeny obtained from an interracial fusion has provided additional evidence of haploidy (Fig. 2) [20].

C. fulvum has been shown to possess a gypsy-class retrotransposon [21], and such elements have been suggested to play a role in pathogenic fungi to permit the observed rapid appearance of new races which can overcome recently bred resistance [22]. The molecular characterization of the *C. fulvum* retrotransposon, named CfT-1, has allowed the mapping of different copies of CfT-1 in the *C. fulvum* genome by Southern analysis, using parasexual progeny. These different loci show a degree of clustering (unpublished results) and the widespread presence of the element in at least six of the eleven chromosomes resolved by PFGE [11].

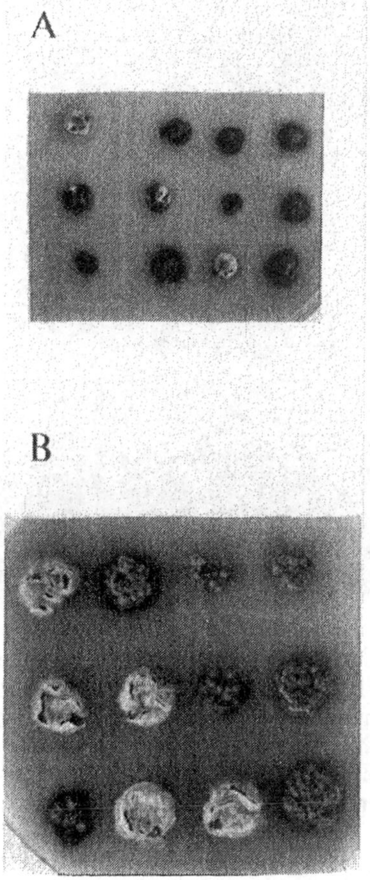

Figure 1 Ploidy analysis in the NOM2453 × NOM2512 parasexual progeny. Panel A shows colonies obtained after transfer to non-selective medium of individual dark green colonies from the fusion plates. Red sectors of different size are indicated in some of the progeny (arrowed). Panel B shows the growth subsequently transferred red and dark green sectors and the absence of sectoring in this stage.

Recently, a telomeric clone has been characterized in our laboratory [12]. This clone has been shown to detect polymorphisms between different *C. fulvum* races. Thus, telomere-derived RFLP markers can be used to test the consistency of this genetic analysis. Telomere RFLP should map exclusively to the end of linkage groups, since they define chromosomal ends.

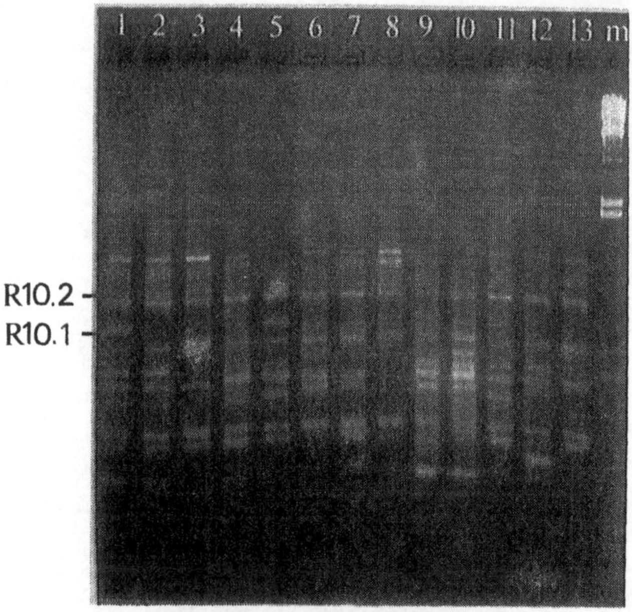

Figure 2 Segregation and ploidy analysis of the NOM2457 × NOM2512 parasexual progeny using RAPD markers. Amplification profile obtained using primer RC09 (5'-GATAACGCAC-3') from parental strains NOM2457 (lane 1) and NOM2512 (lane 2), and from 11 parasexual progeny (lanes 3 to 13). Two RAPDs are obtained (R10.1 and R10.2, indicated to the left) and all possible presence/absence combinations can be seen among the progeny [20].

The study of the segregation of all available (phenotypic, RFLP and RAPD) markers in a 100 progeny obtained from an interracial fusion is being carried out. To date 11 linkage groups comprising 43 markers have been established (Arnau and Oliver, manuscript in preparation).

If the genetic map is aimed at the cloning of genes, then a saturated map must be generated to allow a new marker to show linkage to any of the pre-existing markers. When close linkage is found between a gene of interest and a molecular marker, chromosome walking towards the gene can be used to clone it. The molecular marker is used as a probe to identify clones from a gene bank which contained that DNA region. Restriction analysis of positive clones provides flanking DNA which can then be used as a probe in a new round of hybridization permitting the walking towards the target gene. A problem arises when the molecular marker (i.e., RFLP or RAPD) contains repetitive DNA. In this case, a large number of clones from the gene bank will hybridize to the probe, precluding its use to walk to the gene.

The *C. fulvum* genetic map shows the widespread presence of RAPD markers in all 11 linkage groups identified [20]. It was necessary to characterize these markers to ensure they can be used to initiate chromosome walking. Moreover, single copy RAPD markers are useful in the assignment of linkage groups to separated chromosomes. The strategy followed consisted of the isolation of RAPD bands from agarose gels, and their use for reamplification and as probes in Southern analysis of total *C. fulvum* genomic DNA. Single copy RAPDs should show a simple hybridization pattern (one or a few bands) while multicopy RAPDs yield a complex pattern of hybridization (a large number of bands).

3. PRACTICE

A. Experiments

Ploidy Studies in the Parasexual Progeny Using a Phenotypic Marker

Two *C. fulvum* strains were used in this experiment: Strain NOM2453 is a race 4, hygR (pAN7-1 transformant), red colored isolate and strain NOM2433 is a race 5, met^- isolate [8]. Protoplast formation and fusion, selecting for both prototrophy and resistance to hygromycin B, was carried out as described [10] between the two strains described above. Two types of colonies can be expected to grow in the fusion plates: red colonies, resulted from non- or self-fusing NOM2453 protoplasts and true fusion products. Non- or self-fusing protoplasts of strain NOM2433 cannot grow due to hygromycin selection and the methionine auxotrophy it carries. Dark green (wild type) colonies appeared in the fusion plates and were considered true fusion products. Transfer of these colonies to non-selective medium (rich medium with no hygromycin) to allow for the recovery of haploid progeny resulted in 18–25% of the colonies showing red sectors (Fig. 1). These red sectors are observed where the red allele has segregated as a consequence of haploidization. The size of the red sector is inversely correlated to the timing of the haploidization (i.e., large red sectors are produced by rapidly haploidizing fusion products) (Fig. 1A). Additional evidence for the recessiveness of the red allele is provided by the fact that original red colonies obtained in the fusion plates do not exhibit dark green sectors and neither do red sectors when transferred to fresh non-selective medium (Fig. 1B).

Ploidy Studies in the Parasexual Progeny Using RAPD Markers

As mentioned above, RAPD markers are suitable for ploidy studies of parasexual progeny obtained from genetically uncharacterized fungi due to their dominant nature [20,23].

Polyploid progeny will exhibit all RAPD markers amplified from the two parental strains, while recombinant haploid progeny will only contain a set of RAPDs which should be different in each progeny.

We studied the inheritance of 49 RAPD markers in the parasexual progeny derived from an interracial fusion involving the following *C. fulvum* strains: strain NOM2457 (a race 5 isolate, cnx^-, $mor1^-$) and strain NOM2512 (a race 4 isolate, nia^-, nii^- (nir^-), hygR) [10,20]. A different phenotypic test, based on a two way selection method for either mutant or wild type genotype for the nitrate assimilatory pathway, was originally used to ensure haploid progeny [10]. Random 10-mer primers were screened in a PCR reaction for the amplification of polymorphic bands from NOM2457 and NOM2512. The number of polymorphic bands produced by each useful primer varied between 1–10 [20]. The inheritance of these markers was studied in the progeny and showed additional evidence of haploidy, since no progeny exhibited a full set of RAPDs and that any given RAPD was present and/or absent in at least 10% of the progeny [20].

B. Flow Chart

Phenotypic analysis of the ploidy in the (NOM2453 × NOM2433) parasexual progeny:

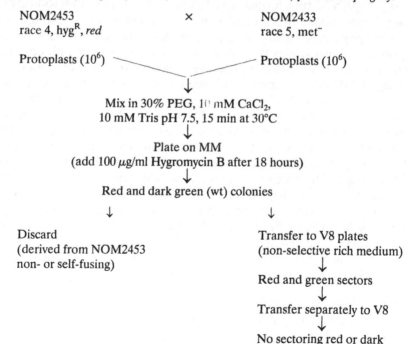

NOM2453 × NOM2433
race 4, hyg^R, *red* — race 5, met^-

Protoplasts (10^6) — Protoplasts (10^6)

↓

Mix in 30% PEG, 10 mM $CaCl_2$,
10 mM Tris pH 7.5, 15 min at 30°C

↓

Plate on MM
(add 100 μg/ml Hygromycin B after 18 hours)

↓

Red and dark green (wt) colonies

↓ (left): Discard
(derived from NOM2453
non- or self-fusing)

↓ (right): Transfer to V8 plates
(non-selective rich medium)

↓

Red and green sectors

↓

Transfer separately to V8

↓

No sectoring red or dark
green colonies

C. Materials Needed and Experimental Procedure

Protoplasts Formation and Fusion

A list of the materials needed for the fusion experiments discussed above is as follows:

Culture conditions and media. *C. fulvum* is grown in V8 juice agar at 23°C and conidia harvested after 12–14 days by washing the surface of the plate with 2–3 ml sterile 10% glycerol. Conidia stocks can be kept at –80°C for long term storage. Mycelium for protoplasts production is obtained by inoculation of freshly harvested conidia in 50 ml liquid Potato Dextrose Broth (Difco) in 250-ml Erlenmeyer flasks and incubation at 23°C for 46 hours in the dark. Mycelium is collected onto sterile cellulose nitrate membranes (5 mm, Whatman). All media are adjusted to pH 6.0 with HCl or NaOH. Protoplast regeneration media were supplemented with 0.8 M sucrose as osmotic stabilizer. Agar (Bitek) was added (2%) for solid media.

Protoplast formation and fusion. Protoplasts were prepared by treatment of filtered mycelia (see above) with either Novozym 234 (7.5 mg/ml) or Glucanex (150 mg/ml) in 1 M $MgSO_4$; 20 mM MES pH 5.8 for 3–4 hours at 28°C, with gentle shaking. Protoplasts were filtered through sterile scintered glass and the filtrate was centrifuged at 1000 rpm. The pelleted mycelial debris is discarded, and two volumes of 1 M NaCl; 20 mM MES pH 5.8 is added to the supernatant containing the protoplasts to allow pelleting. Approxi- mately 10^6 protoplasts of parental each strain are mixed and centrifuged as above. The pellet is resuspended in 1 ml of prewarmed (30°C) 30% PEG 4000; 10 mM $CaCl_2$; 10 mM Tris pH 7.5, and incubated at 30°C for 15 min. 5 ml of the appropriate overlay (OMMNO3, used for the NOM2457 × NOM2512 fusion, is a minimal medium with sodium nitrate as sole nitrogen source, osmotically stabilized as described above; OCM, used for the NOM2453 × NOM2433 fusion is osmotically stabilized complete medium [8]). Hygromycin B (100 μg/ml final concentration) was added as a 5 ml overlay (1% agar OCM; 48°C). The appropriate pre- and post-fusion regeneration controls are carried out by plating aliquots of protoplasts from the strains onto non-selective OCM medium. Plates are incubated at 23°C and colonies are visible after two weeks. Well separated colonies are transferred to an eppendorf tube containing 250 μl 10% glycerol, vortexed and an aliquot is plated on V8 juice agar. The remaining volume is kept at –80°C.

Amplification Conditions, Primer Screening and Characterization of RAPD Markers

Amplification reactions were performed in volumes of 50 μl containing 10 mM Tris-Cl, pH 8.3, 50 mM KCl, 4 mM $MgCl_2$, 0.2 mM primer, 20–100 ng genomic

DNA (see below), 0.2 mM each of dATP, dCTP, dGTP and dTTP (Pharmacia), 100 mg/ml BSA (Sigma), and 1.2 Units of Taq DNA polymerase (Perkin Elmer Cetus or Boehringer). Amplifications were carried out, in either a Perkin Elmer Cetus DNA Thermal Cycler or in a PREM III Thermocycler (LEP Scientific), for 45 cycles of 1 min at 94°C, 1 min at 35°C, 2 min at 72°C, followed by 5 min final step at 72°C. Amplification products (25–45 μl of the reaction mix) were analysed by electrophoresis in 1–1.5% agarose gels (Fig. 2). Fungal DNA was obtained using a Qiagen tip-20 column, as described [10].

Twenty-six 10-mer primers, out of approximately 100 tested, detected a total of 49 polymorphisms between strains NOM2457 and NOM2512. To obtain probes for Southern analysis, chosen RAPDs were purified using GeneClean (BIO 101), and 1–5 ng DNA was reamplified using the original conditions (see above). Probe DNA (10–100 ng) was ^{32}P-labelled using the random hexanucleotide labelling procedure [10]. All blots were washed to high stringency (0.1 × SCC; 0.1% SDS, final wash at 65°C). RAPD markers which showed one or a few hybridizing bands, using at least two different restriction enzymes for the digestion of *C. fulvum* genomic DNA, were considered single copy. RAPDs yielding a complex hybridization pattern were considered multicopy [20].

D. Questions

We have shown conclusive evidence of the haploidy of fusion products derived from two different experiments. Recombination and reassortment of markers is also observed in a 100 progeny derived from the NOM2457 × NOM2512 fusion [10,20], allowing the establishment of 11 linkage groups and suggesting that the *C. fulvum* parasexual cycle can be used for the genetic analysis of this imperfect fungus. However, several problems derive from the observation that 48 of the studied 49 RAPD markers include repetitive DNA [20]. These markers cannot be used to initiate a chromosome walk to a gene of interest as discussed above. Moreover, multicopy molecular markers cannot be used to test the consistency of the genetic map by hybridization of linked markers to PFGE blots, since their sequence is present in a number—if not all—of the *C. fulvum* chromosomes. RAPDs have been shown to be a quick and powerful source of genetic markers [19]. Thus, the genetic analysis of relatively uncharacterized organisms can be carried out using a set of random primers, without any additional genetic knowledge. This can lead to the establishment of a genetic map in which most (or all) the markers include repetitive DNA in these organisms. Like *C. fulvum*, most phytopathogenic fungi studied show evidence for a complex genome with a high level of repetitive DNA [24]. This situation must be taken into account when RAPD markers are used, because amplification results from the presence of inverted repeated sequences [19]. Therefore, the molecular characterization of the markers used in the generation of a map should be carried out to ensure their usefulness for map-based strategies. In lettuce, a gene (Dm) conferring resistance to downy mildew (*Bremia lactucae*) was shown to be linked to a number of RAPD markers, the majority of which represented repetitive DNA [25].

The consistency of a parasexual genetic map must be tested using linked single copy markers as probes to demonstrate their presence in the same electrophoretically separated chromosome.

E. Extensions

Further experiments can be devised to test the usefulness of parasexuality in genetic analysis. First, at least for *C. fulvum*, it is necessary to include single copy RFLP markers in the genetic analysis. A cDNA library is currently being screened and a two positive clones have been mapped (Arnau and Oliver, unpublished). This approach may be required if, like in *C. fulvum*, alternative molecular markers (cosmid clones, RAPDs) are shown to include repetitive DNA.

Second, the use of teolomere RFLP analysis may become a powerful tool since, at least for ascomycetous fungi, telomeric sequences are highly conserved [12]. It is then possible to use a cloned heterologous telomere to demonstrate that the derived RFLP markers do map to the ends of linkage groups obtained in the analysis of the parasexual progeny.

Obviously, the validity of this approach can only be fully tested when a linked gene is cloned by walking from the corresponding marker, and this has not yet been reported. In *C. fulvum*, a recessive morphological mutation (*mor*1) appeared to be linked to a Cft-1 element. The analysis of clones containing Cft-1 flanking regions resulted in the characterization of a RFLP marker, Tfs-1, which maps less than four map units from the mor1 gene (Arnau and Oliver, unpublished). The use of Tfs-1-containing cosmid clones in transformation of the *mor1*$^-$ mutant strain should allow the cloning of the wild type gene. Genetic studies based on parasexual cycles are being used in many economically important fungi [16,26] and may represent a powerful tool for cloning genes for strain improvement.

REFERENCES

1. de Wit, P. J. G. M. (1992). Molecular characterization of gene-for-gene systems in plant-fungus interactions and the application of avirulence genes in control of plant pathogens. *Annu. Rev. Phytopathol. 30*:391.
2. Marmeisse, R., G. F. J. M. van den Ackerveken, T. Goosen, P. J. G. M. de Wit, H. W. J. van den Broek (1994). Disruption of the avirulence gene avr9 in two races of

the tomato pathogen *Cladosporium fulvum* causes virulence on tomato genotypes with the complimentary resistance gene Cf9. *Mol. Plant-Microbe Interact.* (in press).
3. van den Ackerveken, G. F. J. M., J. A. L. van Kan, M. H. A. J. Joosten, J. M. Muisers, H. M. Verbakel, P. J. G. M. de Wit (1993). Characterization of two putative pathogenicity genes of the fungal tomato pathogen *Cladosporium fulvum. Mol. Plant-Microbe Interact. 6*:210.
4. Kenyon, L., B. G. Lewis, A. Coddington, R. Harling, J. G. Turner (1994). Pathogenicity mutants of the tomato leaf mould fungus *Fulvia fulva* (Cooke) Ciferri (syn. *Cladosporium fulvum* Cooke). *Physiol. Mol. Plant Pathol.* (in press).
5. Talbot, N. J. (1990). Genetic and Genomic Analysis in *Cladosporium fulvum* Cooke (syn. *Fulvia fulva* Cooke). PhD Thesis, University of East Anglia.
6. Barr, R., M. L. Tomes (1953). Variation in the tomato leaf mold organism, *Cladosporium fulvum. Am. J. Bot. 48*:512.
7. Talbot, N. J., D. Rawlings, A. Coddington (1988). A rapid method for ploidy determination in fungal cells. *Curr. Genet. 14*:51.
8. Talbot, N. J., A. Coddington, I. N. Roberts, R. P. Oliver (1988). Diploid construction by protoplast fusion in *Fulvia fulva* (syn. *Cladosporium fulvum*): genetic analysis of an imperfect fungal plant pathogen. *Curr. Genet. 14*:567.
9. Oliver, R. P., I. N. Roberts, R. Harling, L. Kenyon, P. J. Punt, M. A. Dingemanse, C. A. M. J. J. van den Hondel (1987). Transformation of *Fulvia fulva*, a fungal pathogen of tomato to hygromycin B resistance. *Curr. Genet. 12*:231.
10. Arnau, J., R. P. Oliver (1993). Inheritance and alteration of transforming DNA during an induced parasexual cycle in the imperfect fungus *Cladosporium fulvum. Curr. Genet. 23*:508.
11. Talbot, N. J., R. P. Oliver, A. Coddington (1991). Pulsed-field gel electrophoresis reveals chromosome-length differences between strains of *Cladosporium fulvum* (syn. *Fulvia fulva*). *Mol. Gen. Genet. 229*:267.
12. Coleman, M. J., M. T. McHale, J. Arnau, A. Watson, R. P. Oliver (1993). Cloning and characterisation of telomeric DNA from *Cladosporium fulvum. Gene 132*:67.
13. Harling, R., L. Kenyon, B. G. Lewis, R. P. Oliver, J. G. Turner, A. Coddington (1988). Conditions for efficient isolation and regeneration of protoplasts from *Fulvia fulva. J. Phytopathol. 122*:143.
14. Roberts, I. N., R. P. Oliver, P. J. Punt, C. A. M. J. J. van den Hondel (1989). Expression of the *Escherichia coli* β-glucuronidase gene in industrial and phytopathogenic filamentous fungi. *Curr. Genet. 15*:177.
15. Oliver, R. P., M. Farman, J. D. G. Jones, K. E. Hammond-Kosack (1993). Use of fungal transformants expressing β-glucuronidase activity to detect infection and measure hyphal biomass in infected plant tissues. *Mol. Plant-Microbe Interact. 6*:512.
16. Debets, F., K. Swart, R. F. Hoekstra, C. J. Bos (1993). Genetic maps of eight linkage groups of *Aspergillus niger* based on mitotic mapping. *Curr. Genet. 23*:47.
17. Bonatelli, R., J. L. Azevedo (1992). A system for increasing variability in filamentous fungi. *Fungal Genet. Newslet. 39*:90.
18. Daboussi, M. J., C. Gerlinger (1992). Parasexual cycle and genetic analysis following protoplast fusion in *Nectria haematococca. Curr. Genet. 21*:385.

19. Williams, J. G. K., A. R. Kubelik, K. J. Livak, J. A. Rafalski, S. V. Tanskey (1990). DNA polymorphisms amplified by arbitrary primers are useful as genetic markers. *Nucleic Acid Res. 18*:6531.
20. Arnau, J., R. P. Oliver (1994). The use of RAPD markers in the genetic analysis of the plant pathogenic fungus *Cladosporium fulvum. Curr. Genet.* (in press).
21. McHale, M. T., I. N. Roberts, S. M. Noble, C. Beaumont, M. P. Whitehead, D. Seth, R. P. Oliver (1992). Cft-1: an letter retrotransposon in *Cladosporium fulvum*, a fungal pathogen of tomato. *Mol. Gen. Genet. 233*:337.
22. de Wit, P. J. G. M., R. P. Oliver (1989). The interaction of between *Cladosporium fulvum* (syn. *Fulvia fulva*) and tomato: a model system in molecular plant pathology, In *Molecular Biology of Filamentous Fungi* (H. Nevalainen, M. Penttila, eds.). *Found. Biotech. Industr. Ferment. Res. 6*:227.
23. Durand, N., P. Reymond, M. Fevre (1993). Randomly amplified polymorphic DNAs assess recombination following an induced parasexual cycle in *Penicillium roqueforti. Curr. Genet. 24*:417.
24. Anderson, P. A., B. A. Taylor, A. Pryor (1992). Genome complexity of the maize rust fungus *Puccinia sorghi. Exp. Mycol. 16*:302.
25. Paran, I., R. W. Michelmore (1993). Development of reliable PCR-based markers linked to downy mildew resistance genes in lettuce. *Theor. Appl. Genet. 85*:985.
26. Durand, N., P. Reymond, M. Fevre (1992). Transmission and modification of transformation markers during an induced parasexual cycle in *Penicillium roqueforti. Curr. Genet. 21*:377.

18

Production and Analysis of Meiotic Mutants in *Coprinus cinereus*

Patricia J. Pukkila
The University of North Carolina at Chapel Hill, Chapel Hill, North Carolina

1. INTRODUCTION

Coprinus cinereus is a small hymenomycete that is very easy to culture in the laboratory. Its rapid life cycle makes it particularly suitable material for a wide variety of genetic investigations. This case study will focus on genetic analyses of meiosis. The details of meiotic chromosome behavior are remarkably similar among eukaryotes (see [1] for review). However, very little is understood concerning the mechanisms that underlie homologue recognition, synapsis, crossing over, and segregation. Genetic tools provide an extremely powerful approach to understanding the detailed steps in such a complex process. Recessive mutants are the easiest to characterize. With a large collection of mutants, the investigator can concentrate on the most interesting complementation groups and utilize the most informative alleles. The goals of this case study are to recover a representative sample of sporeless mutants of *C. cinereus*, and to determine the proportion that exhibit abnormalities in meiotic chromosome behavior. The *AmutBmut* background will be used, since mutagenized variants with defects in spore production can be recognized in *AmutBmut* strains without further genetic manipulation, as described below.

AmutBmut strains have several properties that facilitate the recovery and analysis of developmental mutants. First, they continue to make asexual spores (oidia). Upon germination, a single oidium (containing a single haploid nucleus) will undergo all the morphological changes associated with dikaryon formation. The resulting homokaryotic colony (which contains two identical nuclei in each cell) forms fruit bodies that exhibit normal chromosome behavior during meiosis [2]. Thus, an oidium harboring a recessive meiotic mutant may give rise to a fruit body with defects in basidiospore formation. There is one further property of *AmutBmut* strains that is essential for this analysis. As mentioned above, each cell contains two haploid nuclei that are genetically identical. The fact that nuclear fusion is delayed until the onset of meiosis means that meiosis is not essential for the recovery of a haploid nucleus. Consequently, subsequent genetic analyses need not depend on the recovery of a small number of surviving spores. Instead, any *AmutBmut* isolate with an interesting phenotype can be mated to a conventional haploid strain. If the mutation is recessive, the cross will result in normal fruit bodies. The new mutation can be recovered in a conventional genetic background for complementation analysis.

2. PRINCIPLES

In *C. cinereus*, meiosis occurs during development of the fruit body in cells called basidia. Buller [3,4] drew attention to the synchronous development of basidia in this species, and Lu [5] was the first to exploit the synchrony for analysis of chromosome behavior during meiosis. It is known that the two haploid nuclei fuse immediately prior to meiotic prophase. Chromosomes remain anchored in the nuclear envelope, but also move within the nucleus so that homologues become aligned. Synapsis occurs, and the prominent tripartite synaptonemal complexes (SCs) can be observed. Typically, genetic crossing over results in the formation of 28 chiasmata per meiocyte. These are sufficient to ensure regular disjunction of the 13 homologous chromosome pairs during metaphase I. Following the second meiotic division, the four haploid nuclei in each basidium are drawn to the cell surface, through four projections (sterigmata), and into the developing spores.

There are two approaches that have been utilized to assemble a representative collection of meiotic mutants. Zolan et al. [6] and Valentine et al. [7] took advantage of the fact that haploid strains can be recognized that have recessive defects in the repair of DNA damage induced by ionizing radiation. They found four complementation groups that are also required for the completion of meiotic prophase. A broader approach would be to identify all the genes that, when defective, confer meiotic anomalies but allow normal growth and fruit body formation. These can be recognized easily when homozygous, because viable spores are not formed. However, such mutants would exhibit no obvious phenotype in haploid vegetative cultures. The brute force approach (mutagenizing the haploid, crossing to a compatible strain, back-crossing the progeny to reveal any meiotic anomalies) is particularly labor-intensive in this organism, due to the mating system. The difficulties involved in recognizing recessive mutants whose traits are only expressed in the dikaryotic or diploid state was largely overcome with the development of *AmutBmut* strains [8].

The principles that underlie the generation and recovery of mutations are common to many systems (see [9] for review). A mutagen that generates a broad spectrum of mutations should be chosen. UV exhibits these properties [10], and it is the safest and most convenient mutagen to use in the laboratory. The dose should be chosen so as to result in a convenient proportion of mutant cells while minimizing the number of multiple mutants. The optimal dose is a balance between the yield of mutants and the amount of damage that can be accepted. We use a dose resulting in 1% survival, and we do back-crosses in order to detect if the mutant is monogenic or not. The effective dose can be kept reasonably constant from experiment to experiment by monitoring the percent of cells that die as a consequence of the irradiation, and recovering mutants from cultures with a similar percentage kill. Detailed analysis of the mutants should be deferred until the strain is back-crossed and the segregation pattern is that predicted from a single Mendelian factor.

The number of mutations represented in the collection should be determined by complementation analysis. Any variation in phenotype among alleles of one locus should be noted, especially since some may be more informative than others. The distribution of mutants among their complementation groups can allow estimates of the number of genes that the screen could uncover. If the initial estimates are very large, it can be useful to carry out a rapid secondary screen to allow the most effort to be concentrated on the desired class. Finally, two classes of mutants should result from abnormal meiosis. In theory, spore formation might proceed despite abnormal chromosome behavior, although the resulting aneuploidy would be expected to block germination and/or subsequent growth of the mycelium. Alternatively, spores might fail to form at all. Both types of mutants have been recovered in *C. cinereus*. These principles are illustrated in the following experiments.

3. PRACTICE

A. The Experimental Approach

The goal is to recover a representative sample of sporeless mutants of *C. cinereus* and determine the proportion of these that exhibit abnormalities in meiotic chromosome behavior. Oidia from strain 326 (*AmutBmut*) will be collected and mutagenized to 99% kill. The survivors will be examined, and any that produce sporeless (white) fruit bodies will be retested. Those that repeatedly produce white fruit bodies will be crossed to strain 301 (*A3B1*). Complementation tests will be performed. Representatives from each complementation group will be examined by propidium iodide staining to examine nuclear divisions, and chromosomes will be spread and stained with silver nitrate to examine SC formation. Strains exhibiting meiotic anomalies will be further crossed to strain 301 (*A43B43*), and tetrad analysis will confirm that a single Mendelian factor is segregating. Interesting mutants will be candidates for rescue by DNA-mediated transformation.

B. Materials

For mutagenesis:

1. YMG media (4 g/L yeast extract [Difco], 10 g/L malt extract [Difco], 4 g/L dextrose, 15 g/L agar [Difco]). Note that fruiting is inhibited by some types of agar. Pour plates and fruiting tubes. Disposable 16 × 150 mm tubes with reusable Kaput (Bellco) closures work well. It is critical that the closures be allowed to rest loosely on the top of the tube, *not* pressed down to seal. If the CO_2 tension builds up in the tube, fruiting will be inhibited.
2. Sterile water
3. Sterile syringe plugged with glass wool
4. Germicidal UV lamp
5. Tungsten needle

For propidium iodide staining:

1. NS buffer (20 mM Tris-HCl pH 7.5, 0.25 M mannitol, 1 mM EDTA, 1 mM $MgCl_2$, 0.1 mM $CaCl_2$)
2. 500 mg/mL ribonuclease A (Worthington) in TE (10 mM Tris-HCl pH 7.5, 1 mM EDTA) 3.10 mg/mL propidium iodide (Sigma) in NS buffer
3. Ethanol
4. Fluorescence microscope equipped with rhodamine filter set (e.g., Zeiss 487715)

C. Experimental Procedure

1. Grow the *AmutBmut* strain for 5–7 days at 37°C on YMG plates. Oidia production is reduced relative to wild type, so use three to five plates per experiment.

2. Flood each plate with 5 mL sterile water, and scrape the surface with the 5 mL-pipette tip to release the oidia into suspension.

3. Filter through a glass wool plug in a sterile syringe to remove hyphal debris. Count the oidia using a hemocytometer. Expect about 5 × 10^7/plate (10–100 × less than usually found in wild-type strains).

4. Adjust concentration to 5 × 10^6/mL. Dilute the starting suspension 1:100 and plate 0.1 mL to determine the initial viability (expect about 500 colonies). Irradiate 10-mL aliquots of the undiluted suspension to determine the dose that gives 0.5–1% survival (typically 20–40 sec with a 15-watt GE Germicidal UV lamp G15T8 at 35 cm from the dish). Be certain to remove the lids, and agitate throughout the irradiation period.

5. Incubate in the dark (to prevent photoreactivation) for 24 h at 37°C, and for 12–24 h at 25°C or until the colonies are a convenient size. It is not necessary to keep the plates in the dark after the initial 24-h period.

6. Rescue colonies onto a master plate (20/plate). Inoculate fruiting tubes, and incubate 2 days at 37°C and 1–2 weeks at 25°C under a 16-h light–8-h dark regimen.

7. Expect 0.1–1% of the fruit bodies to be sporeless. Retest any putative mutants defective in basidiospore development (*bad* mutants). Note that other developmental anomalies, such as stalks that fail to elongate and caps that fail to open, are often observed. Many tubes may arrest prior to fruit body formation.

8. Return to the master plate and prepare a stock plate of any confirmed bad mutants. It is convenient to freeze chunks of freshly grown mycelium plus underlying agar in 15% glycerol at –70°C for long-term storage.

9. Cross each mutant to strain 301 (*A3B1*). The *AmutBmut* mycelium will donate nuclei to the monokaryon, which will become dikaryotic. Inoculate a portion of this mated mycelium into a fruiting tube.

10. Rescue 50 spores from each cross. Recover a *A3B1bad* isolate from each mutant for complementation tests (these will produce white fruit bodies again when mated to the original *AmutBmutbad* isolate).

11. Cross each *A3B1bad* isolate with each of the *AmutBmutbad* isolates to test if any mutations fail to complement (produce white fruit bodies).

12. Choose a representative from each complementation group. Cross the *A3B1bad* strain to strain 306 (*A43B43*). Dissect 10 tetrads and confirm that the Bad phenotype segregates 2:2. Recover an *A43B43* bad strain for cytological analysis.

13. Cross compatible strains that are homozygous for the *bad* mutation. When primordia form (1 day before spores are released), monitor nuclear behavior using propidium iodide staining. Remove gill samples, and fix in 70% ethanol/30% NS buffer. If desired, samples can be held at –20°C before analysis. It is convenient to sample after at the start of the 16-h light period, and then at 3-h intervals thereafter. Rinse through three changes of NS buffer, and then treat with RNase A in TE at 37°C for 30 min. Rinse through an

additional three changes of NS buffer. Tear into single basidial layers, and transfer the cell sheet into a small drop of propidium iodide in NS buffer. Remove into fresh drop of propidium iodide on a coverslip, lower a microscope slide onto the drop, and examine by fluorescence microscopy. Initially, basidia with two nuclei (prekaryogamy) and a single nucleus (postkaryogamy) should be apparent. In wild type, basidia that have progressed through prophase (two and four nuclei per basidium) are apparent 9–10 h after the start of the 16-h light period. In several mutants, it is apparent that the basidia arrest with a single nucleus.

14. Monitor chromosome behavior using silver staining. The methods are detailed in [11] (see also [1], this volume, Figure 6).

D. Questions

1. The screen will recognize only those meiotic mutants that exhibit cell-cycle arrest (and thus fail to make spores). Some meiotic mutants might excape cell-cycle arrest and produce dead spores. It is likely that at least some mutant alleles of all potentially interesting genes will exhibit cell-cycle arrest, or is it likely that important genes will be missed by this method?

2. Estimate the minimum number of genes essential for spore formation from the number of complementation groups represented in the mutant collection [12]. Calculate the standard error of your estimate [13].

3. Explain how the above procedures would need to be altered to permit an analysis of bad genes that are linked to either the *A* or the *B* gene.

E. Extensions

It is particularly valuable to obtain multiple alleles, because these often exhibit different phenotypes. For example, marked differences in sensitivity to ionizing radiation, intergenic recombination, and spore viability were observed between the *rad3-1* and *rad3-2* alleles [6].

Having identified genes that are required to complete normal meiosis, it is then of considerable interest to learn about the products they encode. There is a well-developed DNA-mediated transformation system in *C. cinereus* [14]. To date, many genes have been cloned by sib selection from genomic or chromosome-specific cosmid libraries. Future studies will focus on the subset of genes required for homology recognition, since both meiotic and premeiotic homology searches can be analyzed in this system (reviewed in [15]).

REFERENCES

1. Lu, B. C. (1996). In *Fungal Genetics, Principles and Practice* (C. J. Bos, ed.). Marcel Dekker, New York, pp. 119–176.

2. Kanda, T., H. Arakawa, Y. Yasuda, T. Takemaru (1990). Basidiospore formation in a mutant of incompatibility factors and in mutants with arrest at meta-anaphase I in *Corpinus cinereus. Exp. Mycol. 14*:218–226.
3. Buller, A. H. R. (1909). *Researches on Fungi*, Vol. I. Longmans, Green, London, pp. 196–215.
4. Buller, A. H. R. (1924). *Researches on Fungi*, Vol. III. Longmans, Green, London, pp. 118–356.
5. Lu, B. C. (1967). Meiosis in *Coprinus lagopus*: a comparative study with light and electron microscopy. *J. Cell Sci. 2*:529–536.
6. Zolan, M. E., C. J. Tremel, P. J. Pukkila (1988). Production and characterization of radiation-sensitive meiotic mutants of *Coprinus cinereus. Genetics 120*:379–387.
7. Valentine, G., Y. J. Wallace, F. R. Turner, M. E. Zolan (1995). Pathway analysis of radiation-sensitive meiotic mutants of *Coprinus cinereus. Mol. Gen. Genet. 247*: 169–179.
8. Swamy, S., I. Uno, T. Ishikawa (1984). Morphogenetic effects of mutations at the A and B incompatibility factors in *Coprinus cinereus. J. Gen. Microbiol. 130*:3219–3224.
9. Bos, C. J., D. Stadler (1996). Mutation. In *Fungal Genetics, Principles and Practice* (C. J. Bos, ed.). Marcel Dekker, New York, pp. 13–42.
10. Harris, S. D., J. R. Pringle (1991). Genetic analysis of *Saccharomyces cerevisiae* chromosome I: On the role of mutagen specificity in delimiting the set of genes identifiable using temperature-sensitive-lethal mutations. *Genetics 127*:279–285.
11. Pukkila, P. J., C. Skrzynia, B. C. Lu (1992). The *rad3-1* mutant is defective in axial core assembly and homologous chromosome pairing during meiosis in the basid-diomycete *Coprinus cinereus. Dev. Genet. 13*:403–410.
12. Raper, J. R. (1996). *Genetics of Sexuality in Higher Fungi.* Ronald Press, New York, pp. 106–107.
13. Stevens, W. L. (1942). Accuracy of mutation rates. *J. Genet. 43*:301–307.
14. Binninger, D. M., C. Skrzynia, P. J. Pukkila, L. A. Casselton (1987). DNA-mediated transformation of the basidiomycete *Coprinus cinereus. EMBO J. 6*:835–840.
15. Pukkila, P. J. (1994). Meiosis in mycelial fungi. In *The Mycota I* (J. G. H. Wessels, F. Meinhardt, eds.). Springer-Verlag, Berlin, pp. 267–281.

19

A Case Study in Fungal Development and Genetics

Schizophyllum commune

Kirk A. Bartholomew
Pulp and Paper Research Institute of Canada, Pont Claire, Quebec, Canada

Amy L. Marion
Purdue University at Fort Wayne, Fort Wayne, Indiana

Charles P. Novotny and Robert C. Ullrich
The University of Vermont, Burlington, Vermont

1. INTRODUCTION

The filamentous, wood-rotting basidiomycete, *Schizophyllum commune*, exemplifies genetic control of the life cycle common among Basidiomycetes [1]. This extensively studied fungus provides a classic example of complex development culminating in sexual reproduction. The fungus is heterothallic (i.e., genetically obligated to outbreed) and tetrapolar (i.e., four predominant mating types among the progeny). Haploid strains (i.e., homokaryons) that grow in proximity of one another form fusion cells (plasmogamy). Sexual development ensues provided the genetic constitution at the mating-type loci of the potential mates is correct. Full sexual development establishes a fertile dikaryon by converting the uninucleate haploid cells of each mate into binucleate cells, containing one haploid nucleus from each of the two mates. The dikaryon may grow vegetatively for an extended period, but is subject to environmental cues that provoke fruiting body formation, karyogamy, meiosis, and sporulation. Figure 1 portrays the features of this life cycle.

The A- and the B-developmental pathways in concert comprise sexual development. Two complex genetic loci, $A\alpha$ and $A\beta$, regulate the A pathway. These loci are polymorphic, with 9 $A\alpha$ and 32 $A\beta$ polymorphs estimated in the

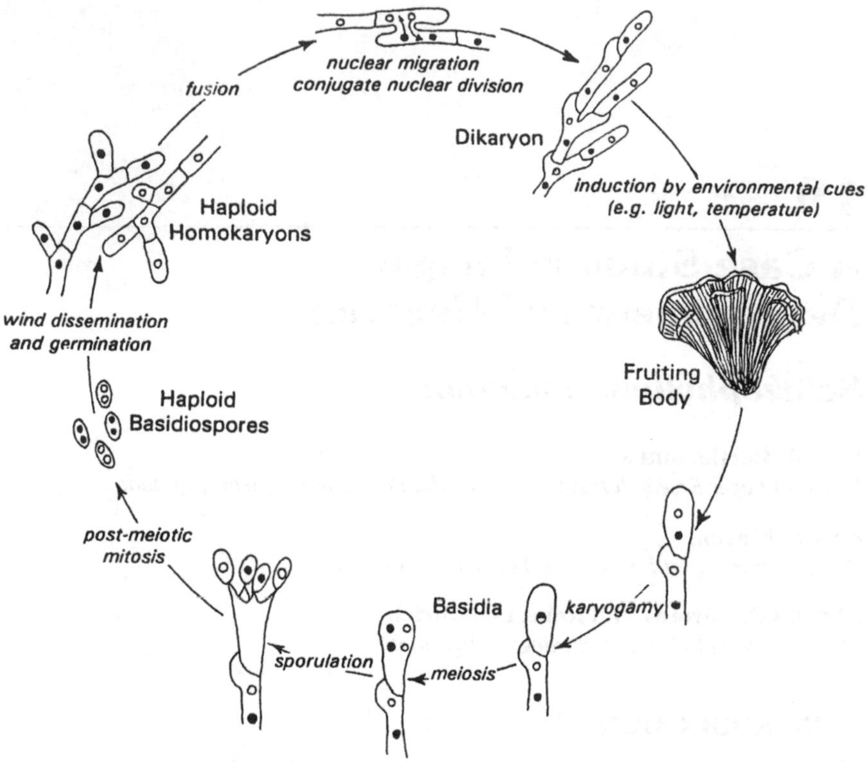

Figure 1 Life cycle of *Schizophyllum commune*. Fusion occurs within each haploid homokaryon, but full sexual development follows fusion of two haploid mates with genetic differences at mating-type loci. Nuclei represented as circles with genetic differences indicated by solid versus open circles.

worldwide population of *S. commune* [2,3]. The mates activate the A pathway provided they differ in polymorphs for either Aα or Aβ [4]. The alternative polymorphs are functionally equivalent (i.e., each fully activates the A pathway provided it is present in a fusion cell with a different polymorph). Bα and Bβ regulate the B pathway in a manner analogous to Aα and Aβ regulation of the A pathway [5]. The B loci are also thought to be genetically complex and have nine polymorphs each [5].

The study of heterokaryons, each differing for polymorphs at a single mating-type locus, permitted assignment of developmental events to the loci (i.e., A or B) that regulate them. The A loci promote pairing of the two nuclei of each heterokaryotic cell, formation of a lateral hyphal appendage called the hook cell, synchronous division of the two nuclei of the heterokaryotic cell, and septation of the hook cell and the hyphal cell. The presence of a hook cell is readily discerned by compound microscopy and serves as an indicator that the A pathway is active.

The B loci regulate enzymatic dissolution of hyphal septa followed by migration of nuclei from the fusion cell throughout the hyphae of the mate. This process is reciprocal, with the haploid cells of each mate becoming converted into a heterokaryotic mycelium. The B loci also regulate an event that takes place during the vegetative growth of the dikaryon; fusion of the hook cell with the penultimate hyphal cell permits the hook cell nucleus to pair with a nucleus derived from the mate, maintaining a 1:1 nuclear ratio in each hyphal cell. Figure 2 details specific events controlled by the A and B regulatory loci.

Event		Locus
Septal dissolution		B
Nuclear migration		B
Nuclear pairing		A
Hook cell formation		A
Conjugate division		A
Cell septation		A
Hook cell fusion		B

Figure 2 Recognized events of the A and B developmental pathways and the regulatory mating-type locus for each. A, activated by either Aα or Aβ; B, activated by either Bα or Bβ.

When differences in the polymorphs of the A and B loci activate both the A and B pathways, the developmental result is the dikaryon with potential to complete sexual development under proper environmental conditions.

Key concepts in fungal development and genetics are these:

1. *S. commune* is an ideal organism for the study of genetics and biochemistry. Features that make *S. commune* particularly amenable to study include a short life cycle with straightforward Mendelian genetics, the production of large numbers of viable spores, rapid growth of homokaryotic and dikaryotic cultures on defined minimal liquid and semisolid media, the ability to store frozen stock cultures, and well-developed molecular genetic and biochemical techniques.

2. These advantages have fostered research on sexual development that emphasizes the following areas: the genetic and biochemical basis of mating type regulation [6,7], cell wall biochemistry [8], the process of fruiting, and the genetics and biochemistry of fruiting [9]. *S. commune* is also well suited for studying proteolysis [10], enzymatic degradation of complex polysaccharides and lignin [11–14], and the nature and activity of excreted polysaccharides [15–19].

2. PRINCIPLES

The experiments in this exercise illustrate the genetic, cytological, and molecular bases of the *self/nonself* recognition phenomenon originally termed incompatibility and today known as mating type. The Mendelian approach illustrates the classical genetic identification of the controlling genetic elements and the morphogenic events visualized during activation of the developmental pathways. The molecular approach involves activation of A-regulated development by transformation with an Aα gene. The methodology in the molecular approach follows the strategy used to elucidate the physical structure of the Aα locus and the biochemical basis for Aα-activated development [6,7].

The classical genetic experiment illustrates the combinatorial action of the independent A and B loci in activating development. Four combinations of A and B activity are possible: 1.) A off, B off, the two mates have identical mating-type genotypes (e.g., A=B=); 2.) A on, B off, the mates differ in Aα and/or Aβ only (e.g., A≠B=); 3.) A off, B on, the mates differ in Bα and/or Bβ only (e.g., A=B≠); A on, B on, the mates differ in A and B loci (e.g., A≠B≠).

The morphological characteristics of these interactions will be distinguished by macro- and microscopic examination (see Practice). These cytological distinctions permit segregation analysis of the four principal mating types among the progeny. Because the Aα and Aβ loci are linked on one chromosome and Bα and Bβ on another, recombinants with new mating types appear as only a few percent of the progeny. Therefore, the classical genetic experiment examines mating of homokaryons, the establishment of the dikaryon, distinguishing the dikaryon from homokaryons and other heterokaryons, fruiting-body formation, sporulation, the collection of meiotic progeny, and the analysis of mating-type segregants and recombinants.

The molecular genetic experiment relies on our understanding of the detailed structure of the Aα locus. The Aα locus consists of two divergently transcribed genes, *Y* and *Z*. Each of the Aα polymorphs has its own *Y* and *Z* alleles (e.g., Aα4 has *Y4* and *Z4*), except for Aα1, which has only *Y1*. Your experiment will show that the A pathway is activated when a *Y* allele of one mating type and a *Z* allele of a different mating type occur in the same cell. The experimental approach transforms the *Z4* allele into an Aα1 homokaryon (containing only the *Y1* Aα gene) and monitors the results by mating the transformants with various tester strains in order to assay the developmental phenotype and hence determine the Aα genotype of the transformants.

3. PRACTICE

A. Mendelian Genetics

The Experimental Approach

The general approach of this experiment is for you to examine the interactions of A and B in all four possible combinations: A=B=, A=B≠, A≠B=, and A≠B≠. The dikaryon will be allowed to undergo fruiting-body formation, meiosis, and sporulation. You will then collect progeny and analyze the segregation patterns of A and B.

The Scheme (Fig. 3)

1. Set up matings illustrating the four possible interactions and observe their macro- and microscopic characteristics.
2. Allow the dikaryon to fruit.
3. Collect progeny and perform mating tests to determine the segregation patterns of A and B.

Materials

S. commune strains (Table 1)

- 30°C Incubator (If not available, use room temperature.)
- Compound and dissecting microscopes
- Transfer lance
- Petri dishes

S. commune growth media (for semisolid media add 15 g agar/L)

- CYM: $MgSO_4$–$7H_2O$, 0.5 g/L; KH_2PO_4, 0.46 g/L; K_2HPO_4, 1.0 g/L; peptone, 2.0 g/L, dextrose, 20.0 g/L; yeast extract, 2.0 g/L
- CYMT: CYM + 4 mM tryptophan

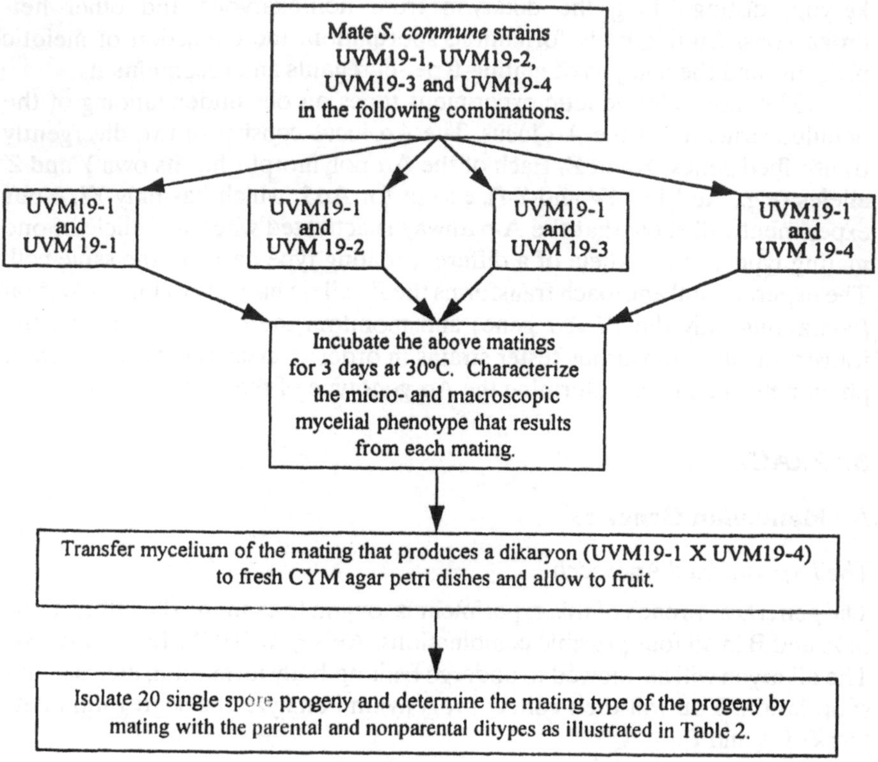

Figure 3 The experimental protocol for the Mendelian genetic experiment. Details of the procedures involved can be found in the Protocols section.

Protocols

Maintenance of *S. commune* strains. Maintain *S. commune* for up to 6 months by culturing on semisolid growth medium in Petri dishes (CYMT for *trp1* strains, CYM for all others). Following growth at 30°C (or room temperature), store the cultures at 4°C [20]. Wrap the edges of the Petri dish with parafilm to store for several months. Frozen stocks of *S. commune* can be maintained for years at –70°C without cryoprotectants. Culture on slants of an appropriate semisolid media in 1-dram, screw-cap glass vials and freeze at –70°C (R. C. Ullrich, unpublished results).

Mating. Set up matings for observation by pairing small pieces (3 mm^3) of mycelial inoculum, from the haploid strains to be examined, in proximity (1–2 mm apart) on appropriate semisolid media. Observe the macro- and

Table 1 *S. commune* strains utilized in the Mendelian and molecular genetic experiments, including their mating types and relevant auxotrophic markers.

S. commune strain	Mating type	Auxotrophic marker
UVM19-1	Aα1β1, Bα2β2	tryptophan (*trp1*)
UVM19-2	Aα1β1, Bα1β1	
UVM19-3	Aα4β6, Bα2β2	
UVM19-4	Aα4β6, Bα1β1	uracil (*ura1*)

microscopic developmental morphologies of these interactions after the cultures have grown at 30°C for 3 days [20].

Fruiting. Following the establishment and identification of a dikaryon (see Mating above and Progeny Analysis below), transfer small amounts of dikaryotic mycelium to fresh semisolid media and incubate the inverted Petri dish at room temperature in incidental light. Fruit bodies should appear at the edge of the colonies in 1–2 weeks, and masses of spores will become obvious to the unaided eye inside the lid of the inverted Petri dish.

Isolation of monosporous progeny. Convenient isolation of single-spore progeny can be accomplished by inverting the Petri dish containing the fruiting culture over a Petri dish of fresh medium for varying lengths of time (try 10 sec initially). Examine the Petri dish with the dropped spores under a compound microscope (≥200× magnification) to determine an appropriate density for the isolation of single-spore progeny. The basidiospores will appear as cylindrical, obliquely apiculate objects (4–7.5 × 1.5–2 μm) on the surface of the agar. Incubate at room temperature (24–48 h). Transfer the germlings to fresh media by observing the culture plate under a dissecting microscope and transferring small pieces of agar containing single germinated spores with a transfer lance. Use care to avoid transferring more than one germling at a time.

Microscopic examination of hyphal morphology and spore density. Examine the microscopic morphology of the hyphae (or the density of dropped spores) by inverting the culture plate on a microscope stage and focusing on the hyphae (looking through the agar) at 200× magnification.

Progeny analysis. The mating type of single-spore isolates can be determined by mating each unknown isolate with four haploid strains carrying the two parental ditypes and the two nonparental ditypes that would result from the independent segregation of the parental A and B mating type loci. The four possible interactions can be distinguished by micro- and macroscopic examination as follows:

A=B= exhibits the phenotype of a monokaryon, good growth of aerial mycelium with no clamp connections present in the hyphae. This type of interaction is designated minus.

A=B≠ exhibits the *flat* morphology. The macroscopic appearance of flat consists of slow-growing mycelium with little or no aerial hyphae. Microscopic examination reveals gnarled, irregular hyphae with bumps and many short hyphal branches forming at right angles to the main hypha. This type of interaction is designated flat.

A≠B= exhibits varying morphologies that cannot reliably be distinguished from the phenotype of the A=B= interaction (see above). This interaction is also designated minus.

A≠B≠ exhibits the full dikaryotic interaction including slightly appressed aerial growth and the unique hyphal appendage called a clamp connection that forms a cell septa. The clamp connection appears as a semicircular bulge visible at the junction of hyphal cells. This interaction is designated plus.

Professor John R. Raper's book *Genetics of Sexuality in Higher Fungi* includes photographs of the mycelial phenotypes described above [1].

The results of the mating tests will reveal a unique pattern identifying each combination of A and B among progeny resulting from a specific mating. See Table 2 for a specific example.

B. Molecular Genetics

Experimental Approach

The objective of this experiment is to demonstrate activation of A-regulated development by transforming an Aα1 homokaryon (*Y1* is its only Aα mating-

Table 2 Details of the mating reactions of progeny from a UVM19-1 × UVM19-4 cross when progeny are mated with the parental and nonparental ditypes that would result from such a cross.

UVM19-1 × UVM19-4	*S. commune* tester strain/mating type			
Mating type of principle progeny	UVM19-1/ Aα1β1, Bα2β2	UVM19-2/ Aα1β1, Bα1β1	UVM19-3/ Aα4β6, Bα2β2	UVM19-4/ Aα4β6, Bα1β1
PDs and NPDs	Mating test reaction (flat, plus, or minus)			
Aα1β1, Bα2β2	Minus	Flat	Minus	Plus
Aα1β1, Bα1β1	Flat	Minus	Plus	Minus
Aα4β6, Bα2β2	Minus	Plus	Minus	Flat
Aα4β6, Bα1β1	Plus	Minus	Flat	Minus

type gene) with plasmid DNA encoding the *Z4* allele of Aα4 strains. Examination of the mating reactions (as done in the Mendelian experiment) of the transformants will reveal altered reactions when compared to the untransformed Aα1 homokaryon. These alterations indicate the activation of A-regulated development.

Transformants are recovered by cotransformation with a selectable marker. Utilizing a *trp1* mutant of appropriate mating type as the transformation recipient, we introduce two plasmid DNAs, one containing *TRP1* and a second containing *Z4*. When the amount of Z DNA is much greater than *TRP1* DNA (e.g., 4:1 molar ratio), the rate of cotransformation may be as high as 80% in *S. commune*. Following selection of *TRP1* transformants on selective media, mating tests will identify cotransformants for *Z4*. These transformants will demonstrate alteration of mating type and activation of A-regulated development.

The mating test is the best way to demonstrate the mating type of the transformant and demonstrate its active A-regulated development. The genotype of the transformation recipient is Aα1 Aβ1 Bα2 Bβ2 *trp1* (*Y1* is at the Aα locus). The objective is the conversion of the recipient to Aα≠1 (i.e., *Y1 Z4*) by transformation. The untransformed recipient mated against the UVM19-2 tester strain (Aα1 Aβ1 Bα1 Bβ1) should give a flat mating reaction (A=B≠). However, up to 80% of the TRP^+ transformants recovered should contain an ectopic integration of *Z4* within their genomes, becoming Aα≠1 (i.e., Aα1 Aβ1 Bα2 Bβ2 *Z4*). The matings of these transformants with the UVM19-2 tester would be A≠B≠ and produce a dikaryon.

The Scheme (Fig. 4)

1. Generate protoplasts from strain UVM19-1.
2. Cotransform the UVM19-1 protoplasts with two plasmids, pZ4 and pTRP1.
3. Select *TRP1* transformants and screen them for the presence of *Z4* by mating tests with UVM19-2.

Materials (in addition to those from the Mendelian genetic experiment)

Plasmid DNAs pZ4 and pTRP1 (prepared from *E. coli* strains UVMD91 (pZ4) and UVMC5 (pTRP1) by alkaline lysis) [21].
Clinical centrifuge
42°C waterbath or heat block
Waring Blendor and semimicro blender cups

The following sterile solutions and materials:

1 M MgOsm: 1 M $MgSO_4$, 20 mM MES, pH 6.3
0.5 M MgOsm: 0.5 M $MgSO_4$, 10 mM MES, pH 6.3
1 M SorbOsm: 1 M Sorbitol, 10 mM MES, pH 6.3

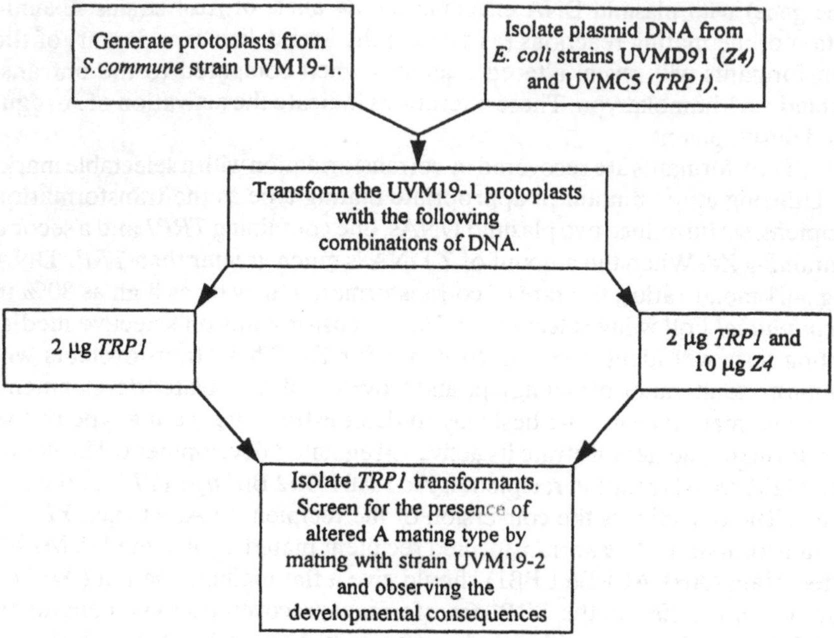

Figure 4 The experimental protocol for the molecular genetic experiment. Details of the procedures involved can be found in the Protocols section.

Novozyme Stock: 25 mg Novozyme 234/mL 0.5 M MgOsm (filter-sterilize and store at –70°C in 0.5-mL aliquots)
SorbOsm/$CaCl_2$: 0.5 M sorbitol, 50 mM $CaCl_2$, 10 mM MES, pH 6.3
50% PEG: 50% polyethylene glycol 4000 in 10 mM MES, pH 6.3 (filter-sterilize and store at –20°C in 0.5-mL aliquots)
Regeneration media: 1× CYM (liquid), 0.5 M sucrose
1% agar in 2× CYM
50-mL disposable centrifuge tubes w/caps (e.g., VWR 21008-146)
15-mL Falcon 2095 conical centrifuge tubes w/caps (e.g., VWR 21008-930)
Sterile transfer pipettes: 7 mL volume, disposable

Protocols

Protoplast generation

1. Inoculate 100 mL of liquid CYMT with the aerial mycelium (scraped from the agar with a spatula or transfer lance) from four semisolid agar

colonies. The colonies should be approximately 1.5 cm in diameter and freshly grown. Blend the inoculated CYMT 1 min in a Waring blendor. Incubate the culture for 30–40 hours at 30°C, with shaking (200 rpm). If the culture becomes viscous, you are growing too dense a culture (i.e., either too much inoculum or too long an incubation).

2. Decant the liquid culture to two 50-mL centrifuge tubes and centrifuge 5 min at 600g.

4. Decant the supernatant and estimate the volume of the loose mycelial pellet.

5. Wash each pellet with 40 mL of 0.5 M MgOsm; centrifuge 10 min, 600g.

6. Decant the supernatant and wash each pellet with 40 mL of 1 M MgOsm; centrifuge 10 min, 600g. (Hint: if the mycelium floats instead of pelleting, simply let the cells sit in the MgOsm for 10 min and centrifuge again.)

7. Decant the supernatant and suspend the pellet in 1 volume 1 M MgOsm (i.e., 10 mL mycelium + 10 mL 1 M MgOsm).

8. Add 0.125 mL Novozyme stock per 10 mL mycelial suspension from step 7 (final concentration = 0.3125 mg Novozyme 234/mL mycelial suspension). (Hint: different lots of Novozyme vary in effectiveness; concentration may have to be altered. Some lots contain deleterious compounds, and a reduced concentration of Novozyme yields a greater numbers of protoplasts.)

9. Incubate at 30°C for 3.5–4 hours with gentle inversion every 30 min.

10. Add an equal volume of sterile dH_2O and centrifuge 5 min, 600 g.

11. Transfer the supernatant containing the protoplasts by pipette to fresh centrifuge tubes (20 mL/tube); add an equal volume of 1 M SorbOsm.

12. Suspend the pellet remaining from step 10 in 20 mL in 0.5 M MgOsm; shake rapidly four or five times and centrifuge 5 min, 600g. This recovers additional protoplasts.

13. Repeat Steps 11 and 12 two times.

14. Centrifuge all the protoplasts suspended in 0.5 MgOsm/SorbOsm from Step 11, 10 min, 600g. Carefully remove all but 1 mL of the supernatant from each tube; suspend the protoplasts by gentle trituration of the protoplasts and the remaining 1 mL of supernatant.

15. Combine the suspended protoplasts and determine the total number with the aid of a hemocytometer. (Hint: a small aliquot of the protoplast suspension may have to be diluted 10–100× in 0.5 M MgOsm to obtain an accurate count.)

16. Centrifuge the protoplast suspension, 10 min, 600 g; pipette off the supernatant and suspend the protoplast pellet in SorbOsm/$CaCl_2$ at a concentration of 10^8 protoplasts/mL. Protoplasts may be used fresh or after storage on ice at 4°C for up to a month or more.

***Schizophyllum* transformation**

1. Prepare the DNA solution as per Fig. 4 and place on ice in a Falcon 2095 tube. DNA (suspended in 114 μL TE or dH_2O) plus 6 μL 1 M $CaCl_2$.

2. Add 200 μL of protoplasts (2 × 10^7) to the DNA solution, mix gently, and incubate on ice 30 min.

4. Slowly add 320 μL 50% PEG. (Hint: let the PEG slide down the side of the tube and collect at the bottom underneath the DNA/protoplast mix.)

5. Incubate on ice 1 min, then mix the two solutions by tapping the tube gently.

7. Add 5 mL regeneration medium, mix gently, and incubate the tube overnight at room temperature.

8. Melt the 2× CYM 1% agar and combine with an equal volume of 1 M MgOsm. Distribute this soft agar overlay to five sterile test tubes prewarmed to 42°C (3 mL/tube) and store in a 42°C waterbath or heat block.

9. Add 1 mL of the regenerated protoplasts to each of the overlay tubes; mix the overlay and regenerated protoplasts thoroughly, and pour each overlay onto a Petri dish containing 25 mL of CYM semisolid medium.

10. Incubate the cultures for 2–3 days at 30°C.

11. As transformed colonies appear, transfer them to CYM semisolid medium. Four transformants may be inoculated equidistantly at the periphery of a 10-cm Petri dish. Allow the transferred inoculum to grow until the colonies are approximately 0.5 cm in diameter. Use this mycelium as the source of inoculum for mating tests and observations of developmental morphology. *Schizophyllum* cultures may be stored at 4°C to retard (but not stop) growth.

Questions

1. Explain the significance of the term "tetrapolar mating type" using your results as an example.

2. Predict the phenotypes of mycelia resulting from mating tests between all pairwise combinations of UVM19-1, -2, -3, and -4.

3. How would you recognize recombinants between Aα and Aβ (or Bα and Bβ) among the progeny of the UVM19-1 × UVM19-2 mating? Did you find any?

4. What would be the pattern of reactions of A or B recombinants when mated with the four tester strains listed in Table 2?

5. How would you prove that the recombinants are in fact recombinants?

6. Why are only a fraction of the *TRP1* transformants isolated in the molecular genetic experiment transformed for A mating type.

7. Does the molecular genetic experiment prove that the minimum requirements for activation of the A-developmental pathway are the presence

of Y and Z protein from different Aα mating types? If not, what experimental approach would you pursue in order to do so?

Extensions

1. Design an experiment to measure the map distance between Aα and Aβ (or Bα and Bβ).
2. Would an Aα null strain (i.e., lacking *Y* and *Z* genes) be useful for analyzing the role of Y and Z proteins? How could an Aα null strain be produced?
3. Consider the molecular nature of Aα mating type. How is the specificity of each polymorph encoded? How could you identify which portion of Y or Z protein encodes the region determining self or nonself recognition?

REFERENCES

1. Raper, J. R. (1966). *Genetics of Sexuality in Higher Fungi.* Ronald Press, New York.
2. Raper, J. R., M. G. Baxter, A. H. Ellingboe (1960). The genetic structure of the incompatibility factors of *Schizophyllum commune*: the A factor. *Proc. Natl. Acad. Sci. USA 53*:1324.
3. Stamberg, J., Y. Koltin (1972). The organization of incompatibility factors in higher fungi: the effects of structure and symmetry on breeding. *Heredity 24*:306.
4. Papazian, H. P. (1950). Physiology of the incompatibility factors in *Schizophyllum commune. Bot. Gaz. 112*:143.
5. Koltin, Y., J. R. Raper, G. Simchen (1967). Genetic structure of the incompatibility factors of *Schizophyllum commune*: the B factor. *Proc. Natl. Acad. Sci. USA 57*: 55.
6. Stankis, M. M., C. A. Specht, H. Yang, L. Giasson, R. C. Ullrich, C. P. Novotny (1992). The Aα mating locus of *Schizophyllum commune* encodes two dissimilar multiallelic homeodomain proteins. *Proc. Natl. Acad. Sci. USA 89*:7169.
7. Specht, C. A., M. M. Stankis, L. Giasson, C. P. Novotny, R. C. Ullrich (1992). Functional analysis of the homeodomain-related proteins of the Aα locus of *Schizophyllum commune. Proc. Natl. Acad. Sci. USA 89*:7174.
8. Wessels, J. G. H., J. H. Seitsma, A. S. M. Sonneberg (1983). Wall synthesis and assembly during hyphal morphogenesis in *Schizophyllum commune. J. Gen. Microbiol. 129*:1607.
9. Wessels, J. G. H. (1992). Gene expression during fruiting in *Schizophyllum commune. Mycol. Res. 96*:609.
10. Lilly, W. W., R. E. Bilbrey, B. L. Williams, L. S. Loos, D. F. Venable, S. M. Higgins (1994). Partial characterization of the cellular proteolytic system of *Schizophyllum commune. Mycologia 86*:461.
11. Paice, M. G., L. Jurasek (1977). Wood saccharifying enzymes from *Schizophyllum commune.* Tappi Conference Papers, Forest Biology, Wood Chemistry Conference, p. 113.
12. Paice, M. G., L. Jurasek, M. R. Carpenter, L. B. Smillie (1978). Production, characterization, and partial amino acid sequence of xylanase A from *Schizophyllum commune. Appl. Environ. Microbiol.36*:802.

13. Paice, M. G., L. Jurasek (1979). Structural and mechanistic comparisons of some β(1→4) glycoside hydrolases. In *Advances in Chemistry Series*, No. 181, *Hydrolysis of Cellulose: Mechanisms of Enzymatic and Acid Catalysis* (R. D. Brown, L. Jurasek, eds.). American Chemical Society, p. 361.
14. Desrochers, M., L. Jurasek, E. Koller (1980). Production of cellulase in a shake-flask culture of *Schizophyllum commune*: optimization of the medium. TAPPI Papermakers Conference Proceedings, Technical Association of the Pulp and Paper Industry, Atlanta, p. 161.
15. Saito, H. T., Ohki, T. Sasaki (1980). A carbon-13 nmr study of polysaccharide gels: molecular architecture in gels consisting of fungal, branched (1-3)-13-β-D-glucans (Lentinan and Schizophyllan) as manifested by conformation changes induced by sodium hydroxide. *Carbohydrate Res. 74*:277.
16. Itoh, W., I. Sugawara, S. Kimura, et al. (1990). Immunopharmacological study of sulfated schizophyllan (SPG). I. Its action as a mitogen and anti-HIV agent. *Int. J. Immunopharmacol. 12*:225.
17. Hotta, H., K. Hagiwara, K. Tabata, W. Itoh, M. Homma (1993). Augmentation of protective immune responses against Sendai virus infection by fungal polysaccharide schizophyllan. *Int. J. Immunopharmacol. 15*:55.
18. Sugawara, I., A. Hirata, W. Itoh (1993). Sulfated schizophyllan as a anti-HIV agent. *Jpn. J. Clin. Med. 51*(suppl):158.
19. Suzuki, T., N. Ohno, T. Yadomae (1992). Activation of the complement system by (1-3)-beta-D-glucans having different degrees of branching and different ultrastructures. *J. Pharmacobio-Dynamics 15*:277.
20. Raper, J. R., R. M. Hoffman (1974). *Schizophyllum commune*. In *Handbook of Genetics*, Vol. 1 (R. C. King, ed.). Plenum Press, New York, pp. 597–626.
21. Lee, S., S. Rasheed (1990). A simple procedure for maximum yield of high-quality plasmid DNA. *Biotechniques 9*:676.

20

Genetics of *Phycomyces*

Arturo P. Eslava and María I. Alvarez
University of Salamanca, Salamanca, Spain

1. INTRODUCTION

The genus *Phycomyces* belongs to the class Zygomycetes, characterized by fusing gametangia which form thick-walled zygospores, by nonflagellated, nonmotile vegetative spores and by a mycelium that is almost completely nonseptate (coenocytic). *Phycomyces* belongs to the order Mucorales, characterized by the development of spores endogenously inside the sporangia. Among the 20 genera of the family Mucoraceae [1], *Phycomyces* is characterized by the enormous size of the sporangiophore and the large, multispored sporangia with a well-defined columella. The large size of the zygospore, the characteristics of the zygophores on both sides of the zygospores (suspensors) and the black appendages surrounding the zygospores are also characteristic features of *Phycomyces*. Examples of some genera belonging to the family Mucoraceae used in research include *Mucor*, *Rhizopus*, and *Absidia*.

A. Life Cycle

Asexual Cycle

The asexual life cycle of *Phycomyces* is shown in Figure 1. The vegetative spores of *P. blakesleeanus* are nonmotile, ellipsoid, multinucleate cells. In the stand-

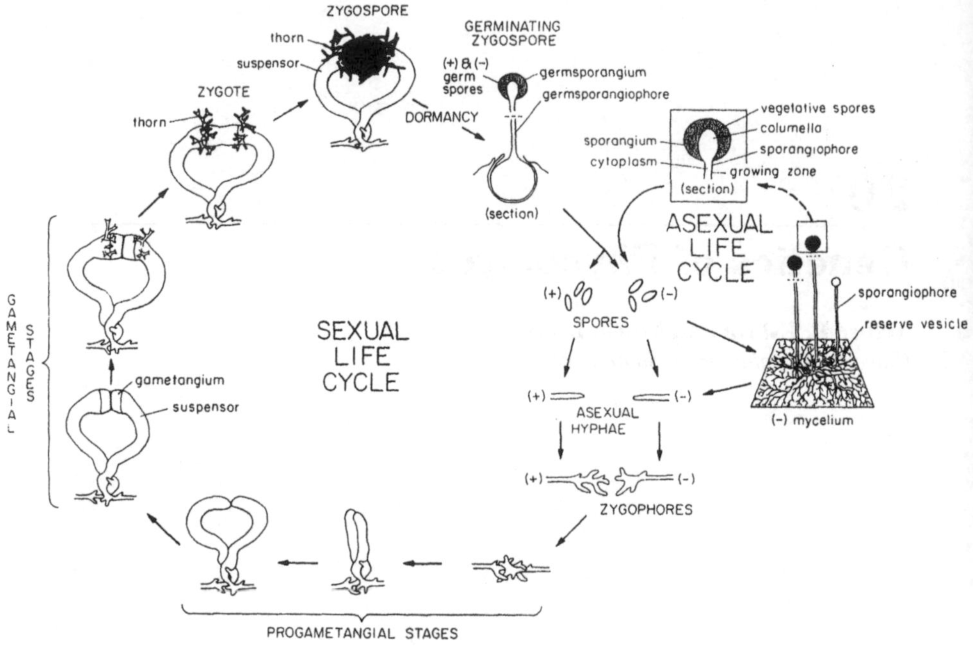

Figure 1 Asexual and sexual life cycles and the principal structures of *Phycomyces blakesleeanus*.

ard wild-type strain NRRL1555, the spore size is about 9 μm long and 6 μm wide on average [2–4]. Only 0.3% of the spores are uninucleate, and about 85% contain three or four nuclei [5]. The *Phycomyces* genome comprises 6, 6 $\times$ 10^7 base pairs [6]. Another study [7] estimates about half that number.

Under suitable conditions the spore germinates. The spore swells and germ tubes (one or two) emerge from the spore to form the first hyphae, which grow and branch rapidly, forming the mycelium. Growth occurs only at the hyphal tips. In minimal medium very few spores germinate. The transition from a dormant to a germinating spore requires an activation, heat shock (15 min at 48°C) being the most commonly used method [8]. No cross walls separate various parts of the mycelium (with some rare exceptions), and the hyphae never anastomose. The mycelia are large coenocytes containing millions of nuclei.

After the mycelium has grown on solid medium for 2 to 3 days, formation of asexual reproductive structures, the sporangiophores, is initiated. They are large, unbranched hyphae which grow upward into the air and form an apical sporangium. The size of the sporangiophores varies depending on environmental conditions. The macrophores are the sporangiophores used in most physiological research. They can reach a length of more than 100 mm, and the diameter is of about 100 μm. The mature sporangia are about 500 μm in diameter and contain some 10^5 spores. The asexual spores are formed by the division of a large mass of cytoplasm into multinucleate portions that develop cell walls. This packaging of nuclei into spores is random [5,9]. The microphores are about 1 mm long, and the corresponding sporangia are about 100 μm in diameter and contain about 1000 spores. They are produced under special environmental conditions. Under the usual laboratory conditions, macrophores are predominantly formed; the term sporangiophores usually refers exclusively to them.

Sexual Cycle

Phycomyces is a heterothallic fungus with two mating types, called (+) and (–) [10], that are indistinguishable except that, when placed near each other in suitable conditions, they form special structures called zygospores [reviewed in 11], which are large, black spheres about 0.5 mm in diameter, surrounded by thorns. The zygospores remain dormant for about 2 months or more depending on the strain and the environmental conditions. The zygospore eventually germinates forming a germsporangiophore whose germsporangium contains about 10,000 germspores. The prefix "germ" is used to designate the products of zygospore germination. The morphology of these structures is similar to that of the asexual cycle (see Fig. 1).

Nearly all nuclei of both mating types that enter the zygospore degenerate, either after fusion or in the initial haploid state. In general, only one diploid nucleus undergoes meiosis and the four meiotic products later undergo successive mitotic divisions [12–14]. The germspores are mainly homokaryotic and are formed, unlike in the asexual cycle, from protospores containing a single nucleus. This nucleus undergoes one or more mitotic divisions. Under appropriate conditions the germspore germinates into a vegetative mycelium and thus enters the asexual cycle. Since a very small proportion of germspores are heterokaryotic, it is assumed that some germspore primordia contain more than one nucleus [3]. When these nuclei are of different mating types, the germspore will give rise to a heterosexual heterokaryotic mycelium which shows a particular deep yellow color and a morphology clearly distinguishable from homokaryotic mycelia [4].

B. Key Notes

Sensory Physiology

The sporangiophores of *Phycomyces* show growth response to several stimuli including light, gravity, wind, chemicals, and the presence of objects in the proximity of the growing zone (avoidance response). The growing zone is about 2 mm long and is located about 0.1 mm below the sporangium. When the stimuli are symmetrically distributed around the sporangiophores, the responses are called mecisms—i.e., changes in growth velocity; asymmetrically distributed stimuli give rise to tropisms—i.e., changes in the direction of growth. Among the different stimulus-response systems of *Phycomyces* the most attractive and studied is the phototropic response. Light is a convenient stimulus because it is quantitative, easy to apply and remove, and available over a long range of wavelengths.

The mycelium of *Phycomyces* also shows responses to light such as initiation of sporangiophores (photophorogenesis) and induction of β-carotene synthesis (photocarotenogenesis). All these properties make *Phycomyces* a good system for the study of intracellular sensory transduction processes.

The developmental steps that have been investigated in this fungus most extensively are the germination of the spores [15], the formation of the sporangiophores (phorogenesis) [16], and the initial stages of sexual development [17]. The study of carotene biosynthesis and its regulation is another main aspect of research in *Phycomyces*. This last line of research is facilitated by the availability of mutants, the use of quantitative complementation techniques, and the development of an in vitro system [18].

Behavioral Genetics

The existence of mutants has facilitated the study of sensory transduction processes, as well as carotenogenesis, differentiation, and sexual development. Mutants with defective phototropism are designated *mad*. The *mad* mutants define nine unlinked genes, *madA* through *madI*. The *madA*, *madB*, *madC*, and *madI* gene products act early in the phototropism signal transduction pathway. The products of genes *madD* to *madG* are thought to regulate cell wall growth. This last point may have to be modified in view of the recent findings of Campuzano et al. (in preparation) which suggest that, by measuring the phototropic action spectrum of a *madF* mutant, these later genes are also involved in photoreception. Mutants in these genes show altered tropisms to several unrelated stimuli (light, gravity, presence of objects in the proximity of the growing zone).

Mutants altered in the synthesis of β-carotene are designated *car*. They can result from a block in the pathway or from an alteration in the regulation of the pathway. Some mutants produce no β-carotene (those altered in the genes *carB*, *carR*, *carRA*), others accumulate small amounts (*carA*, *carC*), and still others produce much greater amounts of β-carotene than the wild-type (*carD*, *carF*, *carS*). The *pic* mutants are defective in the photoinduction of carotenogenesis [19].

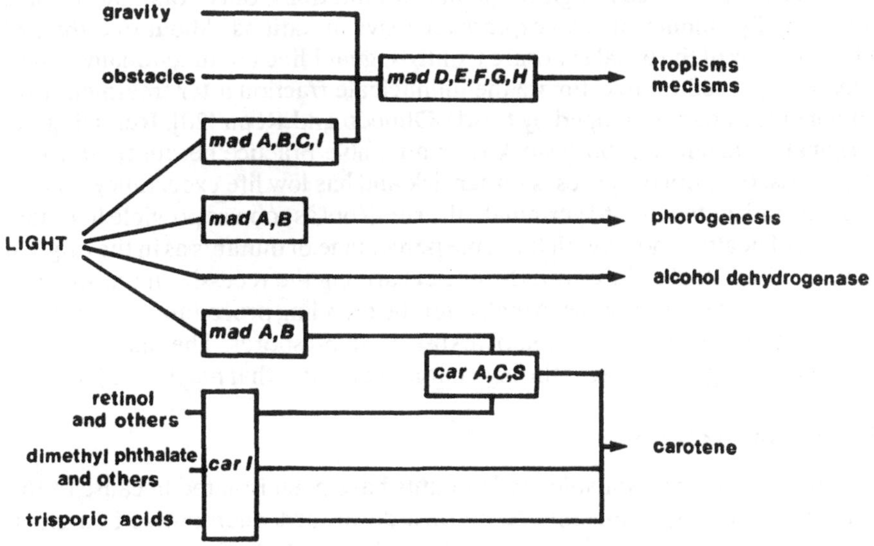

Figure 2 Scheme of sensory pathways of *Phycomyces*. The external stimuli appear on the left, and the responses on the right. Arrows indicate the flow of information. Mutations in the genes listed in boxes hinder the information flow at the place in the pathway. [Adapted from 4 and 21.]

Physiological and genetic studies indicate that stimuli and responses are connected by sensory pathways as shown in Figure 2. The fungus detects the stimuli listed at the left and exhibits the responses at the right. The elements of the sensory pathway, sensors, effectors, and transducers, are named after the genes that control them. By studying mutants with altered behavior, one can identify the genes affecting the operation of each step on the sensory pathway (boxes in Fig. 2). The pathway represents the flow of information, and it is plausible that some genes intervening in the process have not yet been found. A tentative conclusion for the above sequence is that relatively few genes govern the behavior in this organism, and a given gene may govern different processes in different structures of the organism.

2. PRINCIPLES

A. Mutagenesis

The isolation of mutants that carry recessive mutations has relied so far on the killing of most of the nuclei by chemical or physical agents, generally the same agents used to induce a high frequency of mutations. Survivors that become functionally uninucleate can express recessive mutations. About one-third of the survivors of the usual exposure to nitrosoguanidine are functionally uninucleate. A procedure to estimate the uninucleate fraction after treatment with any mutagen was developed by Cerdá-Olmedo and Reau [20]. Recycling the original mutagenized spore stock is an advisable practice because, after mutagenesis, the pool of spores is rather sick and has low life expectancy. Therefore, it is advantageous to germinate the survivors as soon as possible to obtain a stock of healthy spores with the same percentage of mutants as in the original mutagenized stock. Moreover, the nuclei carrying the recessive mutation may be present in heterokaryosis so that, after the recycling procedure, nuclei carrying the mutation may be in homokaryosis in some spores. The main obstacle of the recycling procedure is the high number of plates that need to be handled.

B. Mutant Isolation

The majority of the available *mad* mutants have been isolated because of the inability of their sporangiophores to turn downward when grown in a glass-bottom box, where all light comes from below [21]. Different screening procedures have recently been used resulting in the selection of novel mutant types [22–25].

The isolation of mutants with abnormal carotenoid content (genotype *car*) is carried out by direct inspection of the color shown by mutagenized colonies. There are other types such as auxotrophic, drug-resistant, and developmental mutants [reviewed in 11] which have been studied to a lesser extent.

C. Heterokaryons and Complementation

Because no stable diploids are known in *Phycomyces*, the presence of two genomes in the same cell for complementation tests is brought about by heterokaryosis—that is, the presence of two genetically different types of nuclei in the same cytoplasm. In nature all large *Phycomyces* mycelia are likely to be heterokaryotic to a small extent as a result of spontaneous mutations.

The hyphae of *Phycomyces* do not anastomose spontaneously upon contact; however, this fungus has great capacity for regeneration. The Ootaki method for heterokaryon formation is based on regeneration of grafted sporangiophores [26]; the method of Gauger et al. [27] is based on regeneration of sexual structures into mycelia, and the method of Suárez et al. [28,29] is based on regeneration of fused protoplasts.

The heterokaryons formed by nuclei of the same mating type are morphologically similar to homokaryotic cultures. Heterokaryons formed by nuclei of opposite mating types produce more carotenoids and fewer sporangiophores, and form characteristic structures called pseudophores, which give a grainy aspect to the mycelial surface.

The calculation of nuclear proportions in *Phycomyces* heterokaryons is based on the nuclear behavior which has the following characteristics: 1.) no nuclear divisions while the spores are formed; 2.) the spores are formed by random packaging of the nuclei from the nuclear population of the sporangium; 3.) all parts of the mycelium, including the sporangia, have the same nuclear proportion. This proportion may be calculated [5] from the proportion of the three kinds of spores (heterokaryotic, homokaryotic of one type, and homokaryotic of the other type) that make up the heterokaryon.

The values of biochemical variables or behavioral traits in heterokaryons containing a known proportion of nuclei presenting differences for these variables are obtained as a function of the nuclear proportion. This function, called genophenotypic function by Medina [30,31], describes the dependence of the phenotype on gene dosage. The interest of these functions is that they impose stringent conditions to any hypothesis aimed at explaining the phenomena under study at the molecular level. These conditions may be so stringent as to constitute a very strong argument for a certain hypothesis, as was the case with the study of β-carotene biosynthesis and its regulation in *Phycomyces* [32,33]. The study of behavioral traits such as phototropism has also benefited from this technique [34].

D. Recombination

Phycomyces genetics began early this century [10,35,36], but it has only been in the last 20 years that standard conditions for the germination of the zygospores have been established and clear evidence has been found for the existence of a standard meiotic process in the generation of recombinants [12–14].

The behavior of the nuclei in the zygospore may be deduced from genetic experiments. Hundreds or thousands of haploid nuclei of both sexes enter and form part of the zygospores, which, after a dormancy of 2–3 months, germinate and give rise to a germsporangium containing some 10,000 multinucleate germspores. Though the mechanisms are not known in detail, nearly all nuclei of both mating types that enter the zygospore degenerate, either after fusion or in the initial haploid state. In general, only one diploid nucleus undergoes meiosis; the four meiotic products later undergo successive mitotic division [12–14]. The germspores are, in general, multinucleate and homokaryotic, and a very small proportion may be heterokaryotic (see Sexual Cycle, p. 387).

E. Linkage Analysis

Once the zygospores have germinated, the progeny may be analyzed by two methods (see part c of Fig. 3). In the first method, about 80 colonies from

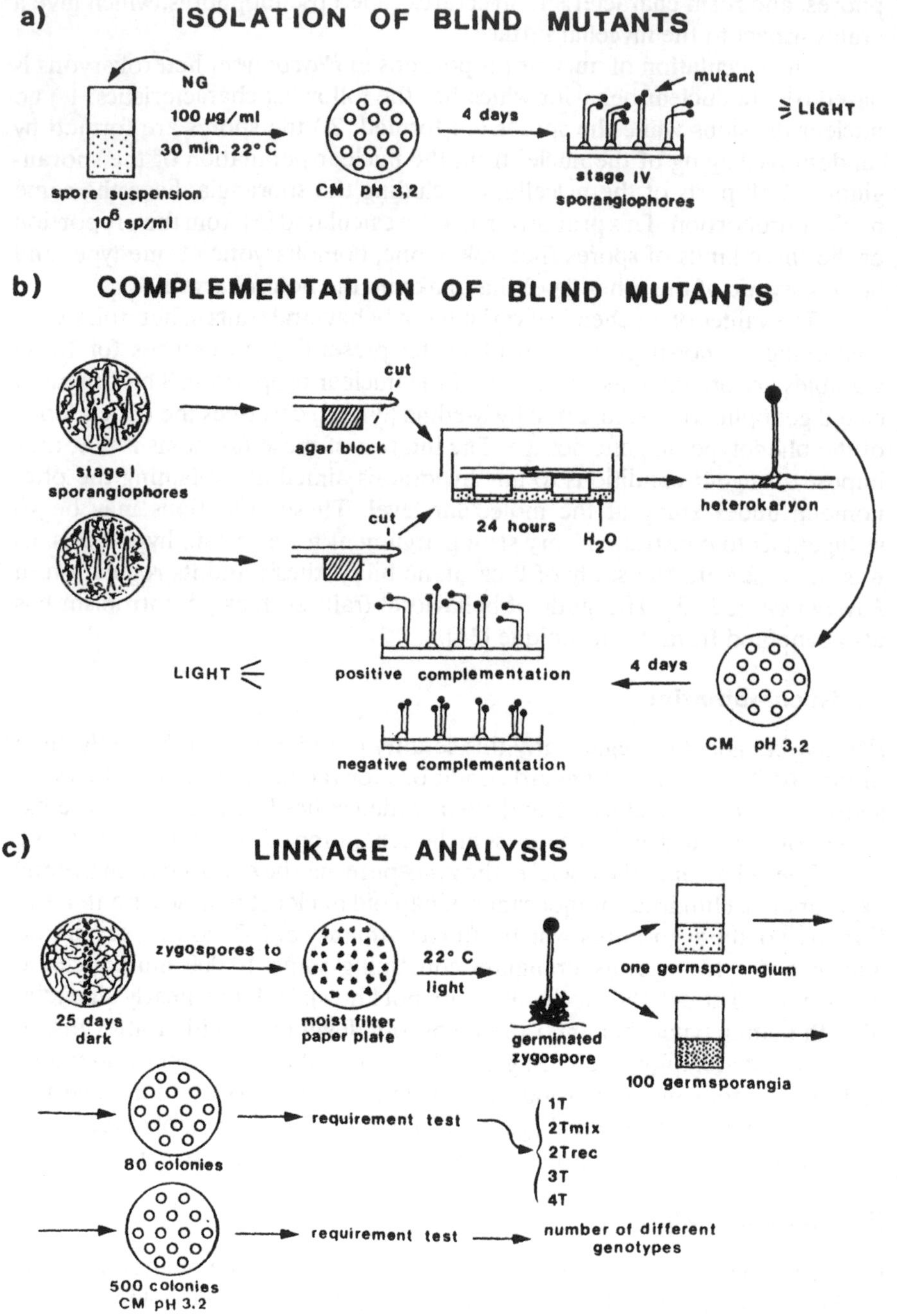
a) ISOLATION OF BLIND MUTANTS
NG
100 μg/ml
30 min. 22° C
spore suspension
10^6 sp/ml
CM pH 3,2
4 days
mutant
stage IV
sporangiophores
LIGHT
b) COMPLEMENTATION OF BLIND MUTANTS
stage I
sporangiophores
cut
agar block
cut
24 hours
H2O
heterokaryon
positive complementation
negative complementation
LIGHT
4 days
CM pH 3,2
c) LINKAGE ANALYSIS
25 days
dark
zygospores to
moist filter
paper plate
22° C
light
germinated
zygospore
one germsporangium
100 germsporangia
80 colonies
requirement test
1T
2Tmix
2Trec
3T
4T
500 colonies
CM pH 3.2
requirement test
number of different
genotypes

germspores from each germsporangium are characterized phenotypically. This is an example of unordered and amplified tetrad analysis. The second method, called random spore analysis, consists of mixing the germspores from about 100 germsporangia and analyzing a large sample of about 500 germspores. The first method is much the more laborious, but provides more information about the meiotic process. Both methods give the linkage relationships among markers, but where linkage is very loose, tetrad analysis may give information unobtainable from random spore analysis. When one is dealing with auxotrophic markers, and particularly when the recombinant classes are rather infrequent, random spore analysis has a very marked advantage [37].

When a pair of markers is studied, tetrads are classified as monotypes (1T) if all the germspores of the sample taken from a germsporangium have the same genotype. If the germsporandium contains two different genotypes, the tetrad is called a ditype (2T). The two genotypes may be reciprocal (2Trec) or not (2Tmix). Similarly, the tetrads are classified as tritypes (3T) or tetratypes (4T) when three or four different genotypes are found in the sample of germspores analyzed [14]. This classification takes into account only the presence or absence of genotypes in the germspore samples and ignores the occurrence of intersexual heterokaryons. With this procedure of scoring, the effects of secondary mechanisms causing disproportion (i.e., sterility of some germspores) are largely eliminated.

A regular meiosis gives rise only to reciprocal ditypes—either a parental ditype (PD) or a nonparental ditype (NPD)—or to a tetratype (TT). When PD and NPD occur with the same frequency, the markers are said to be unlinked; if the PD is significantly more frequent than the NPD, the markers are said to be linked. For linked genes, the distance between the two is calculated from the formula:

$$\text{Percentage recombination} = (\text{NPD}+\text{TT}/2) / (\text{PD}+\text{NPD}+\text{TT}) \times 100$$

Quite often, the expected genotypes do not appear in the sample of germspores taken from each germsporangium. The incomplete tetrads are derived from the complete ones by loss of some of the genotypes, either through an error in the sampling or through some fault in meiosis or in the postmeiotic divisions [13,14]. The parental monotypes are considered as PD, the recombinant monotypes may be counted as NPD, and the tritypes and

---

Figure 3 Flow chart for the isolation of blind mutants in *Phycomyces* and genetic characterization by complementation and recombination. Phototropism is tested with unilateral blue light at a fluence rate of 10^{-6} W m^{-2} and exposure lasting 8 h in darkness. For colonial growth the complete medium (CM) is acidified to pH 3.2.

mixed ditypes may be considered to be TT.One should consider that, in a single meiosis, each allele cannot appear in more than two genetic combinations (unless gene conversion is occurring). Germsporangia are classified as resulting from one meiosis if they are compatible with thus rule; otherwise, the existence of two or more meiosis has to be assumed.

The different arrangement of the genetic material in the wild types may be responsible for the meiotic irregularities. Isogenic strains differing only in mating types show increased viability of the germspores and a regular pattern

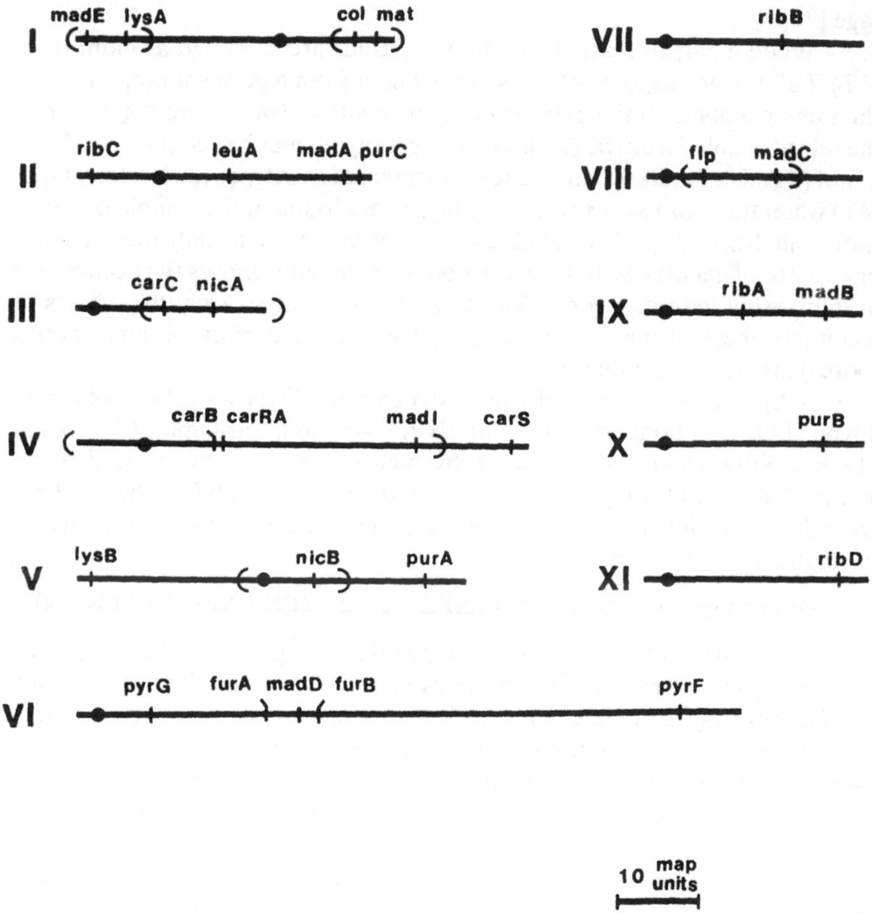

Figure 4 Localization of 29 *Phycomyces* markers into 11 linkage groups. The ordering of the genes within parentheses has not been determined relative to outside markers.

of meiotic segregation [38]. However, the dormancy of the zygospores increases slightly with isogenicity.

Unordered tetrad analysis allows calculation of gene-centromere distances by using an indirect method of estimation of second division segregation frequencies first developed by Whitehouse [39]. By using this method, the distances of several markers from their respective centromeres have been calculated in *Phycomyces* [40,41].

With the random spore analysis method, one should check that the two alleles of each marker appear in the progeny with equal frequencies in order to show the absence of allele-specific selection. The frequency of recombination between two markers is calculated by dividing the number of recombinants by the total.

The number of chromosomes and the number of linkage groups in *Phycomyces* are unknown. The existence of 11 linkage groups (Fig. 4), among a total of 29 markers used, has been established in the initial genetic map of *Phycomyces* [40,42]. This number of linkage groups is a lower limit, since the number of markers analyzed is rather small.

3. PRACTICE

The experiment involves the isolation of mutants of *Phycomyces* with abnormal phototrophism, phenotypic characterization, testing for genetic complementation and mapping. The scheme of the experiment is shown in Figure 3.

A. Materials Needed

Strains

The wild-type NRRL1555 (–), from Northern Regional Research Laboratories, USDA, Peoria, Illinois, is taken as the standard with which other strains are compared. Several other wild types of (+) mating type are also used for crosses, mainly UBC21 (+), from R. J. Baldoni, Botany Department, University of British Columbia, Vancouver, Canada, and A56(+), a strain largely isogenic with NRRL1555 [38]. All strains are named according to the genetic nomenclature proposed for bacteria by Demerec et al. [43]. Prefixes refer to strains isolated at the University of Salamanca (A), the Max Planck Institute for Molecular Genetics in Berlin (B), California Institute of Technology (C), and University of Sevilla (S). *mad* indicates a mutant with abnormal phototropism, and *car* indicates a mutant with abnormal carotene production.

Media

The minimal medium (SIV) contains, per liter, 20 g of D(+)-glucose, 2 g of L-asparagine·H_2O, 0.5 g of KH_2PO_4, 0.5 g of $MgSO_4 \cdot 7H_2O$, 28 mg of $CaCl_2$, 1 mg of thiamine·HCl, 2 mg of citric acid·H_2O, 1.5 mg of $Fe(NO_3)_3 \cdot 9H_2O$, 1 mg of $ZnSO_4 \cdot 7H_2O$, 300 μg of $MnSO_4 \cdot H_2O$, 50 μg of $CuSO_4 \cdot 5H_2O$, and 50 μg of $Na_2MoO_4 \cdot 2H_2O$. This medium is based on modifications by Sutter [44] to the minimal medium proposed by Ødegard [45]. Glucose should be autoclaved separately. The minor ingredients, including $CaCl_2$, may be prepared as 50-fold concentrate. Solid minimal medium is obtained by the addition of agar at 20 g/L. The pH of the medium is about 4.2. A complete medium (SIVYC) contains minimal medium supplemented with 0.1% yeast extract and 0.1% Bacto-Casitone. For colonial growth, the medium is acidified to pH 3.2 by the addition of 1 N HCl after autoclaving (SIVYCA). For phototropism tests, a complete medium (PDA) contain 4% potato dextrose agar (Difco) plus 5 μg/mL thiamine-HCl. For crosses, PDA medium plus 0.1 g/L hypoxanthine is used.

Mutagen

N-Methyl-N'-nitro-N-nitrosoguanidine (NG) is the commonly used mutagen in *Phycomyces*. Several other agents have also been used to a lesser extent. The main mutagenesis procedures used in *Phycomyces* have been described elsewhere [46].

Miscellaneous

A dark room and a light source are needed for the phototrophic response. A tungsten-halogen lamp, cooled by a small fan, may be placed outside the dark room in such a way that the light reaches the dark room through an aperture in the wall which can be covered with a heat filter and a blue filter. A box about 100 cm long, 25–50 cm wide, and 20 cm high may be used instead of a dark room.

The formation of heterokaryons by the method of grafting [26] requires a stereomicroscope, fine tweezers, and iris scissors. This is the method we are following in the experiment. The method of protoplast fusion requires a mixture of commercial lytic enzymes preparation (Novozym 234) rich in chitinase, and a *Streptomyces* preparation, "streptozyme," rich in chitosanase [28,29].

For crosses, a temperature-controlled chamber is needed for keeping the temperature first at 17°C and then at 22°C.

B. Description of the Experimental Procedure

Mutagenesis and Mutant Isolation

Heat-activated spores (15 min at 48°C) of the wild-type strain NRRL1555 at a concentration of 10^6/mL are treated with 100 μg/mL nitrosoguanidine (NG) in 0.1 M citrate-phosphate buffer, pH 7.0, for 30 min at room temperature.

The suspension is shaken occasionally. After three washes in water, aliquots are placed on rich acid medium (SIVYCA) to learn the viability. Under these conditions, the survival ranged from 5% to 15%. The spores are inoculated immediately after mutagenesis on PDA medium (about 200–300 viable spores per plate) or SIVYCA medium (about 50 viable spores per plate). In some cases the mutagenized spores are allowed to complete a full vegetative growth cycle on rich medium (recycling).

After 2 to 3 days of growing under overhead light, the plates are uncovered and placed in the dark room or in the box where they are exposed overnight to unilateral blue light at a given intensity (10^{-6} W m^{-2}, for instance). Wild-type sporangiophores bend toward the light source while mutant sporangiophores grow more or less straight up. Sporangia from the mutant candidates are picked up with wet forceps for retesting by the same procedure. If the sporangiosphores on retesting plates again show the same mutant phenotype, permanent stocks are prepared for further studies.

Complementation

The complementation tests are carried out between the *mad* mutants isolated above and representative strains altered in each of the known genes (*madA–madI*) governing the phototropic response. As can be seen in Figure 3, heterokaryons between two strains are made according to Ootaki [26]. Young sporangiophores (1–2 cm long) from each strain are removed from mycelia and placed on agar blocks, tips facing each other, and grafted by inserting a decapitated tip of one sporangiophore into another. About 24 h after grafting, the grafts are tested for success by observing the formation of new cell wall connecting the two sporangiophore portions. Spores are collected only from the grafts showing a single sporangiophore regenerated at the graft union of a successful graft. Such a sporangiophore is almost always heterokaryotic [26, 47–49].

The spores resulting from the regenerated sporangium (homokaryotic and/or heterokaryotic) are resuspended in a shell vial containing 1 mL sterile distilled water (one sporangium per vial), and after heat activation the spores are plated on rich acid medium at about 50 spores per plate. After 2–3 days, when the sporangiophores appear, the Petri dishes are uncovered, fuzzy first-crop sporangiophores are removed, and only straight, mature sporangiophores about 3–4 cm long are tested.

The plates are placed in a dark room or a test box for 6–9 h to test the phototropic response. The testing light is such that the wild-type clearly bend and the mutants do not. Whenever some of the heterokaryotic sporangiophores bend toward the light, the mutants forming the heterokaryon are scored as a complementary pair. If all the sporangiophores grow straight, the mutations do not complement each other. In this way the blind mutants isolated as above can be said to be affected in one of the known *mad* genes (*madA–madI*) or else they define a new gene or genes.

Crosses

The procedures for setting up the crosses and analyzing the genotypes of the descendants are shown in Figure 3. As an example throughout the section we are going to analyze the cross *furA* (+) × *maD* (–) by mass spore analysis, and the cross *madI* (+) × *nicA carA* (–) by tetrad analysis, where *fur* indicates resistance to 5-fluorouracil, *mad* indicates abnormal phototropism, (+) and (–) are the two mating types, *nic* indicates requirement of nicotinic acid, and *car* indicates abnormal β-carotene synthesis.

The crosses are performed by inoculating a piece of mycelium of the strains of opposite sex at the margins of PDA plates and incubating for 10 days in the dark at 17°C and then for 15 days at 22°C also in the dark (the moment of inoculation is taken as time zero). Then 200–400 zygospores from each cross are harvested individually with tweezers and transferred to Petri dishes containing a moist filter paper at 22°C in continuous light [13,14,40]. The zygospores begin to germinate around 50–60 days or later.

For tetrad analysis, the germspores are collected by picking up ripe germsporangium, from germinated zygospores, in a drop of water between the tips of tweezers and putting them into 1 mL of sterile water. Samples of various sizes from such a suspension are plated on rich acid medium (SIVYCA). Of the colonies formed after 2 or 3 days, a sample of 20–40 are picked at random and transferred to rich medium (SIVYC) plates. The genotype of each isolate is then determined, and the germsporangia are classified as PD, NPD, and TT. In the cross *madI* (+) × *nicA carA* (–) (see Table 1), *madI* and *carA* are linked while the pairs *madI* and *nicA*, and *nicA* and *carA* are unlinked. This is a case of loose linkage between the *madI* and *carA* markers that would be very difficult to show by random mass spore analysis.

Random mass spore analysis was used in the cross *furA* (+) × *madD* (–) (see Table 2). Ripe germsporangia from 100 or more germinated zygospores are pooled in 1–2 mL of water. The germspores from the pool are counted

Table 1 Tetrad analysis of the cross *madI* (+) × *nicA carA* (–)

Markers	PD	NPD	TT	% Recombination
madI and *nicA*	7	9	13	53
madI and *carA*	16	3	10	27
nicA and *carA*	10	8	11	47

Table 2 Analysis of the cross B71 × C149. Genotypes *furA* (+) × *madD* (–)

Shortest dormancy (days)	53
Number of pooled germsporangia	174
Average germspores per germsporangium	1.4×10^4
Germspores viability (%)	30

(a) Genotypes of the progeny

wild type (+)	4	wild type (–)	3
furA (+)	81	*furA* (–)	99
madD (+)	98	*madD* (–)	109
furA madD (+)	2	*furA madD* (–)	4

(b) Occurrence of the two alleles of each gene

	Allele	
Gene	(+) or wild-type	(–) or mutant
Mating type	185	215
furA	214	186
madD	187	213

(c) Linkage relationships

Genes	Parental	Recombinant	% Recombination
furA and *madD*	187	13	3.2
furA and mating type	195	205	51
madD and mating type	198	202	50

under the light microscope, and the average number of germspores per germsporangium is calculated. The viability of the pooled germsporangia is calculated by plating on SIVYCA. When the viability is low, say 5% or less, the cross should not be analyzed. The shortest dormancy, defined as the time elapsed from the day on which mating plates were inoculated to the germination of the first zygospore, is recorded in each cross.

A sample of the germspores from the pool is plated on SIVYCA medium at about 50 viable spores per plate. After 2–3 days about 500 individual colonies are transferred to SIVYC plates and the genotype of each isolate is determined. As can be seen from Table 2, the markers *furA* and *madD* are linked, with a 3% recombination, but they are unlinked to the third marker, the mating type.

C. Problems and Questions

1. The products of a regular meiosis with two unlinked markers may be either tetrads with two types of reciprocal genotypes ($2T_{rec}$) or tetrads with four

different types of genotypes (4T), as long as there is crossing over between at least one of the markers and its centromere. In *Phycomyces*, the germsporangia are classified as monotype (1T), ditype (2T), tritype (3T), and tetratype (4T), depending on whether one, two, three, or four different genotypes, respectively, are found in the sample of germspores analyzed on each germsporangium. Among the ditypes, the two genotypes may be reciprocal pairs, parental or recombinant ($2T_{rec}$), or they may not ($2T_{mixed}$). Table 3 shows the results of the crosses from which samples of germspores from individual germsporangia were analyzed. What can be concluded about the use of isogenic strains?

2. The wild type of *Phycomyces* is yellow because it accumulates the yellow pigment β-carotene. Color mutants have been isolated that are unable to synthesize β-carotene. Among the mutants, some are red, because they accumulate lycopene, a precursor of β-carotene; they are altered in the gene *carR*. Another type of color mutants are white, and are defective in the gene *carA*. They do not accumulate carotenoids, except for traces of β-carotene. Torres-Martinez et al. [50] proposed that the *A* and *R* functions are determined by contiguous DNA segments cotranscribed to a single mRNA. They designate the bifunctional gene as *carRA*. What kind of experiments would you suggest to test this hypothesis?

3. What is the advantage of counting in the sample the presence or absence of genotypes instead of counting the total numbers of the respective genotypes?

4. *Phycomyces* spores are multinucleate. Do you think that this characteristic will influence the mutagenesis procedure?

Table 3 Tetrad analysis of the back-crosses

		Class of germsporangium*					
	Type†	1T	$2T_{rec}$	$2T_{mix}$	3T	4T	$2T_{rec}$ + 4T
UBC21 × H1	P	7 (50)	0 (0)	2 (14.3)	3 (21.4)	2 (14.3)	2 (14.3)
UBC21 × S102	P	24 (61.5)	5 (13)	5 (13)	4 (10)	1 (2.5)	6 (15.5)
C247 × NRRL1555	1	14 (27.5)	11 (21.5)	7 (13.7)	11 (21.5)	8 (15.7)	19 (37.2)
C260 × S102	2	18 (34)	10 (19)	10 (19)	8 (15)	7 (13)	17 (32)
C264 × NRRL1555	3	9 (14)	18 (28)	4 (6)	12 (19)	21 (33)	39 (61)
C268 × S102	4	5 (17)	13 (43)	3 (10)	1 (3)	8 (27)	21 (70)
C268 × H1	4	6 (10)	18 (31)	8 (14)	10 (17)	16 (28)	34 (59)
A11 × S102	8	4 (5.2)	26 (34.2)	2 (2.6)	2 (2.6)	42 (55.3)	68 (89.5)

*The percentage of the total is shown in parentheses.
†P represents parental crosses, and the numbers correspond to back-crosses.

5. What experiments should be performed to learn the number of chromosomes in *Phycomyces*?
6. The number of known genes involved in the phototropic response of *Phycomyces* is rather low. Do you have an explanation for this small number?

D. Extensions

The possibility of making heterokaryons and studying complementation, the availability of mutants, and the existence of an initial genetic map made *Phycomyces*, despite some limitations, an organism to which genetic analysis can be applied to solve questions related with the main lines of research in this fungus: the study of intracellular transduction processes, the biosynthesis of carotenoids and its regulation, morphogenesis, and sexual differentiation.

Basic biological facts such as the moment of meiosis and the chromosome number are still unknown. The reduction of the dormancy period of the zygospores would be an important technical advance in *Phycomyces* genetics. More genetic markers to expand the studies on the genetic map and electrophoretic karyotyping studies by pulsed field gel electrophoresis are needed to allow a reasonably correct estimate of linkage groups in *Phycomyces*. In the related fungus *Absidia glauca* these kinds of studies have been reported [51].

The transformation systems developed in *Phycomyces* are inefficient at present or have not been well characterized [52–56]. An important goal would be the development of simple, efficient, and repetitive methods of transformation with exogenous DNA in order to allow gene cloning by direct complementation. In the related fungus *Mucor circinelloides*, the transformation system is very efficient [57], and gene cloning by direct complementation has been achieved [58,59]. Recently, integrative transformation by homologous recombination, resulting in either additions or else gene replacement and ectopic insertion, has been reported [60]. In another related fungus, *Absidia glauca*, transformation by autonomous replication seems to be the preferential mode of DNA propagation [61,62]. In this organism integrated transformation has also been achieved [63]. The transformation of another member of the Mucorales, *Rhizopus niveus*, has also been reported [64].

There are several areas of research in which *Phycomyces* and/or several other related Mucorales are used to some extent; the initial physiological events of sexual development in Mucorales have been studied in *Phycomyces*, *Blakeslea*, and *Mucor* [17]. Through the mutational analysis of sexual development and the study of the biological, chemical, and metabolic properties of trisporic acids and related compounds, it should be possible to learn more about the molecular basis of sexuality in the Mucorales.

Members of the genus *Mucor* produce a number of extracellular enzymes of industrial application such as acid proteases, amylases, lypases, and cellulases, and display the property of undergoing morphogenetic changes from hyphal to budding yeastlike growth in response to various environmental stimuli [65]. In spite of the advantages of the transformation system in *Mucor circinelloides*, formal genetic analysis in this fungus has not been developed due to the difficulties in utilizing the sexual cycle [66]. However, in *Phycomyces*, the genetics is the most developed among the Mucorales but the transformation system is not efficient. Recently the expression of *Phycomyces* genes in *Mucor circinelloides* has been reported [67,68], and it may be possible to use a heterologous transformation shuttle system between *Mucor* and *Phycomyces* to take advantage of the particular characteristics of both organisms.

Phycomyces and many other organisms exhibit a variety of blue-light responses including photostimulation of β-carotene synthesis, photomovement (phototropism, light-growth response, phototaxis), photocontrol of the intermediary metabolism, and photomorphogenesis. However, the blue-light receptor(s) has not been isolated, and the molecular basis of the light responses is unknown. The only well-characterized blue-light photoreceptors to date are the DNA photolyases involved in DNA repair. Recently, Ahmad and Cashmore [69] reported molecular and genetic evidence that a protein with strong structural homology to DNA lyase may be a photoreceptor in *Arabidopsis*. Probes with the HY4 gene of *Arabipdosis* may be useful to isolate the corresponding gene in *Phycomyces*.

For a review of all aspects of *Phycomyces* research, see the book edited by Cerdá-Olmedo and Lipson (1987). The book edited by Senger (1987) reviews the blue-light responses field in different organisms. Two recent books, by Elliot (1994) and Griffin (1994), cover some aspects of the topics described in this review. An excellent book about the molecular bases of development has been published by Russo et al. (1992). In Volume 214 of *Methods in Enzymology* there are several contributions about photocarotenogenesis in fungi, including *Phycomyces*. Complete publishing information on these books follows.

Cerdá-Olmedo, E., E. D. Lipson (eds.). *Phycomyces*. Cold Spring Harbor Laboratory, Cold Spring Harbor, NY (1987).

Elliott, C. G. *Reproduction in Fungi.* Chapman and Hall, London (1994).

Griffin, D. H. *Fungal Physiology* (2nd edition). Wiley-Liss, London (1994).

Packer, L. (ed.). *Methods in Enzymology.* Vol. 214. *Carotenoids.* Part B. *Metabolism, Genetics, and Biosynthesis.* Academic Press, London (1993).

Russo, V. E. A., S. Brody, D. Cove, S. Ottolenghi (eds.). *Development: The Molecular Genetic Approach.* Springer-Verlag, Berlin (1992).

Senger, H. (ed.). *Blue Light Responses: Phenomena and Occurrence in Plants and Microorganisms.* CRC Press, Boca Raton, Fla (1987).

REFERENCES

1. Hesseltine, C. W., J. J. Ellis (1973). Mucorales. In *The Fungi* (G. C. Ainsworth, F. K. Sparrow, A. S. Sussman, eds.). Academic Press, New York, p. 187.
2. Benjamin, C. R., C. W. Hesseltine (1959). Studies on the genus *Phycomyces. Mycologia 51*:751.
3. Bergman, K., P. V. Burke, E. Cerdá-Olmedo, et al. (1969). *Phycomyces. Bacteriol. Rev. 33*:99.
4. Cerdá-Olmedo, E., E. D. Lipson (1987). Biography of *Phycomyces.* In *Phycomyces* (E. Cerdá-Olmedo, E. D. Lipson, eds.). Cold Spring Harbor Laboratory, Cold Spring Harbor, NY, p. 7.
5. Heisenberg, M., E. Cerdá-Olmedo (1968). Segregation of heterokaryons in the asexual cycle of *Phycomyces. Mol. Gen. Genet. 102*:187.
6. Harshey, R. M., M. Jayaram, M. E. Chamberlain (1979). DNA sequences organization in *Phycomyces blakesleeanus. Chromosoma 73*:143.
7. Dusenbery, R. L. (1975). Characterization of the genome of *Phycomyces blakesleeanus. Biochim. Biophys. Acta 378*:363.
8. Rudolph, H. (1960). Weitere Untersuchungen zur Wärmeaktivierung der Sporangiosporen von *Phycomyces blakesleeanus.* II. Mitteilung. *Planta 54*:424.
9. Swingle, D. B. (1903). Formation of the spores in the sporangia of *Rhizopus nigricans* and *Phycomyces nitens. U.S. Dept. Agric. Bur. Plant. Ind. Bull. 37*:1.
10. Blakeslee, A. F. (1904). Sexual reproduction in the Mucorineae. *Proc. Am. Acad. Arts Sci. 40*:205.
11. Eslava, A. P. (1987). Genetics. In *Phycomyces* (E. Cerdá-Olmedo, E. D. Lipson, eds.). Cold Spring Harbor Laboratory, Cold Spring Harbor, NY, p. 27.
12. Cerdá-Olmedo, E. (1975). The genetics of *Phycomyces blakesleeanus. Genet. Res. 25*:85.
13. Eslava, A. P., M. I. Alvarez, P. V. Burke, M. Delbrück (1975). Genetic recombination in sexual crosses of *Phycomyces. Genetics 80*:445.
14. Eslava, A. P., M. I. Alvarez, M. Delbrück (1975). Meiosis in *Phycomyces. Proc. Natl. Acad. Sci. USA 72*:4076.
15. Van Laere, A. J., J. A. Van Assche, B. Furch (1987). The sporangiospore: dormancy and germination. In *Phycomyces* (E. Cerdá-Olmedo, E. D. Lipson, eds.). Cold Spring Harbor Laboratory, Cold Spring Harbor, NY, p. 247.
16. Corrochano, L. M., E. Cerdá-Olmedo (1991). Photomorphogenesis in *Phycomyces* and in other fungi. *Photochem. Photobiol. 54*:319.
17. Sutter, R. P. (1987). Sexual development. In *Phycomyces* (E. Cerdá-Olmedo, E. D. Lipson, eds.). Cold Spring Harbor Laboratory, Cold Spring Harbor, NY, p. 317.
18. Cerdá-Olmedo, E. (1987). Carotene. In *Phycomyces* (E. Cerdá-Olmedo, E. D. Lipson, eds.). Cold Spring Harbor Laboratory, Cold Spring Harbor, NY, p. 199.
19. Corrochano, L. M., E. Cerdá-Olmedo (1992). Sex, light and carotenes: the development of *Phycomyces. TIG. 8*:268.
20. Cerdá-Olmedo, E., P. Reau (1970). Genetic classification of the lethal effects of various agents on heterokaryotic spores of *Phycomyces. Mutat. Res. 2*:369.

21. Bergman, K., A. P. Eslava, E. Cerdá-Olmedo (1973). Mutants of *Phycomyces* with abnormal phototropism. *Mol. Gen. Genet. 123*:1.
22. Lipson, E. D., I. López-Díaz, J. A. Pollock (1983). Mutants of *Phycomyces* with enhanced tropisms. *Exp. Mycol. 7*:241.
23. Alvarez, M. I., A. P. Eslava, E. D. Lipson (1989). Phototropism mutants of *Phycomyces blakesleeanus* isolated at low light intensity. *Exp. Mycol. 13*:38.
24. Campuzano, V., J. M. Díaz-Mínguez, A. P. Eslava, M. I. Alvarez (1990). A new gene (*madI*) involved in the phototropic response of *Phycomyces. Mol. Gen. Genet. 223*:148.
25. Campuzano, V., P. Galland, H. Senger, M. I. Alvarez, A. P. Eslava (1994). Isolation and characterization of phototropism mutants of *Phycomyces* insensitive to ultraviolet light. *Curr. Genet.* (in press).
26. Ootaki, T. (1973). A new method for heterokaryon formation in *Phycomyces. Mol. Gen. Genet. 121*:49.
27. Gauger, W., M. I. Peláez, M. I. Alvarez, A. P. Eslava (1980). Mating type heterokaryons in *Phycomyces blakesleeanus. Exp. Mycol. 4*:56.
28. Suárez, T., M. Orejas, A. P. Eslava (1985). Isolation, regeneration and fusion of *Phycomyces blakesleeanus* spheroplast. *Exp. Mycol. 9*:203.
29. Suárez, T., M. Orejas, J. Arnau, S. Torres-Martinez (1987). Protoplast formation and fusion. In *Phycomyces* (E. Cerdá-Olmedo, E. D. Lipson, eds.). Cold Spring Harbor Laboratory, Cold Spring Harbor, NY, p. 351.
30. Medina, J. R. (1979). Les fonctions géno-phénotypiques: concept et applications. *Rev. Biomath. 67*:39.
31. Medina, J. R. (1987). Quantitative genetic interactions. In *Phycomyces* (E. Cerdá-Olmedo, E. D. Lipson, eds.). Cold Spring Harbor Laboratory, Cold Spring Harbor, NY, p. 355.
32. De la Guardia, M. D., C. M. G. Aragón, F. J. Murillo, E. Cerdá-Olmedo (1971). A carotenogenic enzyme aggregate in *Phycomyces*: evidence from quantitative complementation. *Proc. Natl. Acad. Sci. USA 68*:2012.
33. Aragón, C. M. G., F. J. Murillo, M. D. De la Guardia, E. Cerdá-Olmedo (1976). An enzyme complex for the dehydrogenation of phytoene in *Phycomyces. Eur. J. Biochem. 63*:71.
34. Medina, J. R., E. Cerdá-Olmedo (1977). Allelic interaction in the photogeotropism of *Phycomyces. Exp. Mycol. 1*:286.
35. Blakeslee, A. F. (1906). Zygospore germinations in the Mucorineae. *Sydowia Ann. Mycol. 4*:1.
36. Burgeff, H. (1915). Untersuchungen über Variabilität und Erblichkeit bei *Phycomyces nitens* Kunze II. *Flora 108*:353.
37. Fincham, J. R. S., P. R. Day, A. Radford (1979). *Fungal Genetics*, 4th Ed. Blackwell Scientific Publications, Oxford.
38. Alvarez, M. I., A. P. Eslava (1983). Isogenic strains of *Phycomyces blakesleeanus* suitable for genetic analysis. *Genetics 105*:873.
39. Whitehouse, H. L. K. (1950). Mapping chromosome centromeres by the analysis of unordered tetrads. *Nature 165*:893.

40. Orejas, M., M. I. Peláez, M. I. Alvarez, A. P. Eslava (1987). A genetic map of *Phycomyces blakesleeanus. Mol. Gen. Genet. 210*:69.
41. Campuzano, V., P. Del Valle, J. I. De Vicente, A. P. Eslava, M. I. Alvarez (1993). Isolation, characterization and mapping of pyrimidine auxotrophs of *Phycomyces blakesleeanus. Curr. Genet. 24*:515.
42. Alvarez, M. I., V. Campuzano, E. P. Benito, A. P. Eslava (1993). Genetic loci of *Phycomyces blakesleeanus*. In *Genetic Maps*, 6th Ed. (S. J. O'Brien, ed.). Cold Spring Harbor Laboratory, Cold Spring Harbor, NY, p. 3.120.
43. Demerec, M., E. A. Adelberg, A. J. Clark, P. E. Hartman (1966). A proposal for a uniform nomenclature in bacterial genetics. *Genetics 54*:61.
44. Sutter, R. P. (1975). Mutations affecting sexual development in *Phycomyces blakesleeanus. Proc. Natl. Acad. Sci. USA 72*:127.
45. Ødegard, K. (1952). On the physiology of *Phycomyces blakesleeanus* Burgeff. I. Mineral requirements on a glucose-asparagine medium. *Physiol. Plant. 5*:583.
46. Cerdá-Olmedo, E. (1987). Mutagenesis. In *Phycomyces* (E. Cerdá-Olmedo, E. D. Lipson, eds.). Cold Spring Harbor Laboratory, Cold Spring Harbor, NY, p. 341.
47. Ootaki, T., A. C. Lighty, M. Delbrück, W. J. Hsu (1973). Complementation between mutants of *Phycomyces* deficient with respect to carotenogenesis. *Mol. Gen. Genet. 121*:57.
48. Ootaki, T., E. P. Fisher, P. Lockhart (1974). Complementation between mutants of *Phycomyces* with abnormal phototropism. *Mol. Gen. Genet. 131*:233.
49. Ootaki, T., T. Kinno, K. Yoshida, A. P. Eslava (1977). Complementation between *Phycomyces* mutants of mating types (+) with abnormal phototropism. *Mol. Gen. Genet. 152*:245.
50. Torres-Martinez, S., F. J. Murillo, E. Cerdá-Olmedo (1980). Genetics of lycopene cyclization and substrate transfer in β-carotene biosynthesis in *Phycomyces. Genet. Res. 36*:299.
51. Kayser, T., J. Wöstemeyer (1991). Electrophoretic karyotype of the zygomycete *Absidia glauca*: evidence for differences between mating types. *Curr. Genet. 19*: 279.
52. Suárez, T. (1985). Obtención de protoplastos y transformación en *Phycomyces blakesleeanus.* Resumen de Tesis Doctorales, Ediciones Universidad de Salamanca, p. 38.
53. Revuelta, J. L., M. Jayaram (1986). Transformation of *Phycomyces blakesleeanus* to G-418 resistance by an autonomously replicating plasmid. *Proc. Natl. Acad. Sci. USA 83*:7344.
54. Suárez, T., A. P. Eslava (1988). Transformation of *Phycomyces* with a bacterial gene for kanamycin resistance. *Mol. Gen. Genet. 212*:120.
55. Arnau, J., F. J. Murillo, S. Torres-Martinez (1988). Expression of Tn-5-derived kanamycin resistance in the fungus *Phycomyces blakesleeanus. Mol. Gen. Genet. 212*:375.
56. Ootaki, T., A. Miyazaki, J. Fukui, et al. (1991). A high efficient method for introduction of exogenous gene into *Phycomyces blakesleeanus. Jpn. J. Genet. 66*:189.

57. Van Heeswijck, R., M. I. G. Roncero (1984). High frequency transformation of *Mucor* with recombinant plasmid DNA. *Carlesberg Res. Commun. 49*:691.
58. Roncero, M. I. G., L. P. Jepsen, P. Stroman, R. Van Heeswijck (1989). Characterization of a *leuA* gene and an ARS element from *Mucor circinelloides. Gene 84*:335.
59. Anaya, N., M. I. G. Roncero (1991). Transformation of a methionine auxotrophic mutant of *Mucor circinelloides* by direct cloning of the corresponding wild type gene. *Mol. Gen. Genet.* 230:449.
60. Arnau, J., P. Stroman (1993). Gene replacement and ectopic integration in the zygomycete *Mucor circinelloides. Curr. Genet. 23*:542.
61. Wöstemeyer, J., A. Burmester, C. Weigel (1987). Neomycin resistance as a dominantly selectable marker for transformation of the zygomycete *Absidia glauca. Curr. Genet. 12*:625.
62. Burmester, A., A. Wöstemeyer, J. Arnau, J. Wöstemeyer (1992). The *SEG1* element: a new DNA region promoting stable mitotic segregation of plasmids in the zygomycete *Absidia glauca. Mol. Gen. Genet. 235*:166.
63. Burmester, A., A. Wöstemeyer, J. Wöstemeyer (1990). Integrative transformation of a zygomycete, *Absidia glauca*, with vectors containing repetitive DNA. *Curr. Genet. 17*:155.
64. Yanai, K., H. Horiuchi, M. Tagaki, K. Yano (1991). Transformation of *Rhizopus niveus* using a bacterial blasticidin S resistance gene as a dominant selectable marker. *Curr. Genet. 19*:221.
65. Orlowski, M. (1991). *Mucor* dimorphism. *Microbiol. Rev. 55*:234.
66. Schipper, M. A. A. (1978). On certain species of *Mucor*, with a key to all accepted species. *Stud. Mycol. 17*:1.
67. Iturriaga, E. A., J. M. Díaz-Mínguez, E. P. Benito, M. I. Alvarez, A. P. Eslava (1992). Heterologous transformation of *Mucor circinelloides* with the *Phycomyces blakesleeanus leu1* gene. *Curr. Genet. 21*:215.
68. Benito, E. P., J. M. Díaz-Mínguez, E. A. Iturriaga, V. Campuzano, A. P. Eslava (1992). Cloning and sequence analysis of the *Mucor circinelloides pyrG* gene encoding orotidine-5'-monophosphate decarboxylase: use of *pyrG* for homologous transformation. *Gene 116*:59.
69. Ahmad, M., A. R. Cashmore (1993). *HY4* gene of *A. thaliana* encodes a protein with characteristics of a blue-light photoreceptor. *Nature 366*:162.

21

Genetic Analysis in the Oomycetous Fungus

Phytophthora infestans

D. S. Shaw
University of Wales, Bangor, Gwynedd, Wales

1. INTRODUCTION

Early systematists included the oomycetes in the group Phycomycetes—the algal-like fungi. Their grouping with the brown and golden algae, in the kingdom Chromista (confirmed by rDNA sequencing [7]) rather than with the true fungi, indicates that they evolved from algal ancestors by loss of the chloroplast. These pseudofungi [5] or straminipilous fungi [6] range from reduced forms with nonfilamentous thalli to those with extensively branched but basically coenocytic hyphae. The asexual sporangia of most genera release heterokont zoospores; sexual oospores result from the interaction of male and female gametangia (oogamy). Many pseudofungi are saprophytes living on decaying animal and vegetable tissue, e.g., water molds; others are pathogens of animals, plants, algae, fungi, or pseudofungi. Plant pathogens range from necrotrophs with a broad host range, through hemibiotrophs, to biotrophs showing high host specificity. The group includes some of the world's most destructive crop pathogens—*Pythium* spp. and *Phytophthora cinnamomi* (root rots); *Phytophthora infestans* (foliage blight); *Plasmopara* spp. and *Bremia lactucae* (downy mildews).

Chromosomal cytology [26,34] and genetic analysis in *P. infestans* [29] and in other *Phytophthora* spp. [28] have shown that nuclei in the vegetative hyphae are diploid and that the female gametangia (oogonia) and male gametangia (antheridia) are the sites of meiosis. A single male gametic nucleus is thought to fuse with one of the many gametic nuclei of the oogonium. The other gametic nuclei degenerate and the oogonium differentiates as a persistent oospore with abundant food reserves. Germination of the oospore gives rise to a sporangium with zoospores or to a mycelium developing directly from one or more germ tubes. Most pseudofungi are homothallic; antheridia and oogonia are formed on a single thallus which can be derived from a uninucleate spore. Others are heterothallic, the thallus being self-sterile and bisexual; a few are dioecious, having oogonia and antheridia on separate thalli.

2. KEY NOTES

P. infestans has few of the traits that make ascomycetes and basidiomycetes attractive models for basic studies in genetics. However, several properties recommend it as a model for studies within the oomycetes. It is a highly destructive and economically important pathogen of both potato and tomato; it evolved in Mexico on wild species of *Solanum*. It is an excellent example of a "fungus" which migrated with its host around the world, in this case in at least two waves, one of which was recent [11]. The study of its evolving populations following migration is an intriguing challenge.

P. infestans is easily cultured on a variety of natural and synthetic media, in contrast to the downy mildew fungi, which are obligate biotrophs. The fungus can be efficiently cloned by isolating single germinating uninucleate zoospores. In contrast, the spores (conidia) of many of the downy mildew fungi are multinucleate (e.g., *Bremia*, *Peronospora*). *P. infestans* is heterothallic with two mating types (A1 and A2). This means that compatible parents can be mated to yield hybrid progeny. Theoretically, reciprocal crosses (female × male and male × female) could be conducted by manipulation of sexual expression. Single parents can be made to self-fertilize when compatible thalli are placed in chemical contact but are physically separated with a porous polycarbonate membrane [19]. A gene-for-gene system seems to control the interaction of single resistance genes (available in a nonisogenic differential series of potato clones) with single complementary avirulence genes in the pathogen [1]. Attempts are being made to map and clone these fungal genes. *P. infestans*, like many phytopathogenic fungi (see Chapter 16), has a large genome and large amounts of repetitive DNA. Crude estimates suggest a size of 2.5×10^8 bp [29], at least 90% of which is repetitive DNA [3]. All other phytophthora spp. examined have smaller genomes [13,21]. *B. lactucae* has a relatively small genome (5×10^7 bp, accurately estimated); even here, 65% is repetitive with a short period pattern of intraspersion with single-copy DNA, a pattern not found in the true fungi [9]. One advantage of a large genome is that meiotic chromosomes can be observed by light [26] and UV [34] microscopy; a disadvantage is that intact, chromosome-size DNA molecules may be too long to resolve using pulsed field gel electrophoresis [2,31]. Useful electrophoretic karyotypes are available for pseudofungi with smaller genomes, e.g., *Pythium* spp. [22], several *Phytophthora* spp. [31], and *B. lactucae* in which there is evidence for B-type chromosomes [8]. Hyphal or zoospore nuclei from isolates of *P. infestans* from Mexico have a basic diploid amount of DNA whereas those of isolates from some other countries show a range of DNA content, suggesting the presence of triploids, tetraploids, and intermediate aneuploids [32,34]. Isolates with twice the diploid amount of DNA have a chromosome number indicating tetraploidy [34]. The extent of ploidy variations in other species and genera is unknown; however, there is evidence of hyperploidy in field isolates of *B. lactucae* from California [14].

In fungal populations where a single mating type of a heterothallic species predominates, variation may be generated by somatic hybridization. Work with RFLP markers indicates that a polyploid pathotype in asexual populations of *B. lactucae* in California is a somatic hybrid of two diploid pathotypes [14]. Somatic "matings" between drug-resistant isolates of *P. infestans* [25,29] and *P. sojae* (syn. *P. megasperma* f. sp. *glycinea*) [20] have yielded unstable recombinants which have yet to be analyzed with well-characterized nuclear and mitochondrial markers.

As recessive mutations are not expressed in diploids, conventional markers for *P. infestans* are few [29]. The mating-type locus is chromosomal but shows non-Mendelian inheritance [18]. Resistance to the systemic fungicide metalaxyl is presumed to be chromosomal [17,27], whereas resistance to streptomycin cosegregates with mitochondrial DNA (mtDNA) [35]. There is evidence that avirulence (virulence) to single resistance genes of potato is determined by single dominant (recessive) genes [1]. Of the many enzymes investigated, glucosephosphate isomerase, peptidase [30], and malic enzyme [24] are polymorphic with many alleles at a single locus; inheritance is usually Mendelian. These have been extremely useful and robust markers for population studies of *P. infestans* [11].

DNA Polymorphism

The mtDNA of *P. infestans* has been sequenced; it is a compact genophore of 38 kb, with few noncoding sequences and no large repeated segments (B. F. Lang, unpublished). A major polymorphism found in many isolates is an insert of approx. 2 kb associated with flanking rearrangements [4]. This type II morph can be further differentiated into IIa and IIb on the basis of a single restriction site; type I, without the insert, can be similarly subdivided into Ia and Ib. Although these four main morphs can be detected on agarose gels following restriction of genomic DNA with certain enzymes [3], primers have now been selected to allow amplification of sequences that include the polymorphisms; diagnostic bands are present after restriction and electrophoresis(G. W. Griffith and D. S. Shaw, unpublished).

As expected in a large genome, most probes derived from a genomic library include repetitive sequences. One of these, RG57 is a multilocus probe, detecting up to 13 independently segregating Mendelian loci [12], which has allowed variation within and between population of *P. infestans* to be defined [10]. Single-copy probes from a genomic library have detected RFLPs which have been used as single-locus markers in mapping studies. Recently, the value of fragments amplified by randomly chosen 10-mer primers (RAPD) have been assessed as genetic markers. The mating-type locus has been located using RAPD markers and a bulked progeny DNA technique [18]. Similar methods are being used to map avirulence loci (R. Maufrand, unpublished). A reliable and stable transformation method has been developed for *P. infestans* using vectors bearing regulatory elements from *B. lactucae* and bacterial selective markers [16]. Cotransformations with mixtures of two plasmids showed linked integration; GUS transformants were pathogenic [15].

3. PRINCIPLES

A detailed genetic analysis in *P. infestans* has only been possible since molecular markers became available. One of the first such studies [2] shows that inheritance is not always Mendelian. The extent of anomolous inheritance needs to be determined. Genetic analysis of single-oospore progeny of a mating of field isolates of *P. infestans* (Ca65 from California and 550 from Mexico) using an isozyme marker, glucosephosphate isomerase, indicated that some of the progeny could be "selfs" of one or the other parent. Subsequent analysis with 19 single-gene RFLP markers confirmed that some "selfs" were actually parental in genotype, indicating a failure of meiosis and fertilization. True selfs, products of automyxis, were also detected. Hybridity in the majority of progeny was confirmed by using a probe homozygous in each parent and heterozygous in hybrid progeny; however, some progeny were partial hybrids that inherited some but not other markers from each parent. Using RFLP markers heterozygous in one parent and homozygous in the other, 48 of the progeny appeared to be diploids inheriting one allele from each parent; a linkage analysis of these detected five small linkage groups.

However, 28 progeny showed a more intense Southern hybridization of one fragment (allele) to a probe than of the other fragment (allele), suggesting three copies of that gene and three copies of the relevant chromosome. Confirmation that some of the progeny carried three alleles of a locus comes from the use of probes identifying two alleles in one parent, a unique allele in the other parent, and all three in some of these progeny. In addition to presumptive trisomics, triploid progeny, inheriting an extra allele at all loci at which this could be detected, were identified. As expected, progeny inherited their mtDNA from either Ca65 (type II) or 550 (type I). Most of the presumptive trisomics inherited mtDNA from Ca65 and also derived their extra allele from that parent. All trisomics inheriting their extra allele from 550 also inherited their mtDNA from 550. If mtDNA is transmitted through the oogonium to each zygote, then the oogonial gametic nucleus was disomic. This suggests that nondisjunctional events tend to occur during oogonial meiosis. Whether these anomalies are a common feature of sexual crosses of *P. infestans* requires clarification. There is obviously a need to choose parents for future genetic work which yield a high proportion of diploid hybrids among the progeny.

4. PRACTICE

A. Experimental Approach

The experiment will investigate the inheritance of mitochondrial and RAPD markers in single-oospore progeny of a mating of diploid parents. Parental, selfed, hybrid, and partially hybrid progeny will be identified. Suitable parents are established as single-zoospore isolates. Hyphal inocula of both parents are placed on streptomycin agar to select for resistant sectors. A stable streptomycin-resistant mutant is mated to the compatible wild-type on nutrient agar. Oospores are harvested and plated to allow germination. Single germinated oospores are isolated, and a progeny of around 100 individuals is established; their streptomycin resistance/sensitivity is assessed. DNA from each parent is amplified with one of a number of 10-mer primers to identify polymorphisms. Inheritance of polymorphic bands in progeny should indicate if they are parental, selfs, hybrids or partial hybrids (see Figure 1).

B. Scheme/Flow Chart

Isolation

↓

Selection of streptomycin resistance

↓

Single-zoospore isolation

↓

Mating and harvest of oospores

↓

Germination of oospores

↓

Establishment of F_1 progeny

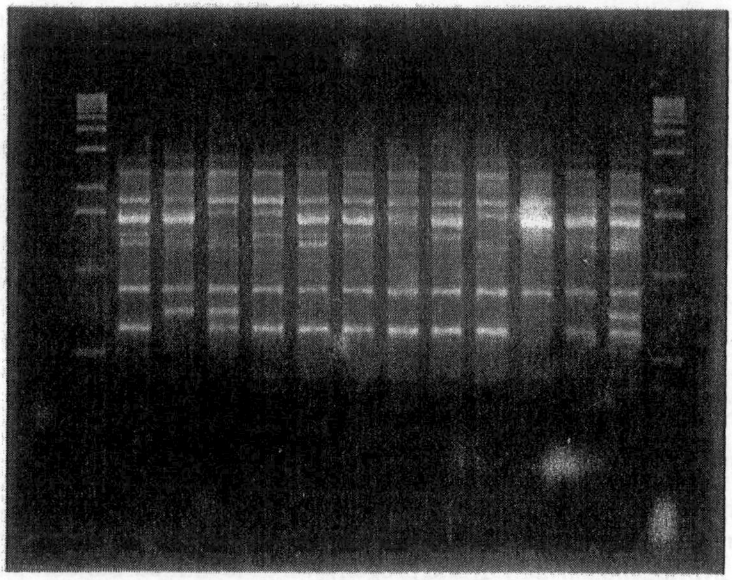

Figure 1 DNA amplified from parents (lanes 2 and 3) and progeny (lanes 4–13) using primer OPG14. The two smallest fragments are clearly polymorphic and are each heterozygous in one parent. Progeny in lanes 4 and 13 are obviously hybrid. Some of the larger fragments also show less obvious RAPD in parents and progeny. Several primers detecting heterozygous loci are usually required to distinguish all hybrids from nonhybrids. Photograph kindly supplied by R. Maufrand.

C. Materials Needed

Isolates

Compatible field isolates from Mexico should make useful diploid parents; other isolates should have the DNA contents of their zoospore nuclei checked by microdensitometry or fluorimetry.

Culture

Rye A agar is used for routine cultivation. Sixty grams of whole rye grains are soaked in 0.5 L water for 36 h. The liquid is retained and the grains are macerated and incubated at 50°C for 3 h. The malted mash is squeezed through nylon mesh or cheesecloth (discard solids); add the retained liquid, 20 g glucose, and

12 g agar. The volume is made up to 1 L before autoclaving. V8-agar is used for mating. One liter contains 200 mL V8 juice (Campbell's), 2 g CaCO3, and 12 g agar.

Incubation

The optimal temperature for growth is 18°C.

D. Experimental Procedure

1. Selection of streptomycin resistance in parental isolates. Having established the minimum concentration that inhibits growth of hyphal inocula on rye agar (probably around 100 μg/mL), inoculate plates of streptomycin agar with plugs of hyphal inocula (four per plate). Incubate and inspect plates at weekly intervals for fast-growing sectors arising from plugs. Isolate single-zoospore germlings from sectors and check that they are stably resistant to streptomycin after growth without the antibiotic. Choose a highly resistant mutant from one or the other parent to act as a mitochondrial marker.
2. Establishment of progeny. Mate resistant × sensitive parents on V8-agar by inoculating a hyphal plug of one parent onto each half of a plate; seal plate with parafilm. Oospores will develop along the midline of contact of the parents within 2 days of meeting. Allow oospores to mature for 3 weeks. Cut out a slab of agar bearing most of the oospores, slice it roughly, and blend in a laboratory homogenizer or hand-held tissue grinder until oospores are freed from agar and subtending hyphae. Wash spores in sterile distilled water (SDW), treat the spore suspension overnight with NovoZym 234 (5 mg/mL), and wash again in SDW. Spores in suspension, uncontaminated with viable hyphae, are plated on 0.8% soft-water agar (approx. 1000 spores per plate) and incubated under continuous fluorescent light. Inspect spores for the formation of germ tubes (with or without terminal germsporangia) each day, cut out single germlings on a block/disc of agar, and transfer each to a plate or Repli-dish of rye agar.
3. Test progeny for inheritance of streptomycin resistance by comparing the growth of each with that of the parents on rye agar, with and without streptomycin.
4. DNA extraction from parents and progeny. Cut out a patch of mycelium (approx. 1 cm^2) and boil for 5 min in 300 μL extraction buffer (0.2 M Tris, 0.25 M NaCl, 25 mM EDTA [pH8], 2% SDS). Add 300 μL phenol, vortex 20 sec, then spin down. Add 5 μL of RNAase (10 μg/mL) to supernatant and incubate for 30 min at 37°C. Add 1 vol chloroform:isoamyl alcohol and spin. Spin aqueous layer again. Precipitate the DNA with 0.5 vol isopropanol and spin down. Wash the pellet with 300 μL ethanol, vacuum dry, and dissolve in 50 μL TE.

5. Characterization of RAPDs markers [23a]. Amplification is performed in volumes of 25 μL overlaid with 25 μL of light mineral oil. The reaction contained 10 mM Tris-HCl (pH 9), 50 mM KCl, 2 mM $MgCl_2$, 1% Triton X-100, 0.5% Tween 20, 100 μM of each dNTP, 0.75 units Taq polymerase (Promega), 5 picamols of 10-mer primer (Operon) and 25–50 ng genomic DNA. Consistent PCR amplifications are produced with a Hybaid Omni-cycler using 35 cycles of 15 sec at 94°C, 30 sec at 35°C, 1 min at 72°C; the first denaturation is for 30 sec and the last extension at 72°C for 3 min. Amplification products (2–3 μL reaction mix) are separated by electrophoresis in 1.6% agarose gels.

5. QUESTIONS AND EXTENSIONS

Initial analysis should allow characterization of the progeny as parental, as selfs (automixis of one parent), or as hybrids, having RAPD bands from both parents. If parents show a polymorphism of mtDNA, the uniparental, mitochondrial inheritance of the streptomycin marker should be verified by testing its cosegregation with the mitochondrial marker [35]. This is most efficiently carried out by amplifying and restricting the polymorphic segments of mtDNA using specific 20-mer primers (G. W. Griffith and D. S. Shaw, unpublished). Further analysis of hybrids for anomalies in chromosomal transmission is best done with RFLP probes. Single-copy probes selected from genomic libraries are available from Bangor or Wageningen (Dr. F. Govers) or can be selected from single-copy RAPD bands (see Chapter 16). Probes identifying two unique alleles of one locus in one or both parents are particularly useful for identifying triploid or aneuploid progeny. Cotransmission of mtDNA and an extra allele from one parent can then be assessed to determine if meiosis in oogonia is particularly prone to aberration. A range of diploid parents from different sources should be analyzed to find out if anomalies are confined to certain crosses (e.g., Californian × Mexican parents) or may be a product of the multinucleate nature of oomycetous gametangia and/or the high plasticity of karyotype as revealed by variations in DNA contents of field isolates.

Inheritance should be analyzed in matings of A1 and A2 isolates from the same geographic area outside Mexico, where isolates are often hyperploid. Are their progeny more anomolous than progeny from Mexican matings? Are hyperploid progeny pathogenic and would they survive in the field?

Matings of the dioecious *Pythium sylvaticum* also show large amounts of unexpected variation [23]. RAPD and RFLP markers show anomalous inheritance, and electrophoretic karyotypes of parents and particularly of progeny are variable, suggesting that nondisjunction and/or translocation at meiosis is common. This genome also seems flexible enough to tolerate aneuploidy.

Phytophthora sojae also has a well-defined gene-for-gene interaction of avirulence loci in the pathogen and resistance loci in the soya bean host. In this homothallic species, "mating" of two parents has produced an F_1 that included about 3% hybrid progeny (identified by RAPD markers). Markers linked to avirulence are being identified in the segregating F_2 progeny [33]. An advantage of homothallic species is that they are unlikely to suffer from the inbreeding noted in second-generation progenies of heterothallic species [29].

ACKNOWLEDGMENT

I wish to thank Remy Maufrand for details of his RAPD method.

REFERENCES

1. Al-Kherb, S. M., C. Fininsa, R. C. Shattock, D. S. Shaw (1995). The inheritance of virulence of *Phytophthora infestans* to potato. *Plant Pathol. 44*:552.
2. Carter, D. A. (1990). DNA polymorphisms as genetic markers in *Phytophthora infestans*. Ph.D. Dissertation, University of London.
3. Carter, D. A., S. A. Archer, K. W. Buck, D. S. Shaw, R. C. Shattock (1990). Restriction-fragment-length-polymorphisms of mitochondrial-DNA of *Phytophthora infestans. Mycol. Res. 94*:1123.
4. Carter, D. A., S. A. Archer, K. W. Buck, D. S. Shaw, R. C. Shattock (1991). DNA polymorphism in *Phytophthora infestans.* In *Phytophthora* (J. A. Lucas, R. C. Shattock, D. S. Shaw, L. Cooke, eds.). Cambridge University Press, Cambridge.
5. Cavalier-Smith, T. (1989). The kingdom Chromista. In *The Chromophyte Algae: Problems and Perspectives* (J. C. Green, B. S. C. Leadbeater, W. C. Diver, eds.). Oxford University Press, Oxford, p. 381.
6. Dick, M. W. (1995). Straminipilous fungi. *CAB* (in press).
7. Förster, H., M. D. Coffey, H. Elwood, M. L. Sogin (1990). Sequence analysis of the small subunit ribosomal RNAs of three zoosporic fungi and implications for fungal evolution. *Mycologia 82*:306.
8. Francis, D. M., R. W. Michelmore (1993). Two classes of chromosome-sized molecules are present in *Bremia lactucae. Exp. Mycol. 17*:284.
9. Francis, D. M., S. H. Hulbert, R. W. Michelmore (1990). Genome size and complexity of the obligate fungal pathogen, *Bremia lactucae. Exp. Mycol. 14*:299.
10. Fry, W. E., S. B. Goodwin, A. T. Dyer, et al. (1993). Historical and recent migrations of *Phytophthora infestans*—chronology, pathways, and implications. *Plant Dis. 77*:653.
11. Fry, W. S., S. B. Goodwin, J. M. Matuszak, L. J. Spielman, M. G. Milgroom, A. Dreuth (1992). Population genetics and intercontinental migrations of *Phytophthora infestans. Annu. Rev. Phytopathol. 30*:107.
12. Goodwin, S. B., A. Drenth, W. E. Fry (1992). Cloning and genetic analyses of two highly polymorphic, moderately repetitive nuclear DNAs from *Phytophthora infestans. Curr. Genet. 22*:107.

13. Hansen, E. M., C. M. Brasier, D. S. Shaw, P. B. Hamm (1986). The taxonomic structure of *Phytophthora megasperma*—evidence for emerging biological species groups. *Trans. Br. Mycol. Soc. 87*:557.
14. Hulbert, S. H., R. W. Michelmore (1988). DNA restriction fragment length polymorphism and somatic variation in the lettuce downy mildew fungus, *Bremia lactucae. Mol. Plant-Microbe Interact. 1*:17.
15. Judelson, H. S. (1993). Intermolecular ligation mediates efficient cotransformation in *Phytophthora infestans. Mol. Gen. Genet. 239*:241.
16. Judelson, H. S., B. M. Tyler, R. W. Michelmore (1991). Transformation of the oomycete pathogen, *Phytophthora infestans. Mol. Plant-Microbe Interact. 4*:602.
17. Judelson, H. S., P. Van West, R. C. Shattock (1994). Towards the isolation of genes determining insensitivity to phenylamide fungicides from *Phytophthora infestans.* In *Fungicide Resistance* (S. Hearney, D. Slawson, D. W. Holloman, P. E. Russell, D. W. Parry, eds.), British Crop Protection Council, p. 167.
18. Judelson, H. S., L. J. Spielman, R. C. Shattock (1995). Genetic mapping and non-Mendelian segregation of mating type loci in the oomycete, *Phytophthora infestans. Genetics 141*:503.
19. Ko, W. H. (1978). Heterothallic *Phytophthora*: evidence for hormonal regulation of sexual reproduction. *J. Gen. Micro. 107*:15.
20. Kuhn, D. N. (1991). Parasexuality in *Phytophthora*? In Phytophthora (J. A. Lucos, R. C. Shattock, D. S. Shaw, L. R. Cooke, eds.). Cambridge University Press, Cambridge, p. 242.
21. Mao, Y. X., B. M. Tyler (1991). Genome organization of *Phytophthora megasperma* f-sp *glycinea. Exp. Mycol. 15*:283.
22. Martin, F. N. (1995). Electrophoretic karyotype polymorphisms in the genus *Pythium*. *Mycologia 87*:333.
23. Martin, F. (1995). Meiotic instability of *Pythium sylvaticum* as demonstrated by inheritance of nuclear markers and karyotype analysis. *Genetics 139*:1233.
23a. Maufrand, R., S. A. Archer, K. W. Buck, D. S. Shaw, R. C. Shattock (1995). The use of RAPD markers in genetic studies of *Phytophthora infestans.* In *Phytophthora infestans 150* (L. J. Dowley, E. Bannon, L. R. Cooke, T. Keane, E. O'Sullivan, eds.). Boole Press Ltd., Dublin, Ireland, p. 55.
24. Mosa, A. A., K. Kobayashi, A. Ogoshi, M. Kato, N. Sato (1993). Isoenzyme polymorphism and segregation in isolates of *Phytophothora infestans* from Japan. *Plant Pathol. 42*:26.
25. Poedinok, N. L., A. V. Dolgova, Y. T. Dyakov (1982). Parasexual process in phytopathogenic fungus *Phytophthora infestans* (Mont) deBary. *Genetika 18*:1423.
26. Sansome, E., C. M. Brasier (1973). Diploidy and chromosomal structural hybridity in *Phytophthora infestans. Nature 241*:344.
27. Shattock, R. C. (1988). Studies on the inheritance of resistance to metalaxyl in *Phytophthora infestans. Plant Pathol. 37*:4.
28. Shaw, D. S. (1988). The *Phytophthora* species. In *The Genetics of Phytopathogenic Fungi* (G. S. Sidhu, ed.). Academic Press, London, p. 27.
29. Shaw, D. S. (1991). Genetics. In Phytophthora infestans, *The Cause of Late Blight of Potato* (N. Robertson, ed.). Academic Press, London, p. 131.

30. Spielman, L. J. (1991). Isozymes and the population genetics of *Phytophthora infestans.* In *Phytophthora* (J. A. Lucas, R. C. Shattock, D. S. Shaw, L. R. Cooke, eds.). Cambridge University Press, Cambridge, p. 231.
31. Tooley, P. W., M. M. Carras (1992). Separation of chromosomes of *Phytophthora* species using CHEF gel electrophoresis. *Exp. Mycol. 16*:188.
32. Tooley, P. W., C. D. Therrien (1987). Cytophotometric determination of the nuclear-DNA content of 23 Mexican and 18 non-Mexican isolates of *Phytophthora infestans. Exp. Mycol. 11*:19.
33. Whisson, S. C., A. Drenth, D. J. Maclean, J. A. G. Irwin (1994). Evidence for out-crossing in *Phytophthora sojae* and linkage of a DNA marker to two avirulence genes. *Curr. Genet.*
34. Whittaker, S. L., R. C. Shattock, D. S. Shaw (1991). Variation in DNA content of nuclei of *Phytophthora infestans* as measured by a microfluorimetric method using the fluorochrome DAPI. *Mycol. Res. 95*:602.
35. Whittaker, S. L., S. J. Assinder, D. S. Shaw (1995). Inheritance of streptomycin and chloramphenicol resistance in *Phytophthora infestans*: evidence for co-segregation of mitochondrial DNA and streptomycin resistance. *Mycol. Res.* (in press).

Author Index

Index of Fungi

Subject Index

Milton Keynes UK
Ingram Content Group UK Ltd.
UKHW020010071024
449327UK00031B/2733

9 780367 448646